Sensors Update

Volume 3

Sensor Technology –
Applications – Markets

WILEY-VCH

Sensors Update

Selected topics from upcoming volumes:

Infrared Sensors

Sensor–Computer Interfaces

Sensors for Process Control

Resonators

Sensor Modelling

Sensor Market: MEMS Marketing

Sensors Update

Edited by
H. Baltes, W. Göpel, J. Hesse

Volume 3

WILEY-VCH

Weinheim · Berlin · New York · Chichester · Brisbane · Singapore · Toronto

Editors:
Prof. Dr. H. Baltes
Physical Electronics Laboratory
ETH Hoenggerberg, HPT-H6
CH-8093 Zürich
Switzerland

Prof. Dr. W. Göpel
Institut für Physikalische und
Theoretische Chemie
Universität Tübingen
Auf der Morgenstelle 8
D-72076 Tübingen, FRG

Prof. Dr. J. Hesse
Carl Zeiss,
ZB „Entwicklung"
Postfach 1380
D-73447 Oberkochen, FRG

e-mail (for orders and customer service enquiries): sales@wiley-vch.de

Visit our Home Page on http://www.wiley-vch.de
or
http://www.wiley.com or www.wiley.co.uk

Other Wiley Editorial Offices

John Wiley & Sons, Inc., 605 Third Avenue,
New York, NY 10158–0012, USA

John Wiley & Sons Ltd, Baffins Lane, Chichester,
West Sussex PO19 1UD, England

Jacaranda Wiley Ltd, Park Road, Milton,
Queensland 4064, Australia

John Wiley & Sons (Asia) Pte Ltd, 2 Clementi Loop #02–01,
Jin Xing Distripark, Singapore 0512

John Wiley & Sons (Canada) Ltd, 22 Worcester Road,
Rexdale, Ontario M9W 1L1, Canada

British Library Cataloguing in Publication Data

A catalogue record for this book is available from the British Library

ISBN 3 527 29433 3

Typset by Thomson Press (India) Ltd, New Delhi
Printed and bound in Great Britain by Bookcraft (Bath) Ltd
This book is printed on acid-free paper responsibly manufactured from sustainable forestation, for which at least two trees are planted for each one used for paper production.

Preface to the Series

The economic realities of productivity, quality, and reliability for the industrial societies of the 21st century are placing major demands on existing manufacturing technologies. To meet both present and anticipated requirements, new and improved methods are needed. It is now recognized that these methods must be based on the powerful techniques employing computer-assisted information systems and production methods. To be effective, the measurement, electronics and control components, and sub-systems, in particular sensors and sensor systems, have to be developed in parallel as part of computer-controlled manufacturing systems. Full computer compatibility of all components and systems must be aimed for. This strategy will, however, not be easy to implement, as seen from previous experience. One major aspect of meeting future requirements will be to systematize sensor research and development.

The rapid development of sensors with computer-compatible output signals has been quite remarkable since its beginnings in the mid 1970s. In addition to the technological advances driving the sensor industry, there has been a boom in new applications and market sectors where new sensor technologies can be applied, thus guaranteeing an increasing demand for sensors and associated instrumentation through the turn of the century.

The aim of *Sensors Update* is to help the reader to stay at the cutting edge of the field. Built upon the 9-volume series "Sensors", it presents an overview of highlights in the field on a regular basis. Coverage includes current developments in materials, design, production and applications of sensors, signal detection and processing, as well as new sensing principles. Furthermore, the sensor market as well as peripheral aspects such as standards are addressed.

Each volume is divided in three sections. 'Sensor Technology' reviews highlights in applied and basic research, 'Sensor Applications' covers new or improved applications of sensors, and 'Sensor Market' provides an overview of suppliers and market trends for a particular sector.

The series editors wish to thank their colleagues who have contributed to this important enterprise whether in editing or writing articles. Thanks are also due to the publishers, Wiley-VCH and their editorial/production staff Dr. P. Gregory, Dr. U. Anton, Dr. J. Ritterbusch and H.-J. Schmitt, and also to W. Greulich for their support in bringing this series into existence.

H. Baltes, Zürich W. Göpel, Tübingen J. Hesse, Oberkochen

Contents

Contributors

H. Becker
Department of Chemistry
Imperial College of Science, Technology and Medicine
South Kensington
London SW7 2AY
United Kingdom

A. Brecht
Institut für Physikalische und Theoretische Chemie
Universität Tübingen
Auf der Morgenstelle 8
D-72076 Tübingen
Germany

U. Dillner
Institut für Physikalische Hochtechnologie IPHT e.V.
Abt. Mikrosysteme/AG Thermosensorik
Helmholtzweg 4
D-07743 Jena
Germany

J. W. Gardner
Department of Engineering
University of Warwick
Coventry CV4 7AL
United Kingdom

G. Gauglitz
Institut für Physikalische und Theoretische Chemie
Universität Tübingen
Auf der Morgenstelle 8
D-72076 Tübingen

M. Glesner
Institut für Mikroelektronische Systeme
Technische Hochschule Darmstadt
Karlstraße 15
D-64283 Darmstadt
Germany

W. Göpel
Institut für Physikalische und Theoretische Chemie
Universität Tübingen
Auf der Morgenstelle 8
D-72076 Tübingen
Germany

U. Grunewald
Carl Zeiss Jena GmbH
Tatzendpromenade 1a
D-07745 Jena
Germany

R. Güttich
Institut für Physikalische Hochtechnologie IPHT e.V.
Abt. Mikrosysteme/AG Thermosensorik
Helmholtzweg 4
D-07743 Jena
Germany

P. Hauptmann
Institut für Prozeßmeßtechnik und Elektronik
Otto-von-Guericke-Universität Magdeburg
Universitätsplatz 2
D-39106 Magdeburg
Germany

T. Hellmuth
Fachhochschule Aalen
Studiengang Optoelektronik
Heinrich-Rieger-Straße 22/1
D-73430 Aalen
Germany

B. Henning
Institut für Automation und Kommunikation
Otto-von-Guericke-Universität Magdeburg
Universitätsplatz 2
D-39106 Magdeburg
Germany

R. Lucklum
Institut für Prozeßmeßtechnik und Elektronik
Otto-von-Guericke-Universität Magdeburg
Universitätsplatz 2
D-39106 Magdeburg
Germany

A. Manz
Department of Chemistry
Imperial College of Science, Technology and Medicine
South Kensington
London SW7 2AY
United Kingdom

J. Müller
Institut für Physikalische Hochtechnologie IPHT e.V.
Abt. Mikrosysteme/AG Thermosensorik
Helmholtzweg 4
D-07743 Jena
Germany

S. Müller-Pfeiffer
Carl Zeiss Jena GmbH
Tatzendpromenade 1a
D-07745 Jena
Germany

T. C. Pierce
School of Engineering
Faculty of Science & Engineering
University of Derby
Kedleston Road
Derby DE22 1GB
United Kingdom

M. Rychetsky
Institut für Mikroelektronische Systeme
Technische Hochschule Darmstadt
Karlstraße 15
D-64283 Darmstadt
Germany

A. Steudel
Institut für Mikroelektronische Systeme
Technische Hochschule Darmstadt
Karlstraße 15
D-64283 Darmstadt
Germany

M. Theisen
Institut für Mikroelektronische Systeme
Technische Hochschule Darmstadt
Karlstraße 15
D-64283 Darmstadt
Germany

K. Toko
Department of Electronic Device Engineering
Graduate School of Information and Electrical Engineering
Kyushu University
Fukuoka 812
Japan

PART 1

Sensor Technology

1 Thermoelectric and Bolometric Infrared Microsensors

J. Müller, U. Dillner, R. Güttich, Institute of Physical High Technology (IPHT), Jena, Germany

Contents

1.1 Introduction

At present, in the middle of the 1990s, there is an enormous renaissance of 'ancient' thermocouple/thermopile infrared radiation sensors invented by Melloni in 1833 and of bolometric sensors first described by Langley in 1881 and formerly often considered to be 'exotic'. The new interest in these sensors stems from the potential of microsystem technologies, which has led to far-reaching improvements of their parameters. In this contribution we review the progress made in recent years (from 1992 to the beginning of 1996) regarding thermoelectric and bolometric infrared microsensors, which belong to the class of thermal infrared microsensors. All detectors of this class have in common that the incident radiation causes a temperature increase of the absorbing medium. In thermoelectric sensors this temperature increase is converted into an electrical signal by means of the Seebeck effect while in bolometric sensors the change of the electrical resistance with temperature is used to get the electrical signal. As microsensors these devices are of small size and their manufacture is based principally on the concepts of microsystem technologies, e.g., silicon chips, micromachining, selective etching techniques, thin-film deposition methods, packaging and interconnection technologies, monolithic or hybrid integration.

The main focus of this article is on thermopile-based infrared sensors (Section 1.2). Concerning bolometric infrared sensors (Section 1.3) the very promising new developments of high-temperature superconducting bolometers are considered in detail (Section 1.3.3). The field of multi-element thermal infrared sensors (Sections 1.2.4 and 1.3.2) is also developing very fast and promisingly. The problem of effective absorber layers, which is a key issue to both thermoelectric and bolometric sensors, is dealt with in a separate section (Section 1.4). Finally a short outlook on trends in infrared sensor applications (Section 1.5) is given.

1.2 Thermoelectric Infrared Microsensor Devices

There has been considerable progress in improving the parameters of thermoelectric microsensors in recent years. This progress is based on the increased use of micromachining, thin-film technology, and photolithographic patterning as well as on the employment of thermoelectrically higher effective materials (cf. Section 1.2.1) or similar ones, which can be more favorably processed by reliable IC technologies and, finally, on the availability of better tools for simulation and modeling of new improved and 'custom-mode' sensor concepts (cf. Section 1.2.2). Besides the work devoted to conventional and electrically calibrated thermopile sensors discussed in detail in Section 1.2.3.1 and 1.2.3.2, much attention has been directed to such issues as *more complex sensor solutions* and thermoelectric *smart sensors*, i.e., sensors with, quite near to the sensor element, hybridly or monolithically interfaced compensating elements, amplifier stages or intelligent controlling circuits (cf. Section 1.2.3.3), or to the development of *multi-element thermopile arrays* (cf. Section 1.2.4) with a larger and larger number of single pixels of smaller and smaller dimensions, which should not affect each other thermally. Moreover, some interesting new thermopile-based sensor solutions have emerged recently that will be reviewed in Section 1.2.5.

1.2.1 Useful Thermoelectric Thin-film Materials

The key component of thermoelectric microsensors is a thin-film thermopile supported by a thin membrane or beam to achieve a low thermal capacitance. Useful materials for both the n-type and p-type legs of the thermopile should have a sufficiently high thermoelectric figure of merit $Z = S^2\sigma\lambda^{-1}$ (S: Seebeck coefficient, σ: electrical conductivity, λ: thermal conductivity). Z has units of inverse temperature and is usually referred to as the dimensionless quantity ZT (T: absolute temperature). Up to now, workers in the field of thermoelectricity have not succeeded in breaking the traditional barrier of $ZT \approx 1$, though this barrier is not based on a theoretical upper limit to the thermoelectric figure of merit. Thus, the best p-type thin-film material is still $Bi_{0.5}\ Sb_{1.5}\ Te_3$ with $ZT = 0.88$ at room temperature, which comes close to $ZT = 0.96$ for the bulk material of this compound semiconductor, cf. [1], whereas the best n-type thin-film material is n-Bi_2Te_3 showing ZT values up to about 0.3 [2]. Contrary to the p-type films, there is the potential to increase ZT in n-type thin films since films of n-type material with the highest bulk ZT at room temperature (doped $Bi_2\ Te_{2.85}\ Se_{0.15}$ with $ZT = 0.9$ [3]) have not yet been realized.

A disadvantage of the above-mentioned highly effective thermoelectric materials is their incompatibility with standard technologies in electronics employing materials such as silicon or GaAs. In particular, silicon and polysilicon films are very interesting materials for thermoelectric applications since they are compatible with the well-established standard IC technologies. As to the figure of merit, however, these films are not the most powerful materials mainly because of their large thermal conductivity (between 15 and 30 W/mK for polysilicon films [4]). In recent years extensive studies of the thermophysical properties of CMOS IC polysilicon layers were carried out at the ETH Zurich [4–7]. The CMOS process-dependence of the thermophysical and transport-related parameters, which was investigated in [6], resulted in ZT values between 0.0038 and 0.021 for n-polysilicon while ZT was about 0.01 for p-polysilicon. Other n-type materials studied recently and employed to build an infrared thermopile sensor [8] are GaAs and $Al_x\ Ga_{1-x}\ As$, respectively, which are compatible with the MESFET technology used, e.g., in producing GaAs power MESFETs. While $ZT = 0.004$ for pure GaAs ($x = 0$) is very small, it can be increased by addition of Al to $ZT = 0.012$ at ($x = 0.15$) and even to $ZT = 0.095$ at $x = 0.45$. The latter relatively high value is due to a maximum in the Seebeck coefficient close to $x = 0.45$ at the change of the direct to an indirect band gap.

Besides the two major groups of useful thermoelectric thin-film materials mentioned – highly effective but IC incompatible materials ($0.1 < ZT < 1$) and modestly effective but IC compatible ones ($ZT < 0.1$) – there has recently been some hope that a third group will emerge in the future, which may be called super-effective materials breaking the barrier $ZT \approx 1$. This optimism has been initiated by a paper of Hicks and Dresselhaus who proposed increasing the thermoelectric figure of merit of certain materials by preparing them in quantum-well superlattice structures [9]. In the paper, calculations were performed on superlattices composed of layers of materials having two-dimensional electronic properties given by infinite potential barriers and zero barrier widths. The resulting ZT increased with decreasing layer thickness of the superlattice multilayers made in a particular orientation. It was calculated that using Bi_2Te_3 in such a superlattice has the potential to yield $ZT \approx 7$. However, other publications [10–12] referring to the paper of Hicks and Dresselhaus do not share their optimistic assessment of the potential of superlattices, but anticipate only moderate improvements in ZT to values between 1 and 2. The main reason for this less optimistic view lies in the more realistic treatment of the superlattice systems including two features that have been neglected in [9], namely, tunneling between quantum wells and the finite thermal conductivity of the material forming the barrier. The experimental work to investigate the thermal and thermoelectric properties of thin films based on the proposed superlattice structures is still in its infancy, but the results are expected to appear in the literature within the foreseeable future. Even a moderate improvement of ZT to, say, 1.5 would be highly welcome by the thermoelectrics community.

1.2.2 Thermal Simulation and Modeling of Thermoelectric Microsensors

To meet special customer requirements concerning several sensor parameters, e.g., responsivity, detectivity, and time constant, as well as to optimize the design of the thermoelectric sensors as to these parameters there is a growing need for thermal modeling and simulation, which can replace numerous technological trial-and-error iterations. Essentially, the thermal simulation includes the calculation of the spatial and temporal distribution of temperature in the sensor chip consisting of a free-standing membrane, at which the heat load is applied, and a silicon rim acting as a heat sink. Once this distribution is known, all the parameters mentioned can be easily deduced. In practice, the temperature field can be calculated analytically or numerically using primarily the finite element method (FEM). Both analytical and numerical methods are based on the solution of the heat-transport equation under appropriate boundary conditions.

Analytical solutions of the heat transport equation often rely on a reduction of the dimension of space due to the special geometry and symmetry of the object to be thermally modeled. In the case of a membrane, the ratio of lateral extent to thickness exceeds unity by at least one order of magnitude. Hence, temperature gradients perpendicular to the membrane plane can be neglected, resulting in a two-dimensional problem. A further reduction to an one-dimensional thermal model is possible if the symmetry of the problem enables a description of the in-plane heat conduction by only one coordinate. In recent years, using Cartesian or cylindrical coordinates, various sensor membranes of thermoelectric microsensors have been modeled in that way [13–15]. The multilayer structure of the membrane has been taken into consideration by a lumped model approach summing up the thermal conductances and capacitances of the individual layers. Heat transfer between the membrane surface and the environment, caused by conduction or convection and by radiation, has been included linearly (Newton's law of cooling). In [15] the model included two additional features: (a) a heat load was applied not only at the boundary opposite the heat sink, mathematically described by a prescribed heat flux boundary condition, but also at the region between this boundary and the heat sink, mathematically described by a homogeneous heat power density; (b) an intermediate boundary was introduced dividing the membrane into two regions of different material properties. At this boundary two boundary conditions ensuring the continuity of temperature and heat flow are valid.

In the cases where the analytical method fails due to more complicated geometrical and load arrangements, commercially available FEM codes (e.g., ANSYS, NASTRAN, ABAQUS) provide a state-of-the-art tool for the thermal modeling of thermoelectric microsensors. Recently FEM was used for the thermal simulation of various micromachined thermoelectric microsensors such as an infrared radiation sensor [16], a planar true r.m.s. multijunction thermal converter [17, 18], and a thin-film absolute radiometer [18]. In Figure 1.1 the FE model of the radiation sensor chip is illustrated. While the membrane is modeled using shell finite elements the silicon

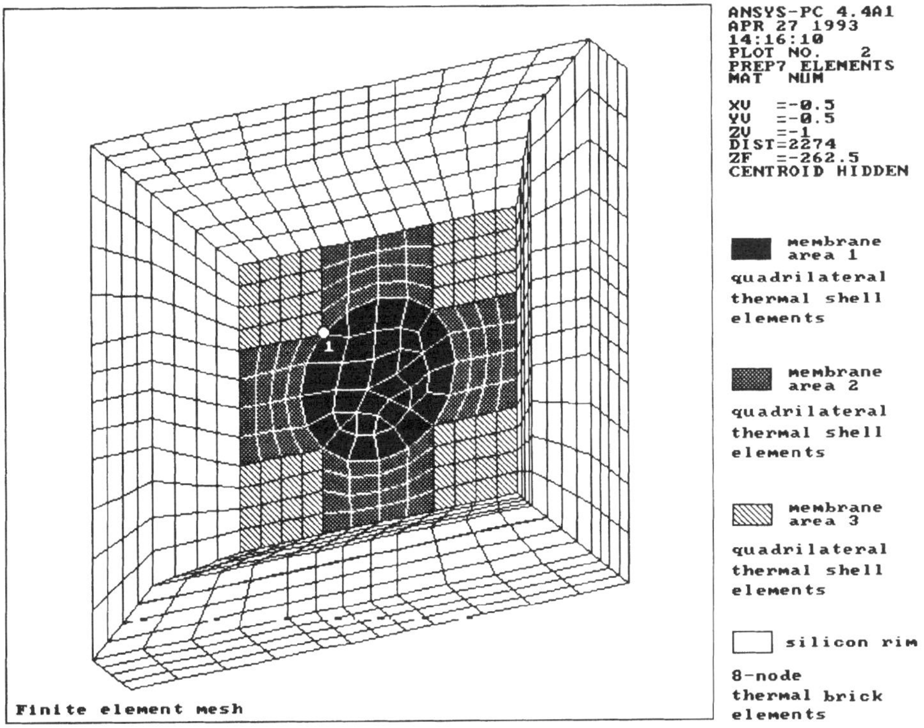

Figure 1.1. Finite element model of the geometry and materials of a radiation microsensor chip, bottom view [16]. The chip is subdivided into three membrane areas and the bulk rim. Each membrane area consists of a stack of functional layers; membrane area 1: substrate, insulation, isothermal collector, black absorber; membrane area 2: substrate, thermopile legs, insulation; membrane area 3: substrate and insulation.

rim is modeled employing brick elements. As the result of the FEM calculation, an irradiance of 1000 W/m^2 as heat load on the thin free-standing membrane causes a temperature increase of about 8.3 K [16] resulting in a responsivity of 56 V/W for the nitrogen filled sensor.

Besides the use of general FEM packages there is a trend to develop special numerical simulation tools for the thermal modeling of microsensors or microstructures. In [19] the program MICROTHERM written for IBM-compatible PCs is introduced, which is based on the solution of the two-dimensional heat conduction equation by means of a finite-difference method. MICROTHERM supplies the 2-D stationary temperature distribution and permits an observation of the temporal development of the temperature field. At the ETH Zurich a special modeling environment SOLIDIS was developed and used, e.g., for the development of new CMOS thermal converters [20, 21].

The often insufficient knowledge on some input parameters continues to be a problem in the thermal modeling of microsensors concerning, typically, important material properties (thermal conductivity, heat capacity, emissivity, etc.) of the functional layers and their dependence on thickness and on quantities, which are clearly influenced by the manufacturing technology (e.g., doping concentrations, grain size, and other structure parameters). In this field of work, progress has been made by investigating these properties for CMOS IC polysilicon materials [4–6].

1.2.3 Single Element Thermopiles

2.3.1 Reliable and Novel Models

In this context, the work has mainly been focused to such objectives as discussed in the following items (a)–(d):

(a) *Increasing of responsivity and detectivity* not only by using more efficient thermoelectric materials (cf. Section 2.1), but also *by integration* of more and more single thermocouples into the receiving area of a thermopile sensor. In comparison to [22], in which the state-of-the-art of thermopile sensors is summarized in a table from the view point at the end of the eighties, in the mid-nineties thermopile sensors have typical parameters given in Table 1.1. Today a highly sensitive thermopile sensor setup usually consists of a circular, square, or rectangular receiving area supported in most cases by a membrane substrate (e.g., consisting of a stress-compensated Si_3N_4 membrane or a SiO_2/Si_3N_4 sandwich, of polyimide, or Mylar), which is freely suspended on the bulk frame (e.g., a micromachined Si wafer or a ceramic carrier) used as a heat sink. The active ('hot') junctions of some dozen single thin-film thermocouples are arranged in the center of the membrane forming the radiation receiving area (most common dimensions are 1 mm^2, 2 mm $\times$ 2 mm, or 0.5 mm in diameter), whereas their reference ('cold') junctions are located on the heat sink. The receiving area is covered by an appropriate absorbing layer (cf. Section 1.4). As a result of radiation absorption, a temperature difference arises between the receiving

Table 1.1. Technical features of current thermoelectric infrared radiation microsensors (mean values according to data sheets of Barnes Eng. Div./EDO Corp., Shelton/USA; Dexter Research Center, Dexter/USA; EG&G Heimann Optoelectronics GmbH, Wiesbaden/D; HL-Planartechnik GmbH, Dortmund/D; Fraunhofer-Institut für Mikroelektronische Schaltungen und Systeme (IMS2), Dresden/D; IPHT e.V., Jena/D; Meggitt/Northwoods Sensors, Manchester/USA; Mitsubishi Yuka, JP; Oriel Corp., Stratfort/USA; Xensor Integration, Delft/NL). A: receiving area or ⊘: its diameter, $\mathcal{R}$: responsivity, τ: time constant, R: resistance, D^*: detectivity.

Note: It is not possible in any case to draw an assessing comparison of the parameters amongst themselves, because all special conditions have to be considered, under which these parameters were determined by the manufacturers, *e.g.*, filling gas and window material.

	Model	No. of elem.	A or Ø (mm² or mm)	$\mathcal{R}$ (V/W)	τ (ms)	R (kΩ)	D^* (10^8cmHz$^{1/2}$/W)
Barnes/EDO		6	1 × 1	15	80	8	1.3
Dexter	M14	12	0.92 × 0.4	16	14	2	1.6
	M5	10	0.5⊘	55	28	3	3.5
	1M	15	1⊘	23.5	32	3	3.0
	2M	48	2×2	19.2	40	10	2.9
	3M	72	3×3	15	50	18	2.5
	6M	59	6⊘	4	60	30	0.5
	S25	20	0.25×0.25	250	12	25	3.0
	S60	44	0.6×0.6	120	12	60	2.0
Heimann	TPS424	40	0.5×0.5	100	25	45	2.0
	TPS434	40	0.5×0.5	48	20	40	0.9
	TPS634	68	1.8×1.8	12	50	15	1.3
HL Planar	TS 8060	80	0.26	60	60	35	1.5
	TS 10060	100	0.5	60	60	50	1.5
IMS2	TP1	60	0.6 × 0.6	210	26	518	1.3
IPHT	TS-72/Kr	72	0.5⊘	280	35	20	7.0
	TS-100/Kr	100	1,13⊘	115	75	30	5.0
	TS-116/Kr	116	1.2×1.2	58	70	62	2.1
	TS 144/Kr	144	2×2	45	100	9	1.5
Meggitt/	PL-82	20	0.8×2.3	15	75	1-5	1.7
North-	PS-20	48	2 × 2	20–40	100	6–15	2.0
woods	S-15	28	1.5×1.5	8–24	100	6–15	1.1
	PC-6	75	6⊘	3–5	200	18–35	1.1
Mitsubishi	MIR100	100	2×2	11.5	100	11	3.0
Oriel		20	0.8 × 4	12	150	5	2.0
		48	2×2	17	75	10	2.0
Xensor	IR3704		1.5×1.2	7.5	25	30	

area and the bulk rim causing a thermoelectric signal voltage on the base of the Seebeck effect. Today typical responsivity values are in the order of 10^0–10^2 V/W according to the size of the receiving area and other parameters, cf., e.g. [23]. Depending on the thermal coupling between the sensor chip and its housing (due to the thermal conductance of the pins [24] and/or the specific filling gas) sensor time constants range from 10^0 to 10^2 ms.

(b) *Cheap batch-production and large scale manufacturing of low-cost sensors.* Therefore, a good choice of appropriate materials and technologies, e.g., for materials having a sufficiently high thermoelectric efficiency and being preferably processable also in well-tried CMOS processes, of appropriate substrate materials, absorber layers, sensor packaging, and test methods for the inspection of special sensor parameters is crucial. Meanwhile, the development of semiconductor thermopile infrared microsensors manufactured on the basis of CMOS technology has reached relevant importance. Silicon–metal thermopiles fabricated by CMOS and micromachining have been described in detail recently [1, 25, 26, 27–29]. A 52-element poly-Si/Al thermopile with an internal resistance of 56 kΩ has a responsivity of 58 V/W, and a time constant of 15 ms [30], a thermopile consisting of 76 elements shows 35.7 kΩ, 33.7 V/W, and 50 ms [31], while a novel sensor based on n-poly Si/p-poly Si reaches 2.5 MΩ, 96 V/W, 31 ms, and $D^* = 4.5 \times 10^7$ cm Hz$^{1/2}$/W [32]. Further examples are reported in [33, 34] on poly Si/Al thermopiles already manufactured commercially by EG&G Heimann Optoelectronics GmbH (cf. Table 1.1) employing a CMOS standard IC process. Finally, in [8] an approach was introduced to make a thermopile sensor set-up by using a GaAs-AlGaAs material system. A responsivity of 145 V/W and $D^* = 4.1 \times 10^7$ cm Hz$^{1/2}$/W was reached. This work attempted to design the thermal resistance of the membrane and the thermopile as high as possible. Therefore, triangular holes were additionally etched into the AlGaAs membrane.

(c) *Novel sensor concepts to solve special measuring problems.* As an example, in a microthermopile consisting of 20 copper–constantan elements, the supporting membrane was completely removed [35]. The active junctions were made of free-standing self-supporting wires (1 μm wide, 0.1 μm thick, 20, 40, or 60 μm long) while the reference junctions were in contact with the bulk substrate. Such a construction results in active junctions of very small mass, which are thermally well isolated to make a particularly quick, but nevertheless sensitive infrared radiation thermopile. The device reaches a time constant of 20 μs and a responsivity (at 10.6 μm) of 2.5 V/W and is suitable as a very good room temperature CO_2 laser radiation detector since the physical dimensions of the free-standing structures are comparable to the wavelength of the radiation being detected and, thus, the structure of free-standing wires can be looked upon as a dipole antenna. A similar device was constructed using free-standing n-GaAs/Al microthermocouples resulting in a responsivity of 2.75 V/W and a time constant of $\sim$250 μs [36]. The GaAs microthermocouples can be fabricated as an array and used as imagers for laser beam diagnostics.

For detection of fire before it has spread significantly a fast response thermopile-based infrared flame sensor has been drafted [37]. The fabrication process for such a sensor uses a combination of both standard and novel microfabrication techniques to produce 8-element thermopile arrays. In this way, thin polyimide membranes, which carry the single active thermocouple junctions, are supported on a silicon frame. The IR absorption is achieved by a careful control of the polyimide thickness and the resistivity of the front metalization (cf. Section 1.4). In this context, the spectral signature of the incident radiation is used to discriminate against sources of false alarms, such as the sun or welding equipment. This is achieved by using an array of such thermopiles, each with a special associated infrared bandpass filter in a single detector unit. The signals obtained from the individual thermopiles are analyzed using an embedded microprocessor control system.

(d) *Optimal matching of the thermal transition behavior* at changing ambient temperature ('bounce') of thermopile sensors and of the instrument, into which the thermopiles were built in as detector [38].

1.2.3.2 Electrically Calibrated Thermopiles

For describing radiometric units and their passing on, standard sensors with high request for large dimensions of the active receiver area, high detectivity, short time constant, uniform spectral responsivity and its homogeneity, good linearity, and very small uncertainties are needed. A new sensor for the absolute measurement of radiation fluxes in the IR range with, in this respect, very advantageous parameters has been developed recently [39, 40] by combining a thin-film thermopile with a meandered thin-film heater localized within the receiving area system. The sensing junctions and legs of the thermopile and the heater are supported on a thin SiO_2/Si_3N_4 free-standing sandwich membrane (8 mm × 8 mm) formed by anisotropic etching of a Si wafer. The BiSb/Sb thin-film thermopile is fabricated using a multilayer technique, i.e., both the legs of each single thermocouple are put one over the other separated by a passivation layer, which enables a higher integration of the thermopile arrangement. The 76 sensing junctions arranged at the periphery of the receiving area of 7 mm diameter are electrically isolated with a photoresistive layer against the heater and the receiving area system, which consists of a thermally equalizing Ag layer covered by an absorbing layer of silver (or gold) black (cf. Section 1.4). The sensor chip is mounted on to a ceramic header and worked into an evacuated and Xe backfilled glass housing equipped with a quartz or CaF_2 window wedge-shaped with an angle of about 1.5°.

The new radiometric sensor stands out for a low resistance (4–5 kΩ to reduce thermal noise), a high responsivity of 10–12 V/W with a low temperature coefficient (-5.4 to $-5.7 \times 10^{-3} K^{-1}$), a time constant of 1 s, NEP of < 5 nW $Hz^{-1/2}$ corresponding to $D^* = 1.25 \times 10^8$ cm $Hz^{1/2}$/W, and a low dependence of the signal voltage on the heating power. The good uniform local responsivity (Figure 1.2) across the receiving area (± 1% within a central area of 5 mm diameter) is to guarantee a good equivalence between responsivity due to irradiation and due to electrical heating and a non-linearity of < 1% below 1 mW. Such a sensor is very suitable as a transfer or secondary standard and radiation detector for absolute radiometry.

1.2.3.3 Sensor Systems with Integrated Electronics

In most thermopile applications, typical output voltages are in the millivolt range only. Therefore, it seems to be advisable to amplify the signal very near the sensor to obtain a larger output signal for the following signal processing stage and to decrease the influences of interfering signals. Some years ago such concepts were realized,

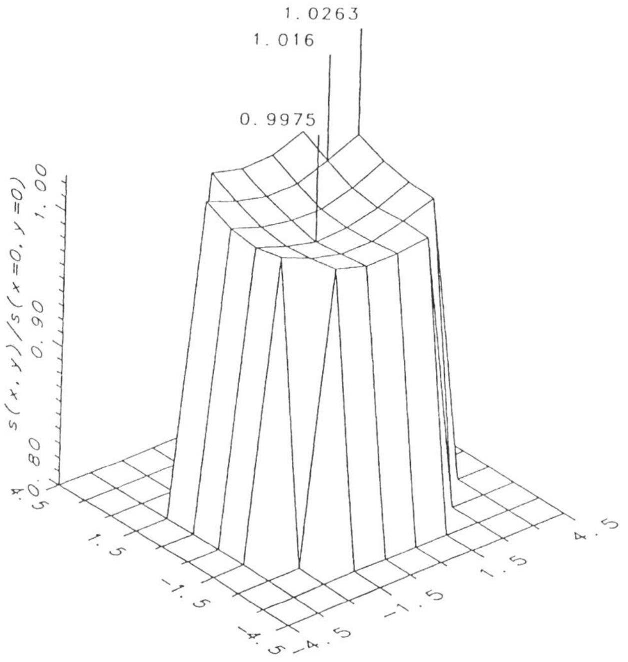

Figure 1.2. Homogeneity of the responsivity across the receiving area of an electrically calibrated thermopile sensor TS-76 from IPHT, Jena [40].

e.g., by a hybrid design [41] for a one-element thermopile sensor or monolithically for a linear array [42].

A very recent approach is an IR sensor system (Figure 1.3) consisting of a p-Si/n-Poly-Si 60-element thermopile with both a monolithically integrated temperature sensor measuring the chip temperature and an amplifier circuitry with programmable amplification for the thermopile signal [43] realized in a CMOS process on SIMOX wafers. The sensor is characterized by a receiving area of 0.36 mm^2, a responsivity (without filter) of about 200 V/W, a time constant of 25 ms, a resistance of about 500 kΩ and a detectivity $D^* = 1.3 \times 10^8\,cm\,Hz^{1/2}/W$. In [43] the influence of the integrated circuits on the sensor signal is discussed in detail. A thermal offset occurs because of the monolithic integration of thermopile and power-consuming circuitry. But, because of its high stability, the resulting offset voltage acts like a constant offset and it is concluded that thermopiles with monolithic integrated circuitry can practically be used in many applications. In contrast to this, in [44] the concept of a thermoelectric sensor module discretely built up on a module board is described (Figure 1.4). After passing an IR filter and mirror optics the IR radiation hits the absorbing area of the thermopile. A preamplifier with adjustable amplification increases the signal level and equalizes variations of responsivity. A reference signal coming from an integrated thermistor (for analog circuit design) or from a reference

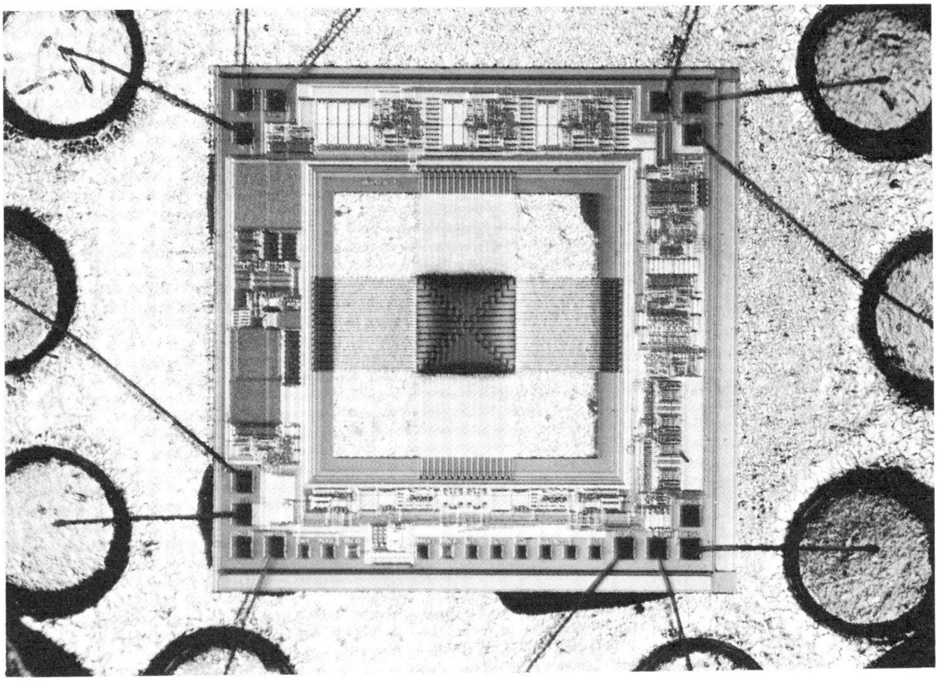

Figure 1.3. Thermopile sensor with monolithically integrated amplifier stage and temperature (by courtesy of IMS2, Dresden).

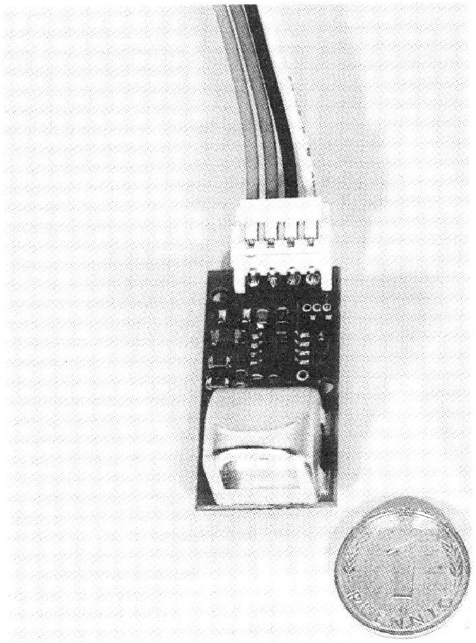

Figure 1.4. Thermoelectric sensor module (by courtesy of EG&G Heimann Optoelectronics GmbH, Wiesbaden).

diode (for digital signal processing) [34] is changed into a compensation voltage, which is subtracted from the amplified sensor signal in the following differential amplifier. The temperature compensation allows for the compensation of the influence of ambient temperature changes between 0 and 180°C to the output. Finally, the temperature-compensated difference output signal can be evaluated in a subsequent control unit.

1.2.4 Multi-Element Thermopiles

Multi-element thermopiles are one- or two-dimensional arrangements of a number of pixels. The arrangements are characterized by pixel size (the assembly of thermocouples to a thermopile as a sensitive area) and pixel pitch (the distance from a center of a sensitive area to the center of the neighboring pixel). Pixel size, pixel pitch, and the noise-limited resolution of the pixels are important as distinguishing features of multi-element sensors.

One-dimensional or linear arrangements are useful as detectors in two important applications: spectrometers (spectral photometers) and line scanned pyrometers. In spectrometers the spectral decomposition is a one-dimensional distribution vs. the wavelength and the resolution (wavelength) is limited by the above mentioned features. Line scanned pyrometers, on the other hand, permit the reflection of a reference source to each pixel for calibration purposes. Two-dimensional or matrix array elements are used with optics in thermal imaging applications. Since they are arranged in the focal plane of these optics the matrix elements are described as focal plane arrays (FPA). The limitation factors are the same as for one-dimensional applications. The aims of modern constructions are a minimization of pixel size and pixel pitch in order to increase the number of working pixels and CMOS compatible solutions of the thermal construction including a combination of the sensor layout with the electrically read-out circuitry.

1.2.4.1 Linear Array Thermopiles

In [45] a linear array was reported on realized in silicon using micromachined dielectric windows and thermopile detector elements formed by thin-film junctions of $Bi_{0.87}\,Sb_{0.13}/Sb$. The array consists of 16 pixels, each having 67 series-connected thermocouples arranged in two staggered 8-element linear arrays. The thermopiles are supported on 1 μm thick diaphragms of Si_3N_4/SiO_2. The detectors provide a responsivity of 70 V/W and a detectivity D^* of $3 \times 10^8 \mathrm{cm\,Hz^{1/2}/W}$. The pixel sizes are 250 μm × 2000 μm and the pitch is 1500 μm; by staggering, the effective pitch becomes 750 μm. Another paper [46] describes a multiplexed line array, which consists of 32 pixels in two staggered 16-element linear arrays. The thin-film thermopiles are made of n^+-polysilicon/Au with 40 series-connected thermocouples supported on 1.2 μm thick diaphragms of SiO_2/Si_3N_4 . The pixels are characterized by a responsivity of 64 V/W and $D^* = 7 \times 10^7 \mathrm{cm\,Hz}^{1/2}/\mathrm{W}$. The pixel sizes are smaller than 400 μm × 800 μm (size of diaphragm windows) and the pixel pitch is 600 μm. The signals are multiplexed by using on-chip enhanced depletion-NMOS circuitry. This array is distributed commercially [47].

Kruse [48] summarized developments in the field of thermal multi-element sensor arrays under the aspects of thermal imaging. In the late 1980s the Honeywell Corporation began developing linear thermoelectric uncooled arrays. This work has been exploited in the SnapShot^{TM} infrared camera being developed by Infrared Solutions Inc. [49] under license from Honeywell. The IR SnapShot^{TM} is a portable, battery-powered IR camera with a mechanical device that scans the two-dimensional scene over the one-dimensional array. The cursor function provides temperature read-out down to individual pixels for instant, detailed review. Specifications of the linear array in the IR SnapShot^{TM} are 120 elements of the size 50 μm × 50 μm, NETD (noise equivalent temperature difference) ≤ 0.2 K with optics f/0.8 (18 mm, 25 mm, 33 mm focal length), and scan in horizontal within a time of 1.44 s.

1.2.4.2 Matrix Array Thermopiles

In the field of two-dimensional arrangements of thermopiles the NEC Corporation has a leading position. In [50] a 128 × 128 pixel array is described showing a pixel size of 100 μm × 100 μm with a fill factor as large as 67% (fill factor means the percentage part of the usable membrane area for the pixel). Each of the thermopile pixels is formed by 32 in series-connected p-type boron-doped polysilicon/n-type phosphorus-doped polysilicon thermocouples. The diaphragm is made of a 450 nm thick silicon dioxide film. Under the diaphragm, there is a hollow isolating the diaphragm thermally. Theses hollows are formed by etching a 1 μm thick sacrificial polysilicon layer. The technology of this sensor array is CMOS-process compatible and the integrated signal read-out circuit contains a vertical CCD for image area, a vertical CCD for storage area, a horizontal CCD, and a floating diffusion scheme output having a source-follower amplifier. The responsivity in the low-frequency region was found to be 1550 V/W (the FPA is packaged in a vacuum housing) and the cutoff frequency is 130 Hz. With frame rates of 120 Hz, clock frequencies for vertical CCD and horizontal CCD of 17 kHz and 2.6 MHz, respectively, and optics of f/1.0, the NETD of a thermal imaging system with this FPA is in the order of 0.5 K. Constructions of uncooled thermoelectric arrays and additional bolometric arrays were patented by NEC [51].

In a recent paper [52] a 32 × 32 pixel array with on-chip address generation and multiplexing is reported on. The pixel size is 375 μm × 375 μm and a pixel consists of 32 and 36 n-poly Si/p-poly Si thermocouples, respectively. The pixel responsivity is 30 V/W. The authors study technological problems of front-undercut and back-etched pixel structures. A special feature of this FPA design is pixel self-testing by using an embedded resistor network forming the front-end of a micro processor-controlled camera system that features automatic per-pixel ranging and gain control.

1.2.5 New Thermopile-based Sensor Solutions

Under the aspect of optimization of responsivity and time constant of a thermopile sensor dependent on its geometrical dimensions, a tunable thermopile radiation microsensor has been suggested in [53] based on the variation of the heat transfer coefficient between the thermoelectric thin-film layer structure and the header of a TO-5 package acting as a heat sink, on which the thermopile chip is mounted. The idea has been realized by an adjusting screw with a small pitch arranged in the center of such a header. Tuning of the screw leads to a variation of the heat transfer coefficient by changing the distance d_1 (cf. Figure 1.5) between the thermoelectric thin-film functional system and the end-surface of the screw and thus results in a variation of responsivity (30–50 V/W) and time constant (25–40 ms).

The idea for another type of tunable micromachined thermoelectric infrared sensor is described in [54]. A sensor with tunable wavelength selectivity for application in infrared spectroscopy as an emission monitor for liquids and gases consists of a tunable interference filter in the form of a thermopile wide-band detector. The applicable spectral bandwidth ranges from 1.5 to 7.5 μm with a resolution of 25 nm over the whole range. The wavelength tuning is carried out by electrostatic forces, varying the voltage at four integrated disk capacitors and, thus, the distance (resonator length) between two mirrors.

Novel thermoelectric sensor devices can be expected since the thermoelectric effect was observed in high-temperature superconductors [55] too. Here, the thermoelectric effect is due to a transverse Seebeck effect. The transverse Seebeck effect is observable firstly because of the anisotropy of $YBa_2Cu_3O_7$ and secondly because it is possible to grow off-axis epitaxial films. Due to a tilt angle between the surface normal and c-axis of YBCO films grown by laser ablation and prepared on a specially cut $SrTiO_3$ substrate, the sequence of metallic CuO_2 layers and less conducting

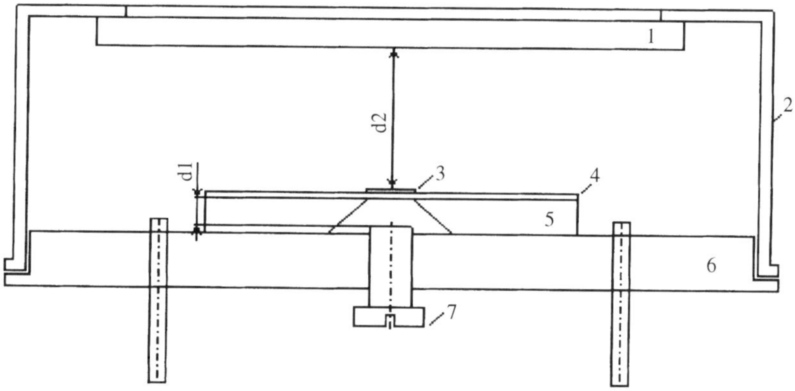

Figure 1.5. Schematic cross-sectional view of a tunable thermopile microsensor: 1–IR window, 2–can, 3–thermoelectric thin-film system, 4–membrane substrate, 5–bulk Si substrate, 6–package header, 7–adjusting screw.

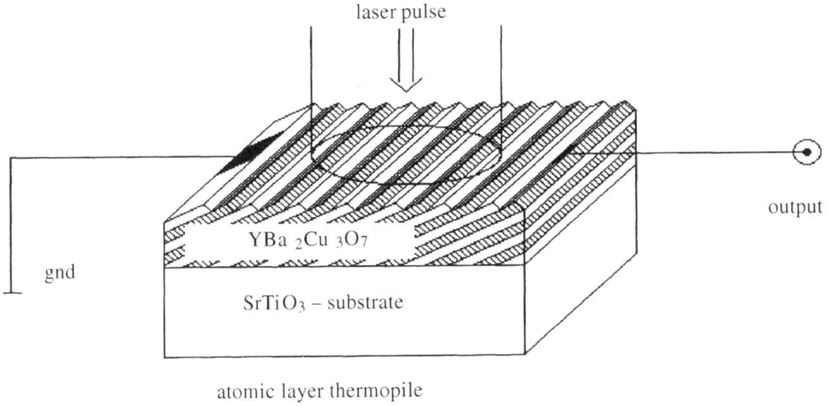

Figure 1.6. Atomic layer thermopile [56]. The thermoelectric voltage is perpendicular to the temperature gradient.

intermediate YBCO layers forms a series connection of some millions of thermocouples of atomic dimensions (Figure 1.6). This thermopile effect can be used for developing detectors for laser radiation with a sensitive area of 10 mm × 10 mm, a film thickness of 300 nm, a responsivity of 1 mV/ W, a time constant of 1–100 ns, and a NEP of 10^{-6} W/Hz$^{1/2}$. Signal pulses up to 200 V and currents of several amperes have already been generated in such YBCO atomic layer thermopiles in response to pulsed laser irradiation at room temperature [56]. Moreover, the partial substitution of Y by Pr causes a dramatic change of thermoelectric response [57]. Specimens with 20% Pr substitution showed voltage signals enhanced by a factor of 20 compared to [56].

1.3 Bolometric Infrared Microsensors

In bolometric infrared microsensors, the absorbed infrared radiation changes the on-chip temperature gradient caused by applying a bias voltage at a conductive film. This temperature change is transduced into an electrical signal voltage using the thermoresistive effect. The strength of the thermoresistive effect is expressed by the temperature coefficient of resistance (TCR), TCR $= \mathrm{d}R/(R\mathrm{d}T)$ (R: ohmic resistance of the conductive layer). The necessity of a bias current associated with $1/f$ noise reducing performance at low frequencies is a disadvantage of bolometers in comparison to the self-generating thermopile sensors. On the other hand, higher detectivities are possible in principle since, contrary to thermoelectric materials where $ZT \approx 1$ remains a barrier to date, there is no comparable barrier for the materials of the bolometric conductive film.

1.3.1 Thin-film and Thermistor Single-Element Bolometers

The typical performance of uncooled non-evacuated thin-film and thermistor bolometers characterized by detectivities D^* up to $1\text{–}2 \times 10^8$ cm $Hz^{1/2}/W$ at time constants of some milliseconds has not been improved remarkably in recent years, but novel techniques and materials have been applied. In particular, porous silicon technology [58] has some advantages in micromachining applications in comparison with the surface micromachining method that are useful concerning the detectivity of thermal sensors since, due to thicker sacrificial layers, higher distances of membranes or bridges to the bulk silicon are feasible (100 μm vs.10 μm). Using this technology bolometers were fabricated consisting of 500 nm thick meandered gold films on 200 nm thick amorphous SiC bridges with an area of 150×300 μm^2 [59]. They showed $D^* = 3.5 \times 10^7$ cm $Hz^{1/2}/W$, a value that was estimated to be tripled by using nickel instead of gold. A similar performance ($D^* = 4.4 \times 10^7$ cm $Hz^{1/2}/W$ at 200 Hz) was achieved by a platinum-film bolometer formed on a SiO_2 membrane of small dimensions (active area 96×96 μm^2) suspended on a silicon substrate [60]. The device was fabricated using CMOS technology for the most part with additional equipment necessary for the Pt film and the NiCr black coating only.

Progress has also been made in the problem of encapsulation of microsensors. At Terumo Corporation in Japan, a micro-airbridge thermistor bolometer with a very small packaging size (2.2 mm × 2.3 mm × 0.8 mm) has been developed for a clinical tympanic thermometer [61]. The package serves as a vacuum microchamber formed by bonding silicon wafers to each other in order to increase the responsivity to 5000 V/W, which exceeds the value at atmospheric pressure by a factor 30.

Concerning new materials for bolometers, the use of high-temperature superconducting materials (cf. Section 1.3.3) at room temperature, i.e., above the superconducting critical temperature, is now under discussion [62, 63]. Here bolometric sensors with detectivities from 10^7 to 10^8 cm $Hz^{1/2}/W$ can be expected in the near future.

1.3.2 Multi-Element Bolometers

In the field of two-dimensional arrangements of bolometers, the Honeywell Corporation has a leading position in design and development, a fact that is confirmed by numerous publications ([64] contains a list of them) and patents for a bolometric FPA [65] and for a complete thermal imaging camera system [66]. Honeywell has developed the so-called bolometer microbridge design (cf. Figure 1.7) representing a thermal construction with microbridges above hollows formed by etching of thin sacrificial layers. The microbridges are the carriers of bolometer structures. Studies of different materials such as VO_2 [48] and polysilicon [65] were performed and protected by patent, respectively. The bolometric construction permits the best reduction in pixel size, a good fill factor, and is, moreover, compatible to the design of CMOS circuitry. At Honeywell different bias modes were studied with the

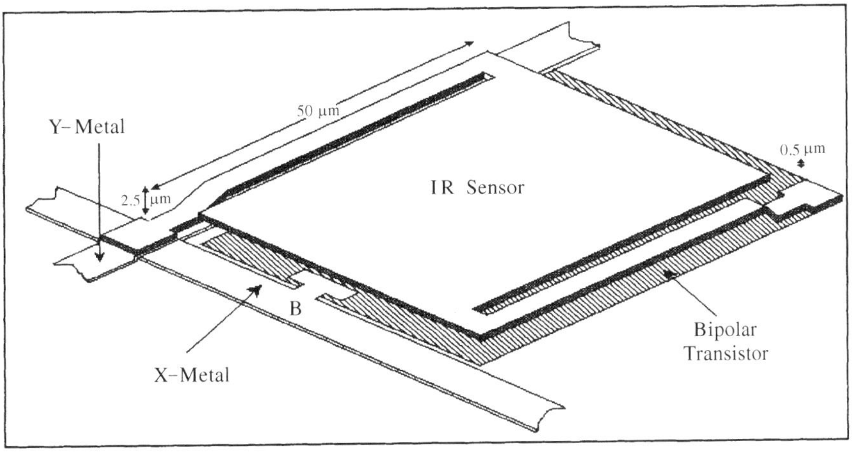

Figure 1.7. Pixel structure of Honeywell's bolometer microbridge.

pulsed bias mode being patented. A vacuum-housed FPA packaging for highest sensitivities was developed and a thermoelectric stabilizer decreasing the great thermal offset and controlling the optimal operating temperature range was integrated. Honeywell has decided not to manufacture arrays and systems. Instead, licenses to the technology are given to other companies, e.g., Hughes and Rockwell [67].

Other authors studied materials such as amorphous, microcrystalline, or polycrystalline silicon thin films prepared by sputtering or chemical vapor deposition [68], nickel-cobalt-manganese oxide compounds with a spinel structure utilized in sputtered film thermistors [69], and platinum [70] evaporated to a thickness of 20 nm and patterned in a meandering shape using the lift-off technique. The adhesion of Pt on Si_3N_4 was improved by a 40 Å thin film of Ti deposited prior to Pt. In [68] three configurations of electrodes to the resistive materials, the so-called gap, sandwich, and the sandwich-gap arrangements, are also presented.

VO_2 as a high-TCR material applied in optical switches has also been studied at the Canadian NOI [71]. Based on this material, they developed a one-dimensional array of 64 pixels of sizes 50 μm × 50 μm [72]. Expecting the cost of IR imagers based on bolometric detector arrays to drop below $10,000 NOI started to fabricate 128 × 128 element bolometric detector FPAs with on-chip electronics in 1996. The NEC group has proposed a 128 × 128 element bolometer array using thin-film titanium [73]. The pixel size is 50 μm × 50 μm and with optics f/0.8 they achieved a NETD of 0.7 K. In the first quarter of 1996, another firm, Amber Engineering, Inc., introduced an uncooled infrared imaging system named 'Sentinel' [74]. This system is based on the uncooled microbridge bolometer FPA developed by Honeywell and shows the following features [75]: 320 × 240 pixels (of size 50 μm × 50 μm), 50 mm f/0.7 optics, thermoelectric thermal stabilization, 30 Hz frame rate, a scene temperature range from −20°C to 150°C, a NETD <0.1 K at 25°C, and RS-170 monochrome (NTSC) or PAL video output.

1.3.3 Cooled Bolometers Including HTSC Bolometers

A bolometer is cooled to obtain higher detectivities D^* in comparison to room temperature operation. Cooling results in a reduction of the two main sources of noise in a thermal sensor, the temperature fluctuation (phonon) noise and the Johnson noise of the resistance, and by cooling a temperature range for the operation of the sensor can be realized where the bolometer has the highest possible TCR. An extremely high TCR can be achieved by using the transition edge between the normal and the superconducting state of a superconducting material. After the discovery of a new class of superconducting materials with a high critical temperature in 1986, this transition was available at much higher temperatures (up to about 90 K) than before. With only liquid nitrogen instead of liquid helium necessary for cooling, the acceptance of superconducting bolometers has clearly risen and much work has been done in recent years to improve the detectivities of high-temperature superconducting (HTSC) bolometers.

The material chosen for the HTSC film has been primarily $YBa_2Cu_3O_{7-x}$ (YBCO) epitaxially grown on various substrates (mainly sapphire or silicon) typically covered with lattice-matched buffer layers. In order to fabricate high-performance HTSC bolometers, it is advantageous to use thin substrates with a very low heat capacity. As early as 1991, a remarkably high detectivity of 4.2×10^9 cm Hz$^{1/2}$/W at 10 Hz was achieved by Verghese *et al.* [76] using a YBCO film on buffered sapphire lapped down to a thickness of 20 μm. To date, a better result was only reported by Brasunas and Lakew [77], their record performance HTSC bolometer showing $D^* = 6 \times 10^9$ cm Hz$^{1/2}$/W at 4 Hz and a time constant of 65 ms. The active area of the bolometer was 0.92 mm^2. Brasunas and Lakew used a sapphire substrate, too, thinned to a thickness of 25 μm and covered by a buffer layer of CeO_2. The YBCO and buffer layers were made by laser ablation.

Other researchers employed different substrates such as $SrTiO_3$ [84], ZrO_2 [78], silicon nitride membranes [79], and silicon [80–83] with an epitaxial yttria stabilized ZrO_2 (YSZ) buffer layer. The YBCO microbridges (19 mm × 30 μm) on 0.5 mm thick $SrTiO_3$ substrates [84] showed $D^* = 2.3 \times 10^8$ cm Hz$^{1/2}$/W at 2 Hz while the YBCO microbolometers (2.5 μm × 2.5 μm) on ZrO_2 substrates [78] had very fast response times of about 10 μs. Their detectivity can be calculated from the responsivity (975 V/W) and noise voltage (1.5 nV/Hz$^{1/2}$) given in [78] to be about 1.5×10^8 cm Hz$^{1/2}$/W at 1000 Hz. Detectivities higher than 10^9 cm Hz$^{1/2}$/W were achieved in two other papers, one by Johnson *et al.* [79] reporting $D^* = 1.1 \times 10^9$ cm Hz$^{1/2}$/W for polycrystalline arrays on silicon nitride membranes and another paper by Neff *et al.* [83] who used $GdBa_2Cu_3O_{7-x}$ instead of YBCO yielding a maximum $D^* = 3.8 \times 10^9$ cm Hz$^{1/2}$/W, which is the best performance for a HTSC bolometer on a Si substrate. Other YBCO bolometers using Si substrates showed smaller detectivities, e.g., $D^* = 6.4 \times 10^7$ cm Hz$^{1/2}$/W at 10 Hz (NEP $= 7 \times 10^{-10}$ W/Hz$^{1/2}$) for a bolometer on a 25 μm thick substrate with a typical detector area of 0.2 mm^2 [80] and $D^* = 8 \times 10^6$ cm Hz$^{1/2}$/W (NEP $= 7 \times 10^{-9}$ W/Hz$^{1/2}$) for a bolometer with a typical detector area of 150 μm × 2 mm [82]. Moreover, in the latter work, different substrates were used instead of silicon. With a YSZ substrate

(200 μm × 1.8 mm, 0.2 μm thick) $D^* = 1.1 \times 10^8$ cm $Hz^{1/2}/W$ at 165 Hz was obtained [82].

Silicon substrates are especially attractive from a technological perspective since cost-effective batch fabrication as well as monolithic integration of signal processing electronic circuitry can be realized by using silicon wafer technology. Therefore, high-D^* HTSC bolometers based on this well-established technology will have a better chance of replacing other thermal infrared sensors or quantum detectors in numerous technical applications than other types of HTSC bolometers. From this point of view and also because of their good D^*, the bolometers described in [83] are very promising. They were fabricated on 1 μm thick single crystalline (100)-oriented Si membranes fully suspended to the surrounding silicon frame. The active area was 0.85 mm × 0.85 mm. Magnetron sputtering was employed to deposit the epitaxial $GdBa_2Cu_3O_{7-x}$ superconducting film of 50 nm thickness. The same thickness was used for the YSZ buffer layer. Figure 1.8 illustrates a comparison of the performance between these bolometers and the record performance devices published in [77]. While the maximum detectivity of the HTSC bolometers on sapphire substrates is about twice as high compared with the maximum D^* of those on Si substrates, the latter ones have much shorter response times (0.4 ms *vs.* 65 ms). Moreover, these silicon-based bolometers have a great potential for an improvement of their performance. If the present fully suspended Si membrane is replaced by a membrane structured in a meandering line with only one suspension point, thus achieving maximal thermal isolation, future D^* values will reach about 10^{10} cm $Hz^{1/2}/W$ with a change in time constant to 7 ms as thermal budget simulations have revealed [85]. The results justify the expectation that these sensors will compete successfully with the present $Cd_{1-x}Hg_xTe$ based quantum detectors, which operate at about the same temperatures as the HTSC devices.

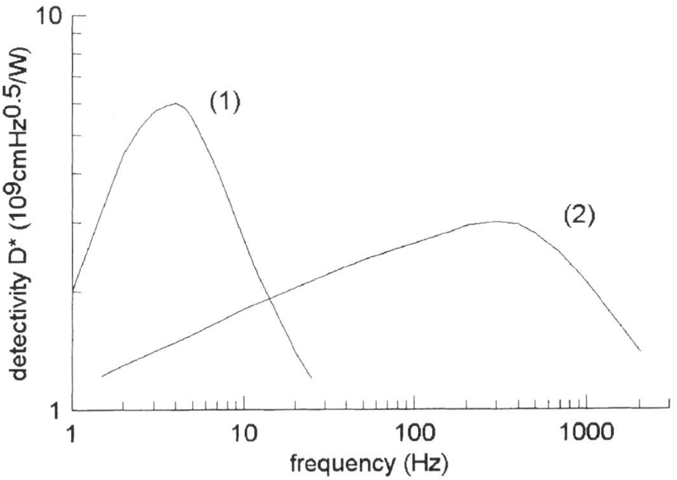

Figure 1.8. Detectivity D^* as a function of frequency for HTSC bolometers showing the highest performances to date (1) on sapphire [77] and (2) on silicon [83] substrates.

1.4 Absorber Layers for Thermal Infrared Microsensors

In contrast to quantum detectors, thermal radiation sensors have the well-known advantage of a constant wavelength-independent spectral responsivity to radiation in a very broad spectral range from the UV to the far infrared if the surface of the sensor functional layer system is able to absorb the incident radiation accordingly uniformly. In the past, several solutions have been proposed for this purpose. A detailed description of the different kinds of well-tried absorbers of optical and infrared radiation is given in [86].

On the one hand, there are some cavity absorbers (e.g., with an absorbance A of 0.99998) and cone absorbers (e.g., with $A = 0.99988$) [86], which are preferably used for precision radiometers; on the other hand, for thermal infrared microsensors, mainly disk absorbers covered by an efficient absorbing layer are currently used. Refs. [87] and [38] offer new summarizing overviews on absorbing layers. Generally one has to distinguish between paints and lacquers [86] (usually used for low-cost sensors, only), metal blacks (porous films), and thin metal films, which are now widespread assembled as interferometric ('quarter-wavelength') structures showing an efficient absorbance in a small spectral range only.

The most effective mechanism for wavelength-independent radiation absorption is high porosity [87]. Such a structure leads to a very high absorption coefficient from UV to FIR since particles with sizes much smaller than the wavelength of radiation generally absorb and diffract it. Layers deposited by galvanic electroplating are characterized typically by dendritic growth. To obtain an absorption better than 95%, e.g., a Pt film of 1.5 mg/cm^2 is needed. A more common procedure for obtaining porous layers is the evaporation of a metal (mainly Ag, Au, Bi, Pt) in a low-pressure atmosphere of nitrogen, argon or another inert gas. Due to the collision of the evaporated atoms with the gas molecules, the metal atoms loose their kinetic energy and are slowed down to thermal speed. Thus the atoms stick at their point of impact with the surface and set up a conglomerate structure of fine ($\sim$25 nm) needles. At present, silver and gold blacks are employed most frequently. Silver films ($A > 99\%$ at 80 $\mu g/cm^2$) show a higher absorbance at the same mass as gold films ($A > 99\%$ at 500 $\mu g/cm^2$). In order to keep the time constant of the sensor low, the mass of the absorbing layer has to be small. In such cases, the application of silver black is especially advantageous. Recently, silver blacks have been examined in detail [88] and it was found that they are a very eligible absorber in the wavelength range from $\sim$200 nm up to $\sim$200 μm. An observed very small reflectance peak near 322 nm is theoretically explained in [88]. In spite of their ideal absorption features, unfortunately metal blacks also have disadvantages. Due to their fragility and low robustness to mechanical or chemical influences, they are patterned by evaporation through punched masks and they also cause technical problems during the mounting process. Therefore, especially for large-scale manufacturing, alternative solutions are of interest, which provide absorbing layers of higher robustness easy to be patterned.

In thermal microsensors based on CMOS IC technology combined with post-processing bulk or surface micromachining, an absorber consisting of a more rugged thin-film sandwich (SiO_2/Si_3N_4, Al, SiO_2) is used. For certain spectral bands no additional material enhancing IR absorption is needed since the passivation layer provided by the CMOS process shows strong absorption bands between 8 and 14 μm [89]. The CMOS process also allows the inclusion a metal layer in the absorbing area underneath the passivation layer for better absorption and thermal equalization. Thus, the incoming radiation is reflected at the metal mirror and passes the absorbing layer twice. Moreover, the spectral absorbance can be increased by structuring such absorber layers [26, 43, 90, 91], too, e.g., in the 8–14 μm band to 32% instead of 25% for unstructured layers [91].

Another class of IR wavelength absorber layers known for many years (cf., e.g., [87, 92]) are metal thin films. In a very narrow range of their sheet resistance ($R_{\square opt} = Z_0/2 = 188.5\ \Omega/\square$) freely suspended metal films can absorb independent of wavelength up to 50% of the incident radiation over the whole IR range above about 5 μm. Z_o is the free space impedance of the vacuum, $Z_0 = (\mu_0/\epsilon_0)^{1/2} = 377\ \Omega$. Such metallic films are deposited by evaporating or sputtering, e.g., bismuth or gold (optimum gold thickness: 17 nm $\sim$ 34 $\mu g/cm^2$) [87]. Prepared on the solid surface of a sensor material, the absorbing properties of a metal layer are, however, strongly influenced by the dielectric function of this material. In [93] a thermoelectric sensor is described where a 400 nm thick silicon layer is used as a support for an absorbing silver film with $R_\square = 150\ \Omega/\square$ resulting in a nearly 50% wavelength-independent absorbance. Besides it has been shown in [93] that the absorbance of a thin metal film broad-band absorber structure is strongly influenced by the refraction index n of the used substrate so that an appropriate selection of the substrate is essential (for substrates with $n \leq 3$ the broad-band absorbance is smaller than 50%).

Enhancing the absorbance of thin metal films to values > 50% is not possible for the whole IR region, but for selected smaller spectral regions. The results of several such investigations have been reviewed in [38]. A theory of absorption in thin-film thermal IR sensors was introduced in [94] for the case where the heat-sensitive (bolometric) component of the sensor is an integral component of a three-layer absorber stack. It is predicted in [94] that a maximum absorption of more than 95% can be achieved when the sheet resistance of an additional rear surface reflector film (e.g., 100 nm NiCr) is $\leq 10\ \Omega/\square$ and the resistance of a front surface film is in the range $300–600\ \Omega/\square$ (e.g., 5 nm Pt or 25 nm Au).

The concept of an 'up to 100% absorption' near a wavelength λ_0 by metal-dielectric-metal (so called 'quarter-wavelength') structures is reexamined in [95]. A metal film (e.g., Ni or Au) at the top is impedance matched to free space, so that 50% of the incident radiation is absorbed and 50% transmitted with none reflected. The transmitted part goes then through a quarter-wave dielectric layer and is reflected by the lower metal film (e.g., Pt or Au) and destructively interferes, giving zero reflection and 100% absorption for λ_0 in an ideal case. In [96] such a quarter-wavelength structure concept is discussed in order to optimize sensitivity and speed of a large bolometric sensor array with $> 128 \times 128$ pixel elements. Therefore, the requirement of a low thermal mass is considered in particular. This concept uses surface micromachining as the method of choice. The structure consists of a 200 nm Au (on Cr) reflector, an air (or vacuum) gap of 2.9 μm produced by removing a polyimide

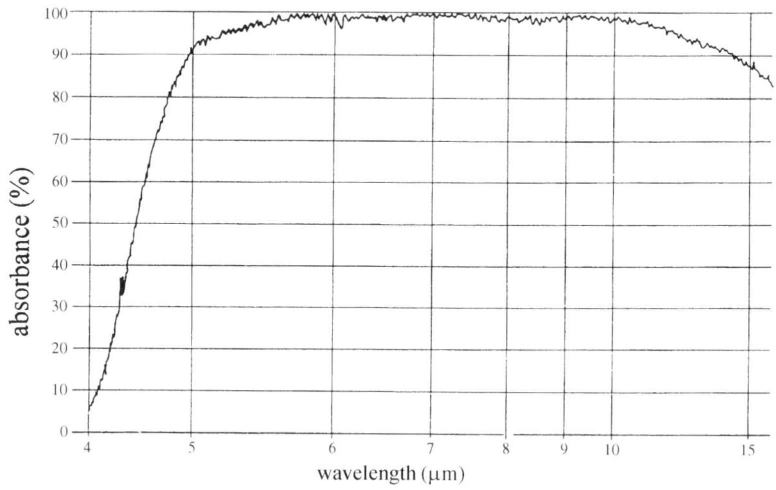

Figure 1.9. Absorbance of an interferometric absorber structure from the IPHT, Jena [38].

sacrificial layer, a bottom dielectric layer (SiN_x), the bolometric metal film of sheet resistance $R_{\square} = 300–400\,\Omega/\square$ (e.g., 5 nm $MoSi_2$) and a top dielectric layer. Good agreement between experimental and theoretical absorbances was obtained in the full wavelength range 2–22 μm with a 95% to ~100% absorbance in the 8–12 μm range.

Finally, Figure 1.9 shows the absorbance of a similar composed interferometric structure, which was developed at the IPHT [38] and which has been used there within a serial production of thermopile sensors for some years. This interferometric absorber structure is extremely resistant to environmental influences and can be handled without any problems. Besides, it is usable up to temperatures of about 180 °C.

Consequently, there is much indication that interferometric structures are today's most promising absorber layers concerning a sensor with a selective sensitivity, e.g., in the 7–15 μm region, needed not only for such widespread customer applications as fire alarms or intruder detectors, but also for sophisticated remote temperature measurement devices.

1.5 Outlook: Trends in Infrared Microsensor Market and Application

The development of thermoelectric and bolometric infrared microsensors has reached a high technological level. The main characteristics such as detectivity or NEP will approach closer and closer to their theoretical limits in the future. Consequently, their distance to the superior cooled quantum radiation detectors is decreasing markedly while their attractivity is increasing on account of the possibility of low-cost mass production and the unnecessity of cooling. Further specialization will be the way of future developments of these sensors. Improved device layouts and material combinations are expected to result in better performance of these devices.

Thermopiles are particularly suited for simple fabrication of IR sensors. There are many applications for such devices ranging from all non-contact measurements of surface temperature (e.g., pyrometry within the scope of industrial process monitoring and controlling) and gas analysis in medical, automotive-exhaust emission control, and ecological equipment to temperature-controlled microwave cooking, toasters, ovens, and other heating systems and, finally, also to security systems such as intrusion and fire alarms as well as presence monitors. In such cases, the improved performance and reliability together with the reduced cost associated especially with micromechanical silicon thermopile sensors open the door to many high-volume sensor applications. In diverse scientific fields thermal infrared microsensors become useful tools, too, especially for in-situ and in-line monitoring of CO_2 laser beam diagnostics and in handy infrared spectrometers built for outdoor use as emission monitors for liquids and gases. Some sophisticated models of thermopile and bolometric sensors are used in space experiments.

The development of thermal FPAs has shown that their pixel sensitivity is far less than that of photon detection arrays. A large thermal FPA could exhibit equal or better sensitivity than a small photon detection FPA. The availability of compact thermal imagers without scanning optics on the basis of cooled FPAs is a commercial reality. Some manufacturers offer moderately priced (less than $50,000) off-the-shelf cooled infrared FPA imagers. In the foreseeable future, truly low-cost uncooled thermal imaging instruments should be available. The SnapShot™ and 'Sentinel' devices are the first examples of infrared cameras equipped with uncooled multi-element sensors, the latter without mechanical scanner. The result of price reduction is that thermal imagers move from the military to the marketplace.

1.6 References

[1] Baltes, H., Moser, D., Völklein, F., in: *Sensors–A Comprehensive Survey*, Göpel, W., Hesse, J., Zemel, J. N. (eds.), vol. 7: *Mechanical Sensors*; Weinheim: VCH, 1994, Chapter 2.

[2] Boyer, A., Cissé, E., *Materials Science and Engineering, B* **13** (1992), 103–111.

[3] Svechnikova, T. E., Maksimova, N. M., Konstantinov, P. P., Alekseeva, G. T., Korzhuev, M. A., *Inorganic Materials*, **29** (1993), 1442–1445.

[4] Paul, O. M., Korvink, J., Baltes, H., *Sensors and Actuators, A* **41–42** (1994), 161–164.

[5] von Arx, M., Paul, O., Baltes, H., *Sensors and Actuators, A* **46–47** (1995), 428–431.

[6] Paul, O., von Arx, M., Baltes, H., in: *Transducers 95 Digest of Technical Papers*, Royal Swedish Academy of Engineering Sciences, Stockholm (1995), pp. 178–181.

[7] Völklein, F., Baltes, H., *J. Microelectromech. Syst.*, **1** (1993), 193–196.

[8] Dehé, A., Fricke, K., Hartnagel, H. L., *Sensors and Actuators, A* **46–47** (1995), 432–436.

[9] Hicks, L. D., Dresselhaus, M. S., *Phys. Rev., B* **47** (1993), 12727–12731.
[10] Sofo, J. O., Mahan, G. D., *Appl. Phys. Lett.*, **65** (1994), 2690–2692.
[11] Broido, D. A., Reinecke, T. L., *Phys. Rev., B* **51** (1995), 13797–13800.
[12] Whitlow, L. W., Hirano, T., *J. Appl. Phys.*, **78** (1995), 5460–5466.
[13] Elbel, T., Lenggenhager, R., Baltes, H., *Sensors and Actuators, A* **35** (1992), 101–106.
[14] Völklein, F., Baltes, H., *Sensors and Actuators, A* **36** (1993), 65–71.
[15] Dillner, U., *Sensors and Actuators, A* **41–42** (1994), 260–267.
[16] Dillner, U., in: *Proceedings of TEMPMEKO '93*; IMEKO Prague: (1993), pp. 284–288.
[17] Dillner, U., Keßler, E., Klonz, M., Thomes, C., in: *Proceedings of Micro System Technologies '94*, H. Reichl, A. Heuberger (eds.); VDE Verlag GmbH Berlin: (1994), pp. 773–782.
[18] Dillner, U., Keßler, E., Müller, J., in: *Tagungsband der 2. Chemnitzer Fachtagung Mikrosystemtechnik*; Technical University of Chemnitz-Zwickau (1995), pp. 293–299.
[19] Kriegl, W., Steiner, P., Folkmer, B., Lang, W., *Sensors and Actuators, A* **46–47** (1995), 637–639.
[20] Jeaggi, D., *Ph.D. Thesis*, Zurich (1996).
[21] Korvink, J. G., Baltes, H., in: *Sensors Update*, Baltes, H., Göpel, W., Hesse, J. (eds.), **vol. 2**; Weinheim: VCH, (1996), pp. 181–209.
[22] Hartung, C., Müller, J., in: *Selected Papers on Temperature Sensing: Optical Methods*, Lucier, R. D. (ed.); SPIE Milestone Series MS 116, Bellingham, Washington, USA (1995), pp. 123–128.
[23] Müller, J., Elbel, T., Poser, S., Völklein, F., *Feingerätetechnik*, **40** (1991), 8–9.
[24] Sun, T., Zhou, N., *J. Infrared Millim. Waves*, **13**(2) (1994), 157–160.
[25] Lenggenhager, R., Baltes, H., Elbel, T., *Sensors and Actuators, A* **37–38** (1993), 216–220.
[26] Müller, M., Gottfried-Gottfried, R., Kück, H., Mokwa, W., *Sensors and Actuators, A* **41–42** (1994), 538–541.
[27] Lenggenhager, R., *Ph.D. Thesis*, Zurich (1994).
[28] Lenggenhager, R., Jeaggi, D., Malcovati, P., Duran, H., Baltes, H., Doering, E., in: *IEDM '94 Technical Digest*; Piscataway: IEEE (1994), pp. 531–534.
[29] Baltes, H., Paul, O., Jeaggi, D., in: *Sensors Update*, Baltes, H., Göpel, W., Hesse, J. (eds.), vol. 1; Weinheim: VCH, (1996), pp. 121–142.
[30] Lang, W., Kühl, K., in: *2. Symp. Mikrosystemtechnik*; Regensburg: OTTI-Technologie-Kolleg (1992), pp. 138–143.
[31] Lu, J., Xiong, B., Lin, C., *Sensors and Actuators, A* **35** (1993), 217–220.
[32] Baltes, H., Paul, O., *Sensors and Materials*, **8** (1996), 409–421.
[33] Schieferdecker, J., Quad, R., Holzenkämpfer, E., Schulze, M., *Sensors and Actuators A* **46–47** (1995), 422–427.
[34] Schmidt, W., *Laser Focus World* **31** (1995) (Aug), 77–79.
[35] Srinivas, T. A. S., Timans, P. J., Butcher, R. J., Ahmed, H., *Rev. Sci. Instrum.*, **64** (1993), 3602–3611.
[36] Chong, N., Srinivas, T.A.S., Ahmed, H., in: *Transducers 95 Digest of Technical Papers, Late News Contr.*; Royal Swedish Academy of Engineering Sciences, Stockholm (1995), pp. 57–58.
[37] *Infrared Flame Detection,* Technology Brief of THORN EMI Central Res. Labs. Ltd., Hayes, UK (1993).
[38] Müller, J., Dillner, U., Güttich, R., Keßler, E., Poser, S., Ratz, P., *Wiss. Z. Techn. Univ. Dresden*, **43**(6) (1994), 15–19.
[39] Keßler, E., Müller, J., Völklein, F., Pawlak, M., Möstl, K., Brandt, F., *Proc. SPIE*, **1712** (1992), 132–141.

[40] Keßler, E., Holstenberg, H.-C., Pawlak, M., Möstl, K., Metzdorf, J., in: *Innovationen in der Mikrosystemtechnik*, **vol. 26**, *Thermoelektrische Mikrosensoren*; Teltow: VDI/VDE, (1995), pp. 107–128 and to be published in: *Proceedings of TEMPMEKO '96*, Turin: (1996).
[41] Elbel, T., Just, H.-J., Poser, S., Wecke, N., *VDI Berichte*, **982** (1992), 63–70.
[42] Choi, I. H., Wise, K. D., *IEEE Trans. Electron Devices*, **33** (1986), 72–79.
[43] Müller, M., Budde, W., Gottfried-Gottfried, R., Hübel, A., Jähne, R., Kück, H., in: *Transducers 95 Digest of Technical Papers,* vol. 2; Royal Swedish Academy of Engineering Sciences, Stockholm, (1995), pp. 640–643.
[44] Schieferdecker, J., Quad, R., Schulze, M., in: *Proceedings of Sensors – Technology and Application*; Bad Nauheim (1994) pp. 585–590.
[45] Schnelle, W., Dillner, U., Poser, S., *VDI Berichte*, **982** (1992), 261–264.
[46] Baer, W. G., Najafi, K., Wise, K. D., Toth, R. S., *Sensors and Actuators, A* **48** (1995), 47–54.
[47] *Preliminary Data Sheet of 32-Element-Silicon Thermopile Linear Array*; Olching, Germany: Laser Components GmbH.
[48] Kruse, P. W., *Photonics Spectra*, **29** (3) (1995), 103–108.
[49] *Data Sheet of IR SnapShot^{TM}*, USA: Infrared Solutions, Inc.
[50] Kanno, T., Saga, M., Matsumoto, S., Uchida, M., Tsukamoto, N., Tanaka, A., Itoh, S., Nakazato, A., Endoh, T., Tohyama, S., Yamamoto, Y., Murashima, S., Fujimoto, N., Teranishi, N., *Proc. SPIE*, **2269** (1994), 450–459.
[51] Tanaka, A., NEC Corp., Patent EP 0 613 185 A1 (1994).
[52] Oliver, A. D., Baer, W. G., Wise, K. D., in: *Transducers 95 Digest of Technical Papers*, vol. 2; Royal Swedish Academy of Engineering Sciences Stockholm: (1995), pp. 636–639.
[53] Elbel, T., Poser, S., Fischer, H., *Sensors and Actuators, A* **41–42** (1994), 493–496.
[54] Rossberg, D., *Sensors and Actuators, A* **46–47** (1995), 413–416.
[55] Lengfellner, H., Krumb, G., Schnellbögel, A., Betz, J., Renk, K. F., Prettl, W., *Appl. Phys. Lett.* **60** (1992), 501–503.
[56] Renk, K. F., Betz, J., Zeuner, S., Lengfellner, H., Prettl, W., *Physica C* **235–240** (1994), 37–40.
[57] Habermeier, H.-U., Jisrawi, N., Jäger-Waldau, G., in: *Applied Superconductivity 1995, Proceedings of EUCAS 1995,* vol. 2, Dew-Hughes, D. (ed.); Bristol, Philadelphia: IOP Publishing Ltd., (1995), pp. 1023–1026.
[58] Steiner, P., Lang, W., *Thin Solid Films* **255** (1995), 52–58.
[59] Lang, W., Steiner, P., Schaber, U., Richter, A., *Sensors and Actuators, A* **43** (1994), 185–187.
[60] Shie, J.-S., Weng, P. K., *Sensors and Actuators, A* **33** (1992), 183–189.
[61] Kudoh, T., Ikebe, S., Satow, H., Komatsu, K., Kimura, M., *Sensors and Actuators, A* **55** (1996), pp. 13–17.
[62] Neff, H., Laukemper, J., Hefle, G., Burnus, M., Heidenblut, T., Michalke, W., Steinbeiss, E., *Appl. Phys. Lett.*, **67** (1995), 1917–1919.
[63] Shan, P. C., Celik-Butler, Z., Butler, D. P., Jahanzeb, A., *J. Appl. Phys.*, **78** (1995), 7334–7339.
[64] Wood, R. A., *Proc. SPIE*, **2020** (1993), 322-329.
[65] Liu, M. S., Haviland, J. S., Yue, Ch. J., Honeywell Corp., Patent US 5 260 225, (1993).
[66] Wood, R. A., Honeywell Corp., Patent WO 94/00950 (1994).
[67] *Data Sheet of 240 × 320 Silicon Micro-Bolometer Uncooled Focal Plane Array*; Goleta, USA: Hughes SBR; *Data Sheet of U 3000 Uncooled Microbolometer Infrared Sensor*; Anaheim, USA: Rockwell Int. Corp.

[68] Liddiard, K. C., Unewisse, M. H., Reinhold, O., *Proc. SPIE*, **2225** (1994), 62–71.
[69] Baliga, S., Rost, M., Doctor, A., *Proc. SPIE*, **2225** (1994), 72–78.
[70] Eriksson, P., Andersson, J. Y., Stemme, G., *Journ. of Micro Electro Mechanical Systems*, **6** (1997), 55–61.
[71] Jerominek, H., *NOI Bulletin*, **3** (1) (1992), 4.
[72] Jerominek, H., *NOI Bulletin*, **6** (2) (1995), 1–3.
[73] Tanaka, A., Matsumoto, S., Tsukamoto, N., Itoh, S., Endoh, T., Nakazato, A., Kumazawa, Y., Hijikawah, M., Gotoh, H., Tanaka, T., Teranashi, N., in: *Transducers 95 Digest of Technical Papers*, vol. 2; Royal Swedish Academy of Engineering Sciences, Stockholm (1995), pp. 632–635.
[74] Stout, A., *Photonics Spectra*, **30**(3) (1996), 86–88.
[75] *Data Sheet of Sentinel*; Goleta, USA: Amber Engineering, Inc.
[76] Verghese, S., Richards, P. L., Char, K., Sachtjen, S. A., *IEEE Trans. Magn.*, **27** (1991), 3077–3080.
[77] Brasunas, J. C., Lakew, B., *Appl. Phys. Lett.*, **64** (1994), 777–778.
[78] Li, K., Johnson, J. E., Aker, B. W., *J. Appl. Phys.*, **73** (1993), 1531–1533.
[79] Johnson, B. R., Ohnstein, T., Han, C. J., Higashi, R., Kruse, P. W., Wood, R. A., Marsh, H., Dunham, S. B., *IEEE Trans. Appl. Supercond.*, **3** (1993) 2856–2859.
[80] Li, Q., Fenner, D. B., Hamblen, W. D., Hamblen, D. G., *Appl. Phys. Lett.*, **62** (1993), 2428–2430.
[81] Rice, J. P., Grossman, E. N., Rudman, D. A., *Appl. Phys. Lett.*, **65** (1994), 773–775.
[82] Tong, F. M., Ravindra, N. M., Ganapathi, L., Giles, S., Rao, R., *Infrared Phys. & Technol.*, **36** (1995), 1053–1058.
[83] Neff, H., Laukemper, J., Khrebtov, I. A., Tkachenko, A. D., Steinbeiss, E., Michalke, W., Burnus, M., Heidenblut, T., Hefle, G., Schwierzi, B., *Appl. Phys. Lett.*, **66** (1995), 2421–2423.
[84] Huang, C. J., Chang, C. Y., Tseng, T. Y., *J. Appl. Phys.*, **72** (1992), 5786–5791.
[85] Berg, M., Laukemper, J., Neff, H., Steinbeiss, E., Michalke, W., Burnus, M., Heidenblut, T., *IEEE Trans. Appl. Supercond.*, **5** (1995), 2435–2438.
[86] Möstl, K., in: *Absolute Radiometry*, F. Hengstberger (ed.); New York: Academic Press, (1989), Chapter 2.
[87] Lang, W., Kühl, K., Sandmaier, H., *Sensors and Actuators, A* **34** (1992), 243–248.
[88] Neff, H., Henkel, S., Sass, J. K., Steinbeiss, E., Ratz, P., Müller, J., Michalke, W., *J. Appl. Phys.*, **80** (2) (1996), 1058–1062.
[89] Lenggenhager, R., Baltes, H., Peer, J., Forster, M., *IEEE Electron Device Lett.*, **13** (1992), 454–456.
[90] Müller, M., Köhler, R., Gottfried-Gottfried, R., Kück, H., Mokwa, W., Vogt, H., in: *Proceedings SENSOR 93,* vol. III, ACS Organisations GmbH Nürnberg: (1993), pp. 163–170.
[91] Schneeberger, N., Paul, O., Baltes, H., in: *Transducers 95 Digest of Technical Papers*, vol. 2; Royal Swedish Acedemy of Engineering Sciences Stockholm (1995), pp. 648–651.
[92] Hadni, A., Gerbaux, X., *Infrared Physics*, **30** (1990), 465–478.
[93] Bauer, S., Bauer-Gogonea, S., Becker, W., Fettig, R., Ploss, B., Ruppel, W., *Sensors and Actuators, A* **1993**, 37–38, 497–501.
[94] Liddiard, K. C., *Infrared Physics*, **34** (1993), 379–387.
[95] Monzón, J. J., Sánchez-Soto, L. L., *Applied Optics*, **33** (1994), 5137-5141.
[96] Eriksson, P., Andersson, J. Y., Stemme, G., *Physica Scripta*, **T54** (1994), 165–168.

2 Fuzzy Logic and Neuro-Systems Assisted Intelligent Sensors

M. Theisen, A. Steudel, M. Rychetsky, M. Glesner, Darmstadt University of Technology, Department of Electrical Engineering and Information Technology, Institute of Microelectronic Systems, Karlstraße 15, D-64283 Darmstadt

Contents

2.1 Fuzzy Systems

2.1.1 Motivation

Fuzzy logic is a superset of conventional Boolean logic where we have the concept of *partial truth* instead of just the two truth values 1 (`TRUE`) and 0 (`FALSE`). It was developed in 1965 by Lotfi Zadeh at the University of California (Berkeley) to model the uncertainties which are inherent to the human language [1].

Today fuzzy logic is used in all areas of information and automation technology where it is useful to model appropriate behavior by concepts of natural language [2]. In expert knowledge, diagnosis, control, analysis, and classification systems it is often much more convenient to use terms like 'high', 'medium', and 'low' instead of crisp concepts like 'greater than 60%', 'between 40% and 60%', and 'less than 40%'.

The decision to use fuzzy methods for the construction of technical systems is made whenever there is (fuzzy) expert knowledge available or complex dependencies have to be modeled transparently. Fuzzy systems allow one to build up an easy human–machine interface and can reduce the expense when developing complex non-linear systems.

In the following sections basic mathematical and set-theoretic operations for fuzzy sets and fuzzy rule-based systems are shown. More comprehensive foundations can be found in [3, 4]. Fuzzy rule-based systems are discussed in [5, 6] and fuzzy controllers in [7, 8].

2.1.2 Foundations

2.1.2.1 Fuzzy Rule-based Systems

Fuzzy rule-based systems implement algorithms by `IF-THEN` rules which are defined in terms of the human language (see Figure 2.3).

For the evaluation of a fuzzy rule the 'real-world' inputs are converted into similarity measures of certain fuzzy sets. This process is called *fuzzification*. The results of the rules is then calculated by *forward chaining*. This process is often referred to as *inference*. Afterwards the results of the different rules are combined to a fuzzy main result by *composition*. Most applications call for crisp output values. Thus, the fuzzy result of the rule-based system has to be *defuzzified* (Figure 2.1).

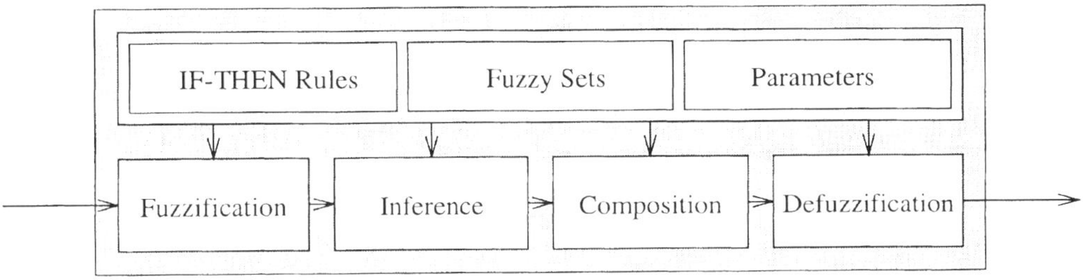

Figure 2.1. The parts of a fuzzy rule-based system.

2.1.2.2 Rule-base

The rule-base contains the knowledge which is necessary for the description of the system [9]. Fuzzy rule-based systems contain rules of the form

$$\text{IF } < \text{premise} > \text{ THEN } < \text{conclusion} > .$$

The premise is an expression that results from the comparison between the input x_i and the fuzzy set $\tilde{A}_j$ or a combination of these comparisons by the operation $\wedge$ (conjunction) or $\vee$ (disjunction). The order of evaluation is determined by the common conventions but can always be modified using brackets.

A conclusion is usually the assignment of a fuzzy value $\tilde{B}_l$ to the fuzzy output $\tilde{Y}_k$.

2.1.2.3 Fuzzy Sets and Linguistic Expressions

The linguistic terms in `IF-THEN` rules are represented by fuzzy sets [1]. Classical (crisp) sets A are defined by their characteristic function μ_A over a domain $\mathscr{X}$.

$$\forall x \in \mathscr{X}, \quad \mu_A(x) = \begin{cases} 1 & \text{if } x \in A \\ 0 & \text{if } x \notin A. \end{cases} \tag{2.1}$$

To represent a fuzzy set $\tilde{A}$, the range of values of the characteristic function is mapped on to the unit interval [0, 1]. The characteristic function becomes the *membership function* $\mu_{\tilde{A}}$.

$$\mu_{\tilde{A}}(x) : \mathscr{X} \to [0, 1]. \tag{2.2}$$

The borders of the interval are the same as with crisp sets (2.1).

$$\forall x \in \mathscr{X}, \quad \mu_{\tilde{A}}(x) = \begin{cases} 1 & \text{if } x \text{ is completely contained in } \tilde{A} \\ 0 & \text{if } x \text{ is completely not contained in } \tilde{A} \\ (0, 1) & \text{else.} \end{cases} \tag{2.3}$$

Linguistic expressions can have very different semantic meanings [10]. They can be used to represent a degree of satisfaction, necessity, certainty, or limitation. Figure 2.2 shows examples for different membership functions of crisp and fuzzy sets.

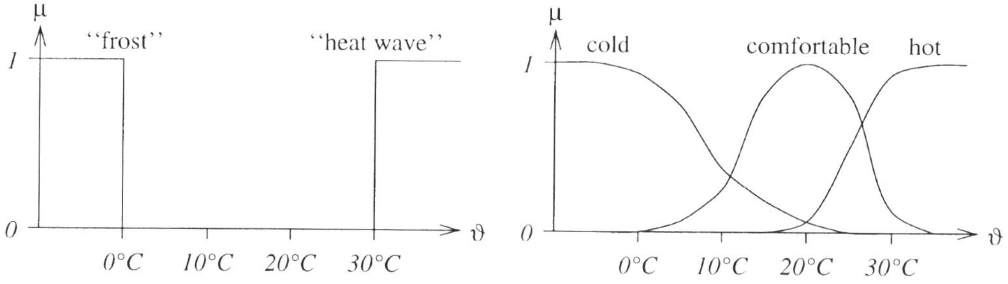

Figure 2.2. Membership functions of crisp and fuzzy sets.

The union of the definitions for the membership functions, the rule-base, and additional parameters is called the *knowledge base*. The knowledge of a rule-base can be acquired by eliciting experts or it can be generated by a computer using some form of automatic learning strategy.

2.1.2.4 Operations on Fuzzy Sets

For operations in fuzzy rule-based systems the membership value of an element $x \in \mathscr{X}$ in the fuzzy result set is unambiguously determined by the membership values of element x belonging to the fuzzy sets the operation is performed on. That is, the membership value of elements belonging to the result can be expressed as a function of the input fuzzy sets:

$$\mu_{F(\tilde{A},\,\tilde{B})}(x) = f(\mu_{\tilde{A}}(x), \mu_{\tilde{B}}(x)). \tag{2.4}$$

The most commonly used operations on membership values are the triangular norms (*t-norms*), $\wedge_t$ and the triangular conorms (*t-conorms*). A function $\wedge_t$: $[0, 1]^2 \rightarrow [0, 1]$ is called a t-norm if the following holds:

$$\text{zero element} \quad 0 \wedge_t 0 = 0 \tag{2.5}$$

$$\text{unit element} \quad a \wedge_t 1 = 1 \wedge_t a = a \tag{2.6}$$

$$\text{monotonicity} \quad a \leq c, b \leq d \Rightarrow a \wedge_t b \leq c \wedge_t d \tag{2.7}$$

$$\text{commutativity} \quad a \wedge_t b = b \wedge_t a \tag{2.8}$$

$$\text{associativity} \quad (a \wedge_t b) \wedge_t c = a \wedge_t (b \wedge_t c). \tag{2.9}$$

For each t-norm $\wedge_t$ there is a corresponding *t-conorm* $\vee_t$:

$$a \vee_t b = 1 - ((1 - a) \wedge_t (1 - b)). \tag{2.10}$$

The strongest t-norm is the minimum and the weakest t-conorm is the maximum:

$$a \wedge_t b \geq \min(a, b) \tag{2.11}$$

$$a \vee_t b \leq \max(a, b). \tag{2.12}$$

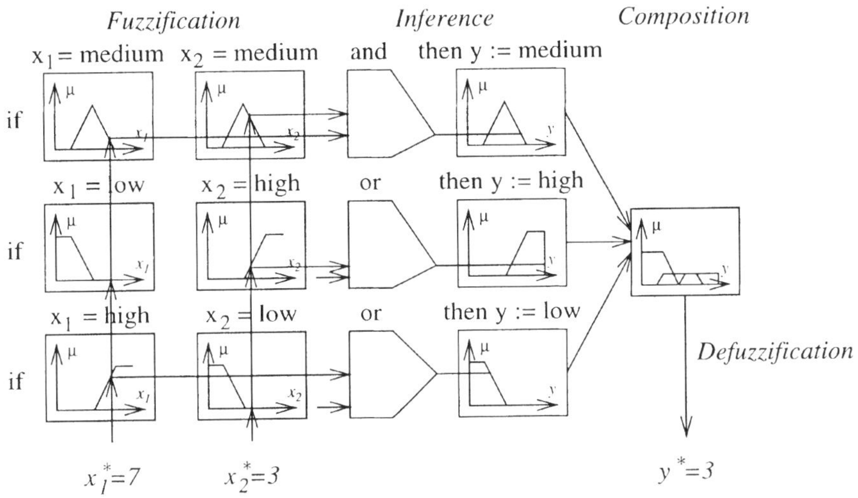

Figure 2.3. Data flow in a fuzzy rule-based system.

Minimum, maximum and one mixed form are the only t-norms and t-conorms which are idempotent [11]:

$$\begin{aligned} a \circ a = a \Leftrightarrow a \circ b &= \min(a,b) \\ \vee\, a \circ b &= \max(a,b) \\ \vee\, a \circ b &= \begin{cases} \min(a,b) & \text{if } a,b \geq x \\ \max(a,b) & \text{if } a,b \leq x \\ x & \text{else} \end{cases} \end{aligned} \tag{2.13}$$

Mathematical derivations for minimum and maximum can be found in [12, 13]. Besides, minimum and maximum are well-suited for practical purposes because they can be evaluated very efficiently.

Figure 2.3 shows how the different parts fuzzification, inference, composition, and defuzzification work together in a fuzzy rule-based system.

2.1.3 Driving Support for Vehicles

Driving a vehicle with a trailer in the reverse direction is a difficult task for a beginner. When the trailer comes to a position where the angle between the longitudinal axes of the vehicle and the trailer cannot be increased even with the maximum angle of the steering wheel in reverse driving, then the driver has to change gear and drive forward in order to avoid further bending and damage of the truck (critical angle). Experienced drivers are able to react early enough by decreasing the angle of the

Figure 2.4. The truck and trailer model.

steering wheel when approaching the critical bending angle. This expert knowledge of continuous reverse driving can be formulated using fuzzy rules and can be integrated in an automatic driving support [14]. The driver's steering proposal and the current angle between truck and trailer are measured by two inexpensive sensors. The fuzzy rule-based system is feasible to represent the expert knowledge and calculate an appropriate front wheel angle for the truck.

2.1.3.1 Set-up

For practical reasons the real-time reverse driving support was implemented and tested on a model truck (52.2 cm long, 18 cm width, and 23.5 cm height) and a trailer (67.8 cm long, 18.7 cm width, and 20 cm height) shown in Figure 2.4.

The steering proposal of the driver is transmitted from the remote control unit (RC) to the microcontroller as a pulse width modulated signal (Figure 2.6). The angle between truck and trailer is measured by a simple circular potentiometer which is wired as a voltage divider. Its voltage is proportional to the angle and is fed to the microcontroller. The output signal of the driving support is converted to a pulse width modulated signal and is used to control a servo that influences the angle of the truck's front wheels. The controller board (Figure 2.5) contains the microcontroller μPD78310A by Motorola which is used for the conversion of the pulse width modulated signals, the processing of the potentiometer voltage, and the serial transmission of programs and data. Furthermore the board contains a 1-kByte EEPROM chip and two 32 kByte RAM chips for program code and data.

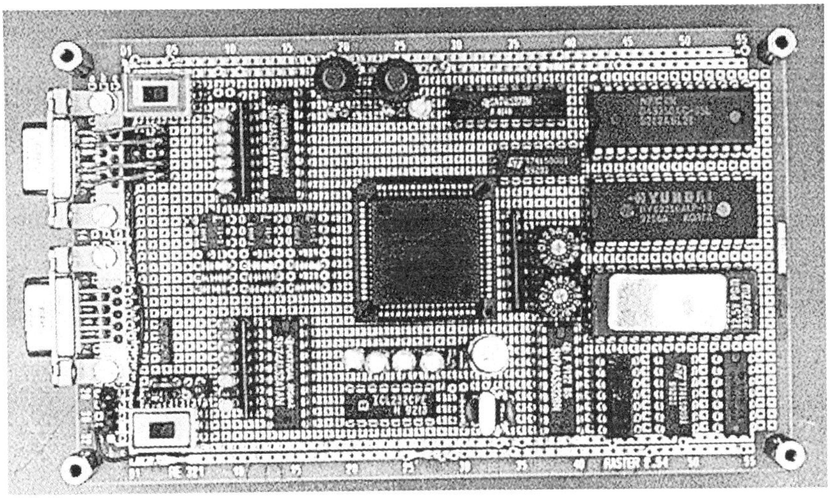

Figure 2.5. The microcontroller board of the driving support.

2.1.3.2 Fuzzy Rule-based System for Driving Support

The fuzzy rule-based system for the reverse driving support was developed using the FPL (*fuzzy programming language*) of Togai InfraLogic. The source code was edited and complied on a standard PC. Afterwards the machine code was transmitted to the controller board via a serial interface.

The controller board has two inputs (steering proposal and angle of trailer) and one output (angle of the front wheels of the truck).

For simplicity the membership functions for input and output signals are piecewise linear functions. The steering proposal has the same semantic meaning as the angle of

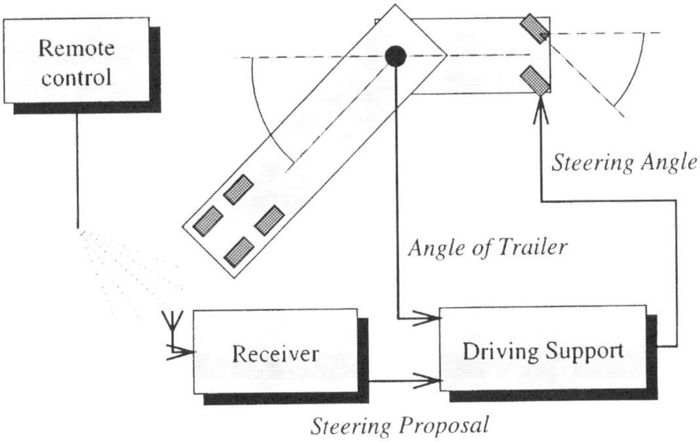

Figure 2.6. Schematic of the driving support.

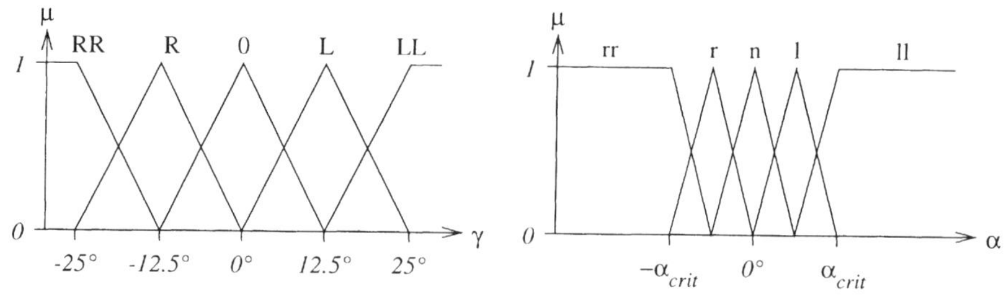

Figure 2.7. Fuzzy sets and membership functions of the driving support.

		Angle of Trailer				
		rr	r	n	l	ll
Steer-	RR	RR	RR	RR	L	LL
ing	R	RR	RR	R	L	LL
	0	RR	RR	0	LL	LL
Pro-	L	RR	R	L	LL	LL
posal	LL	RR	R	LL	LL	LL

Figure 2.8. Rule-base of the driving support.

the front wheels and so the same set of membership functions is used. The range of values is about $-25°$ to $+25°$. Angles of approximately $0°$ are represented by the fuzzy set 'straight ahead' (0) which has its maximum value at $0°$ (see Figure 2.7, left). The angle between truck and trailer is also split up into five different membership functions (Figure 2.7, right). The critical fuzzy sets (rr and ll) are parametrized by the estimated critical angle α_{crit} which was determined experimentally. The membership values for angles larger $\pm\,\alpha_{\text{crit}}$ are equal to one.

All $5 \cdot 5 = 25$ possible combinations between the input values and the corresponding rules can be seen in Figure 2.8. When the angle between truck and trailer is critical then the output angle of the driving support has to steer extremely right or left, respectively (first and last column of Figure 2.8). On the other hand, when the angle between truck and trailer is small then the direction of the front wheels of the truck should correspond to the steering proposal (middle column of Figure 2.8). The remaining columns between these extremes are interpolated.

2.1.3.3 Results

The real-time implementation of the fuzzy controller allowed us to perform several interesting experiments where the fuzzy support for reverse driving of the truck with a long trailer proved to be very efficient.

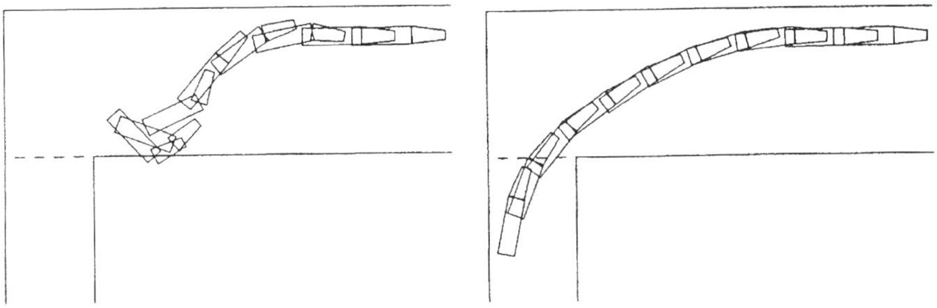

Figure 2.9. Reverse driving without (left) and with (right) the fuzzy support.

One experiment was the competition between an inexperienced driver without the fuzzy steering support and an inexperienced driver with the fuzzy steering support. In most cases the driver without support ended up in a deadlock (Figure 2.9, left) whereas the driver with fuzzy steering support was able to reverse without problems (Figure 2.9, right). Beginners as well as driving experts were able to improve their reverse driving performance significantly. In all case deadlocks between truck and trailer could be avoided.

The results of the experiments with the model truck can easily be utilized in real trucks.

The project showed that a very inexpensive sensor (circular potentiometer) together with intelligent signal processing (fuzzy rule-based system) can very well yield excellent results.

2.2 Introduction to Neural Networks

Neural networks can in general be used to solve the following problems of high interest: pattern classification, clustering and categorization, function approximation, prediction and forecasting, control and optimization tasks [15]. In this section we want to give a short overview of the concepts of neural networks. We will summarize some results concerning the capabilities and limitations of some neural network models.

2.2.1 Neural Networks Overview

Figure 2.10 shows a simple model for an artificial neuron (the perceptron introduced by Rosenblatt in 1958 [16]). It has a biological motivation but is simplified and abstracted from the original model. This mathematical neuron computes a weighted sum of its input signals x_i; $i = 1, \ldots, n$. After the sum a so-called activation function is applied.

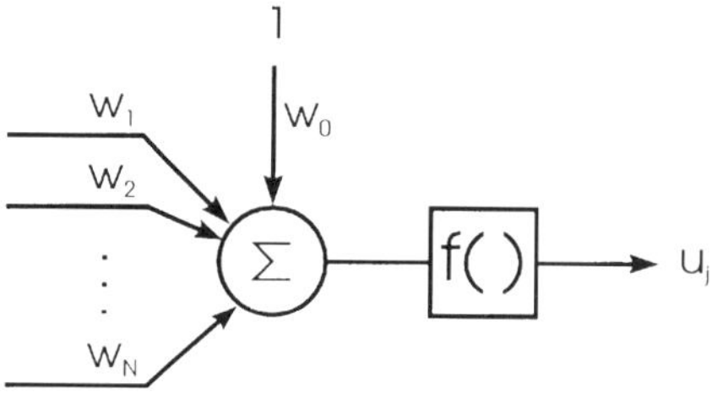

Figure 2.10. The perceptron.

Mathematically the neuron performs the following function:

$$u_i = f\left(\sum_{i=0}^{N} w_i x_i\right) \tag{2.14}$$

where $f(x)$ is the activation function. This can be for example the logistic (or sigmoid) function

$$f(x) = \frac{1}{1 + \mathrm{e}^{-(x)}}, \tag{2.15}$$

the tangens hyperbolicus

$$f(x) = \tanh\,(x), \tag{2.16}$$

a threshold function

$$f(x) = \begin{cases} -1 & x < 0 \\ 1 & x \geq 0 \end{cases} \tag{2.17}$$

or the identity function. To move the threshold of the activation function a bias has to be introduced. This is easily done by an additional input of one. The first neuron model was proposed by McCulloch and Pitts [17] in 1943 based on the threshold activation function. In most applications the sigmoid function is used, especially in multilayer perceptron nets which are discussed later.

Neural networks can be seen as weighted directed graphs in which neurons are nodes and reside on directed edges which form the connections [15]. Based on the connection configuration, neural nets can be split into feed-forward nets (the graph has no loops) and recurrent nets (loops occur because of feedback connections). Recurrent nets are, in contrast to feed-forward nets, dynamical systems. Examples of feed-forward nets are the multilayer perceptron and RBF networks. An example for a recurrent net is the Kohonen feature map.

Important for a neural net is the capability to learn. Therefore a learning algorithm has to be defined for each architecture. This algorithm has to optimize the parameters (weights) of the net by optimizing a target or error function on the training patterns. There are three main learning paradigms: supervised, unsupervised, and hybrid. In supervised learning a correct result is given for every input pattern. In contrast, unsupervised learning does not require a correct answer. The output of the net is learned by self-organization.

2.2.2 Network Architectures

To show the great variety of architectures we briefly introduce some of them by sketching their architecture and learning paradigms. At first we will present the multilayer perceptron which has its main focus on classification and function approximation. In the second subsection the radial basis function net is introduced, a more localized concept. Finally the architecture of the Kohonen feature map, which is an example for unsupervised learning, is shown. There are many other architectures which cannot be shown here, for example nets specialized for time series prediction like Elman or FIR nets or growing structures like the Cascade-Correlation Learning family. But it should be easy to understand these concepts with the basic knowledge given here (for more detailed introductions see [18] or [19]).

2.2.2.1 Multilayer Perceptron (MLP)

Figure 2.11 shows a typical three-layer perceptron with two hidden and one output layer. In general, a standard L-layer feed-forward (the inputs nodes are not counted as a layer) consists of an input stage, $(L-1)$ hidden layers, and an output layer. The units are successively fully or locally connected in a feed-forward fashion. There are no connections between units in the same layer and no feedback connections between layers.

This type of network, also called a multilayer perceptron (MLP), is the most popular class of artificial neural nets. It has been proven that an MLP network can form arbitrarily complex decision boundaries and can represent any Boolean function, if the network size is adequate [20]. Through the development of the backpropagation algorithm (see Algorithm 1) for determination of weights the multilayer perceptron has become the most popular neural network type.

Each neuron in an MLP net is of the perceptron type described above:

$$u_i = f\left(\sum_{i=0}^{N_{l-1}} w_{l,j,i}\, u_{l-1,i}\right) \tag{2.18}$$

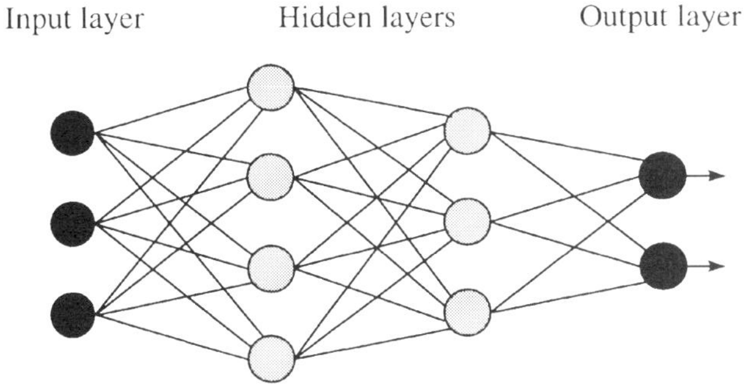

Figure 2.11. Architecture of a typical multilayer perceptron.

where $u_{l,j}$ is the output of the jth neuron of Layer l, $w_{l,j,i}$ is the weight, which connects the ith neuron of the layer $l-1$ with the jth neuron in the layer l, and N_{l-1} is the neuron count in layer $l-1$. The activation function $f(\cdot)$ is usually the sigmoid function which has a special property of its derivative:

$$\frac{\mathrm{d}f(x)}{\mathrm{d}x} = f(x)(1-f(x)). \tag{2.19}$$

With the given model only a simple forward calculation of the net is possible. The backpropagation algorithm now introduces a gradient descent based method to learn the weights. This is done by a backward propagation of the error produced by the actual output through the net. In fact a squared error cost function is minimized:

$$J(W) = \frac{1}{2}\sum_{p=1}^{P}\sum_{q=1}^{N_L}(u_{L,q}(x_p) - d_q(x_p))^2 \tag{2.20}$$

where J is the error measure, $u_{L,q}$ is the output of neuron q at the layer L (which has N_L nodes in this layer) and d_q is the desired output for this layer. The backpropagation algorithm takes the *steepest descent* to change the weights. They are changed by the following value:

$$\Delta w_{l,j,i} = \mu \sum_{p=1}^{P} \left.\frac{\delta J_p(W)}{\delta w_{l,j,i}}\right|_{W(k)}. \tag{2.21}$$

where J_P is the summed squared error over a pattern P depending on the actual weights (stored in the weight matrix W with the components w) and μ is a learning parameter which determines the learning speed. It can be a problem to choose the right μ because a too-big learning parameter can result in instabilities and a too-small parameter will result in very slow convergence. Using the chain rule, the gradient of the hidden layers can be calculated from the output layer:

$$\frac{\delta J_p(W)}{\delta w_{l,j,i}} = \frac{\delta J_p(W)}{\delta u_{l,j}} u_{l,j}(1-u_{l,j})u_{l-1,i}. \tag{2.22}$$

Thus we have changed the gradient based on weights to a gradient based on the output of this layer. This new gradient can be written as a function of the gradient in the next layer to the output:

$$\frac{\delta J_p(W)}{\delta u_{l,j}} = \sum_{m=1}^{N_{l+1}} \frac{\delta J_p(W)}{\delta u_{l+1,j}} u_{l+1,m}(1-u_{l+1,m})w_{l+1,m,j}. \tag{2.23}$$

With a boundary condition for the gradient on the output layer:

$$\frac{\delta J_p(W)}{\delta u_{l,j}} = u_{L,j}(x_p) - d_j(x_p). \tag{2.24}$$

Starting with equation (2.24), equations (2.22) and (2.23) can be used iteratively to calculated the gradient in (2.21) for each layer.

Algorithm 1 Backpropagation

```
Initialization of weights with small random values
repeat
   repeat
      Chose one training pattern including the target vector
      FEED-FORWARD calculation
      GRADIENT CALCULATION for all layers (backward)
      if on-line mode then
         Change weights with the calculated Δw_{w,j,i}
      else
         Store the calculated Δw_{w,j,i}
      end if
   until all patterns used;
   if batch mode then
      Change weights with the stored values
   end if
until termination condition reached;
```

In Algorithm 1 the general structure of backpropagation is shown. The gradient descent is repeated until a termination condition is reached, e.g. the summed squared error falls under a certain threshold.

There are many extensions for backpropagation. For example the learning speed can be improved by the usage of second-order methods or a momentum term which produces a smoother weight correction by adding a small amount of the last weight change to the actual weight change. However there are regularization techniques to avoid over-fitting (the nets learn to reproduce training data perfectly, but perform very badly on new data, the so-called test set).

2.2.2.2 Radial Basis Function (RBF) Networks

The receptive fields in the cerebral cortex are well-known structures. Based on these structures the basic principles of *Radial Basis Function Networks* (*RBFN*) were deduced [21]. These are shown in Figure 2.12.

The output of an artificial receptive field is computed by:

$$\omega_i = R_i(\vec{x}) \quad i = 1, 2, \ldots, H, \tag{2.25}$$

where H is the number of receptive fields, x is the input vector and R_i is the output value of the receptive field. Typically the Gauss function

$$R_i(\vec{x}) = \exp\left(-\frac{||\vec{x} - \vec{c}_i||^2}{\sigma_i^2}\right) \tag{2.26}$$

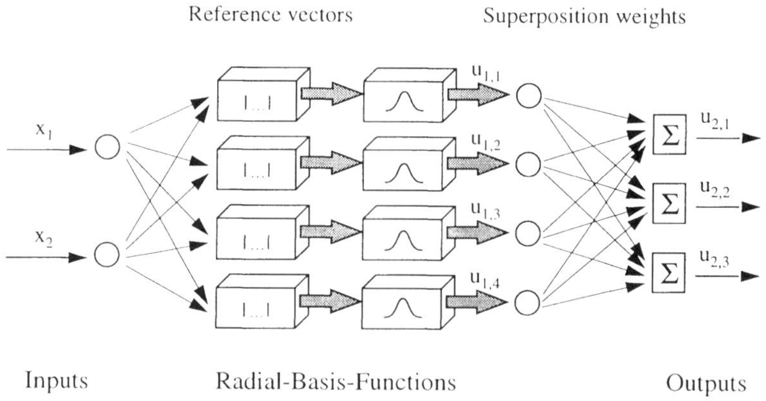

Figure 2.12. The Radial Basis Function network.

is chosen for R_i. Thus R_i has its maximal value at $\vec{x} = \vec{c}_i$. The output of the network shown in Figure 2.12 is computed by:

$$f(\vec{x}) = \sum_{i=1}^{H} f_i w_i = \sum_{i=1}^{H} f_i R_i(\vec{x}) \tag{2.27}$$

where f_i represents the strength of the receptive field i. If additional lateral connections between the receptive fields are added then f can be computed as

$$f(\vec{x}) = \frac{\sum_{i=1}^{H} f_i \omega_i}{\sum_{i=1}^{H} \omega_i} = \frac{\sum_{i=1}^{H} f_i R_i(\vec{x})}{\sum_{i=1}^{H} R_i(\vec{x})}. \tag{2.28}$$

In theory, the RBF network, like the MLP, is capable of forming an arbitrarily close approximation to any continuous nonlinear mapping [19]. The main difference between MLP and RBF is their behavior in regions without data. RBF nets tend to produce a zero output because of their localized responses. However MLP nets extrapolate the data.

Normally learning in RBF networks is split into two stages: learning in the hidden layer, followed by learning in the output layer. Kernel positions and kernel widths are estimated using a unsupervised clustering algorithm like k-means. The normalization parameters σ_i^2 are obtained later. As they represent the spread of the data associated with each node, it is common to make them equal to the average distance between the cluster centers and the training patterns, that is [19]:

$$\sigma_i^2 = \frac{1}{M_i} \sum_{\vec{x} \in \Theta_i} ||\vec{x} - \vec{c}_i||^2 \tag{2.29}$$

where Θ_i is the set of training patterns (M_i is the number of patterns in Θ_i) grouped with cluster centers $\vec{c}_i$. After the learning process of the first layer the linear output layer is typically trained using the Least Mean Squares (LMS) algorithm. This initial solution can be refined by a supervised gradient-based algorithm.

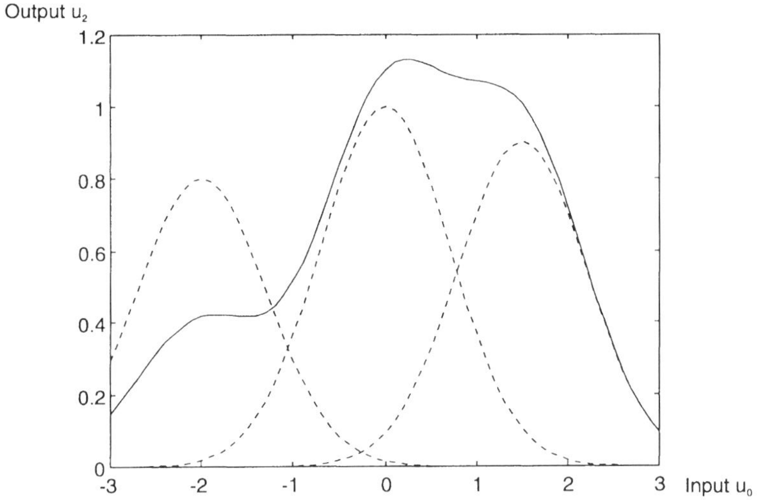

Figure 2.13. Functional approximation with Radial Basis Functions.

2.2.2.3 Kohonen's Self-organizing Feature Map

Self-organizing feature maps are single-layer networks trained with an unsupervised learning algorithm [22]. They have the property of topology preservation, similar to the feature maps in the cortex of higher animals. Figure 2.14 shows the basic architecture.

It consists of an array (e.g. two- or three-dimensional) of neurons, each connected to all n input nodes. For training the weight vector $\vec{w}_j = (W_{1j}, \ldots, w_{nj})$ of neuron j is compared to the input vector $\vec{x} = (x_1, \ldots, x_n)$ using a metric (for example the

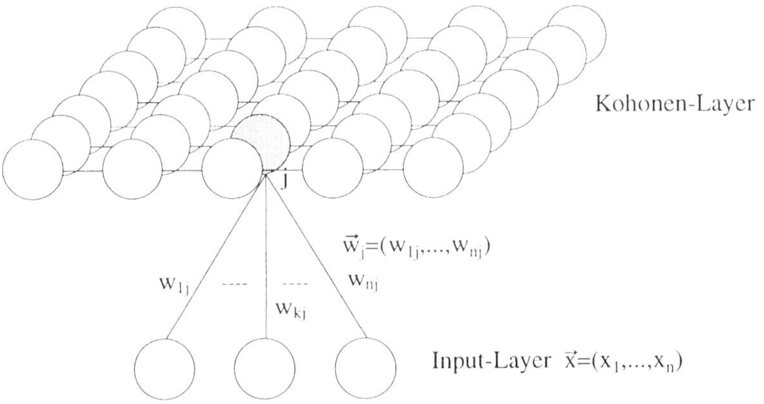

Figure 2.14. Kohonen feature mapping from a square region of the plane on to an array of output units.

Euclidian distance). The neuron c wins, if it has a weight vector $\vec{w}_c$ that is most similar to the input vector $\vec{x}$:

$$||\vec{x} - \vec{w}_c|| = \min_j(||\vec{x} - \vec{w}_j||). \tag{2.30}$$

Only the neurons in a spatial neighborhood of the winning unit are adjusted. The shape and size of the neighborhood may vary over time. Usually one begins with two-thirds of the network size and shrinks the region size over time (e.g. exponentially decreasing). Weights are updated according to the following learning rule:

$$\vec{w}_j(t+1) = \vec{w}_j(t) + \nu(t)h_{cj}(t)[\vec{x}(t) - \vec{w}_j(t)] \tag{2.31}$$

where $h_{cj}(t)$ is the neighborhood function (neighborhood kernel) and $\nu(t)$ is a time-dependent learning rate (which is monotonically decreasing and in the interval [0; 1]). For the neighborhood function one can write:

$$h_{cj}(t) = h(||\vec{r}_c - \vec{r}_j||, t) \tag{2.32}$$

where $\vec{r}_c$ and $\vec{r}_j$ are vectors to the positions of neurons c and j. With increasing distance of the two units $h_{cj}(t) \to 0$ applies. A good neighborhood function is for example the normalized Gaussian.

Kohonen's feature map can be used for projection and analysis of multivariate data, density approximation, and clustering [15]. It has been successfully applied in the areas of speech recognition, image processing, robotics, and process control.

2.3 Neuro-fuzzy Systems

2.3.1 Motivation

In many industrial applications control-technical problems are solved by expert systems. Those systems use human expert knowledge. If this knowledge has to be integrated in such an expert system it is possible that the following problems arise:

1. The expert knowledge is incomplete.
2. The problems are solved intuitively.
3. The rules are built up with many conditions.
4. The rules have to be optimized.

With the help of neural networks these problems can be solved because these are able to extract information from given data sets in order to complete the knowledge. The disadvantage of neural networks is that

1. The knowledge which is included in the network cannot be interpreted by the user and therefore the network is not transparent.
2. No additional human knowledge can be integrated into the system.

It is possible to realize a fuzzy system in the form of a neural network. With the help of such a combined system the knowledge of the fuzzy system can be optimized and completed by the concept of the neural network.

2.3.2 Fuzzy Models in Neuro-fuzzy Systems

In this chapter fuzzy models will be introduced which can be applied in neuro-fuzzy systems. The most popular type of a fuzzy-controller is the Mamdani controller with the standard *center of gravity* method of defuzzification. The flow of information in this controller is shown in Figure 2.15 [23]. Besides that there is the model of Takagi–Sugeno–Kang (TSK) [24]. The difference between both models is that there only a singleton is used for the conclusion as shown in Figure 2.16 and not a fuzzy set with a membership function. At the end the results have to be defuzzified. For that purpose the result of each rule is weighted with the strength of the rule and then the average of these weighted values is computed. For the TSK model there are two types of rules. A first-order rule has the following form

$$\text{if} \quad I_1 \quad \text{is} \quad A \quad \text{and} \quad I_2 \quad \text{is} \quad B \quad \text{then} \quad f_{i,j} = c_{i,j},$$

whereas a second-order rule is of the following type

$$\text{if} \quad I_1 \quad \text{is} \quad A \quad \text{and} \quad I_2 \quad \text{is} \quad B \quad \text{then} \quad f_{i,j} = p_1 I_1 + p_2 I_2 + c_{i,j}.$$

The output for both types is

$$\theta_i = \sum_i f_{i,j} k_j^*,$$

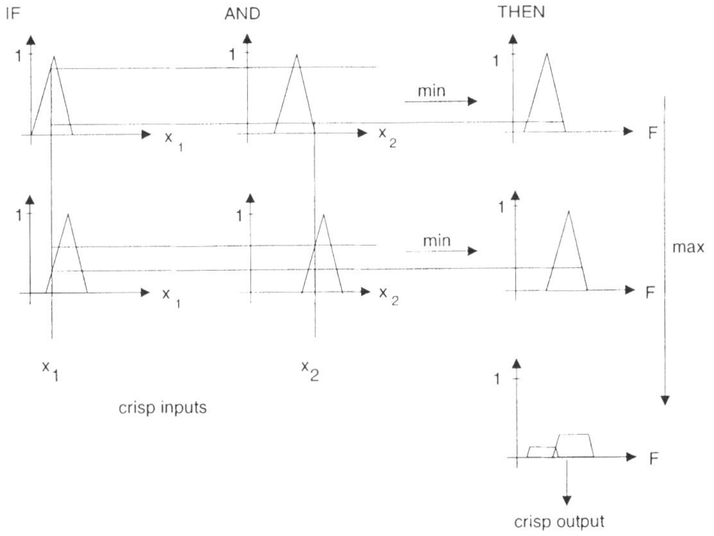

Figure 2.15. Mamdani controller.

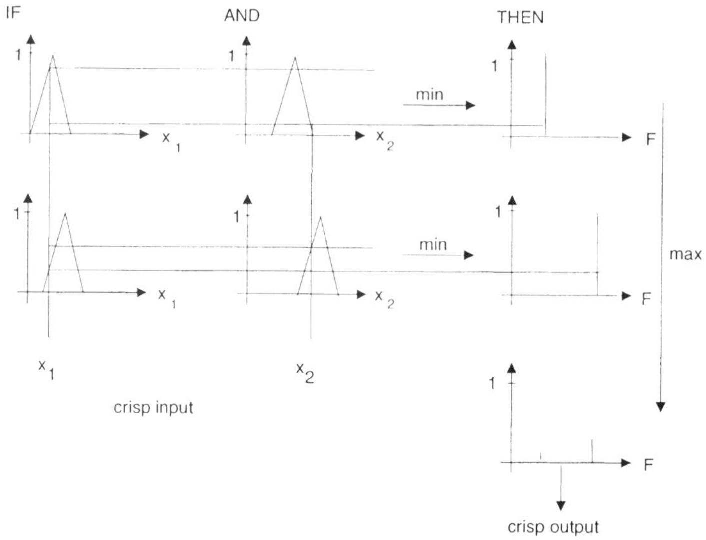

Figure 2.16. TSK-controller.

with

$$k_j^* = \frac{k_j}{\sum_i k_i}.$$

i is the index of the outputs and j of the rules, A and B are the fuzzy sets, $c_{i,j}, p_1$ and p_2 are constants, and k_j^* is the normalized rule strength.

A further model is the FuNeI-model [25]. Its rules are of the following form

$$\begin{aligned} \text{if} \quad I_1 \quad \text{is} \quad A \quad \text{and} \quad I_2 \quad \text{is} \quad B \quad \text{then} \quad f_{i,j} &= c_{i,j}^A \\ \text{if} \quad I_1 \quad \text{is} \quad C \quad \text{or} \quad I_2 \quad \text{is} \quad D \quad \text{then} \quad f_{i,j} &= c_{i,j}^O \\ \text{if} \quad I_1 \quad \text{is} \quad E \quad \text{then} \quad f_{i,j} &= c_{i,j}^S \\ \theta_i &= \text{sig}\left(\sum_j f_{i,j} k_j\right), \end{aligned}$$

where $c_{i,j}^A$, $c_{i,j}^O$ resp. $c_{i,j}^S$ are the weights of the and-, or- resp. simple-rules and where sig is the sigmoidal function. If $c < 0$ the rule has to be interpreted as an if – not rule.

The fourth type to be introduced here is the classificator model [25]:

$$\begin{aligned} &\text{if} \quad I_1 \quad \text{is} \quad A \quad \text{and} \quad I_2 \quad \text{is} \quad B \quad \text{then} \quad O_1 \\ &\text{if} \quad I_1 \quad \text{is} \quad C \quad \text{or} \quad I_2 \text{ is} \quad D \quad \text{then} \quad O_2. \end{aligned}$$

It is possible to have several rules for one class, but one rule can never have several classes as output.

All models can be used for data analysis, especially the last one. Because of the good ability of function approximation the first three are very well suited for control problems.

2.3.3 Fuzzy Systems as Pure Feed-forward Neural Networks

The basic modules of a neural network are the single neurons. In such a combined system the neurons are used to build the membership functions. As shown in Figures 2.15 and 2.16 the results have to be defuzzified. For this process a special neural network can be applied which will be introduced in Section 2.3.3.4. At the end of this section two special methods for the extraction of the rules from a data set will be discussed.

2.3.3.1 Rule-neuron

The model of the neuron of McCulloch and Pitts is shown in Figure 2.10. Thus the value of the output of this neuron is calculated by

$$\theta_i = a\left(\sum_{j=1}^{N} I_j w_j + B_i\right),$$

where a is the activation function. This can be a linear, sigmoidal or tangens-hyperbolical function. In a generalized way the representation of a neuron looks like

$$\theta_i = a(f(B_i, w_{i1}I_1, w_{i2}I_2, \ldots)).$$

Here again a is the activation function and f calculates the net input value. In the previous formula f is the function of summation.

In many neuro-fuzzy systems (e.g. FuNeI) the basic neuron is of the generalized type. Here the linear or sigmoidal function is used as activation function. Since the learning methods are often based on gradient methods derivable activation functions are chosen. Additionally the conjunction and disjunction are no longer implemented by min- and max-functions, but by the *softminimum-* (SMIN) or the *softmaximum-* (SMAX) function, where

$$\text{SMIN}\ (I_1, \ldots, I_n) = \frac{\sum_{i=1}^{n} I_i\, e^{-jI_i}}{\sum_{i=1}^{n} e^{-jI_i}}$$

$$\text{SMIN}\ (I_1, \ldots, I_n) = \frac{\sum_{i=1}^{n} I_i\, e^{-jI_i}}{\sum_{i=1}^{n} e^{-jI_i}}.$$

If the network converges j tends to ∞. Thus in the generalized description of neurons f is chosen as SMIN resp. SMAX, a as the linear or sigmoidal activation function, $B_i = 0$ and $w_{i1} = w_{i2} = 1$.

2.3.3.2 Membership Functions

The membership functions which have to be trained separately can be generated with the help of a multilayer perceptron network [25]. If it is profitted by scaling, shifting

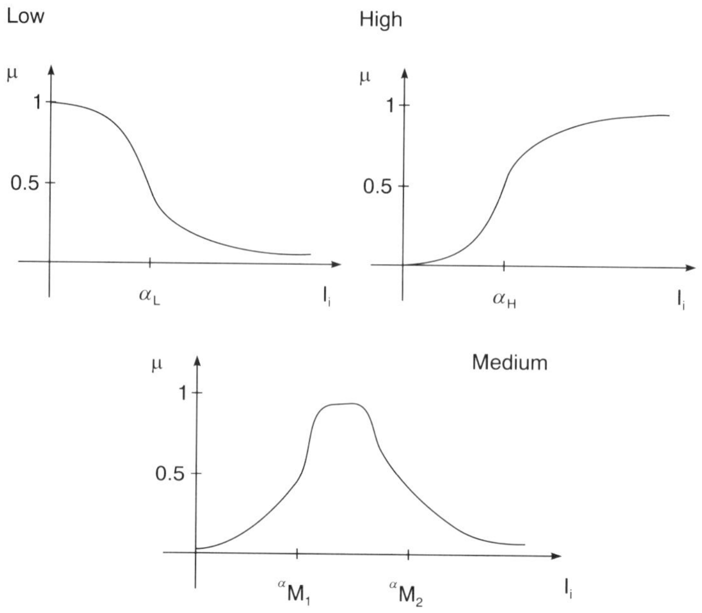

Figure 2.17. Membership functions.

and reflection all membership functions can be deduced from the two basic types for *low* and *high*. For the purpose the two special sigmoidal functions

$$a_1[I_i, C, \alpha] = \frac{1}{1 + \mathrm{e}^{-C(I_i - \alpha)}}$$
$$a_2[I_i, C, \alpha] = \frac{1}{1 + \mathrm{e}^{C(I_i - \alpha)}}$$

are used, i.e. that a_1 is the reflection of a_2 in the y-axis. C is always positive and defines the steepness of the sigmoidal function: if C tends to ∞ the sigmoidal function converges against the step function. α describes the shifting of the sigmoidal function in the x-direction.

Thus the membership functions shown in Figure 2.17 can be described in the following way:

- Low: $L = a_2[I_i, C_L, \alpha_L]$;
- High: $H = a_1[I_i, C_H, \alpha_H]$
- Medium: in this case there are two different methods:

1. As shown in Figure 2.18 a sigmoidal function of type *low* and one of type *high* is generated with the help of two neurons. A third neuron computes the minimum of

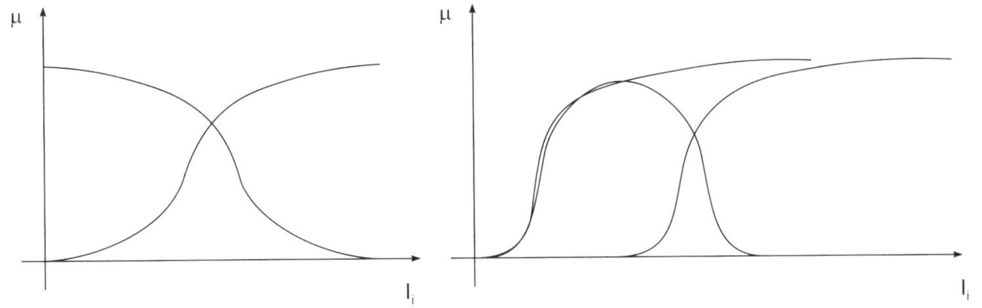

Figure 2.18. Calculation of the membership function *middle*.

both functions:

$$\begin{aligned} M_1 &= a_1[I_i, C^*_{M_1}, \alpha^*_{M_1}] \\ M_2 &= a_2[I_i, C^*_{M_2} \alpha^*_{M_2}] \\ M &= \min\{M_1, M_2\}, \end{aligned}$$

where

$$\alpha^*_{M_1} < \alpha^*_{M_2}.$$

2. Two shifted sigmoidal functions of type *high* are built and then one is subtracted by a third neuron from the other one (see Figure 2.18):

$$\begin{aligned} M_1 &= a_1[I_i, \tilde{c}_{M_1}, \tilde{\alpha}_{M_1}] \\ M_2 &= a_1[I_i, \tilde{c}_{M_2}, \tilde{\alpha}_{M_2}] \\ M &= \sum\{M_1, -M_2\}, \end{aligned}$$

where $\tilde{\alpha}_{M_1} < \tilde{\alpha}_{M_1}$.

2.3.3.3 *Membership Functions for the Conclusions of the Rules*

Normally the values at the output of a rule of a fuzzy system are continuous. For reasons of simplification it is supposed here that they are discrete, but the following can be easily generalized for continuous ones. Let U be the set of all possible normalized output values:

$$U = \{U_1, \ldots, U_n\} \quad \text{with } 0 \leq U_i \leq 1, \quad \forall i : 1 \leq i \leq n.$$

Each value of U has a membership value. The set of these values is described by:

$$\mu = \{\mu_1, \ldots, \mu_n\}.$$

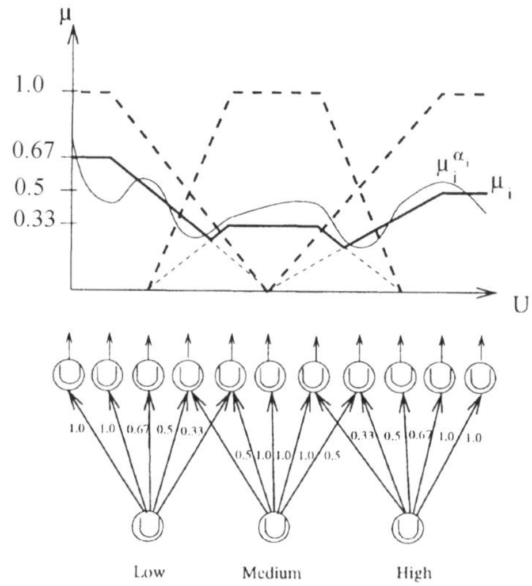

Figure 2.19. Computation of the membership functions for the conclusions.

In Figure 2.19 it is shown how the membership functions for the conclusions of the rules are computed by a two-layer neural network [25]. The neurons of the input layer represent the linguistic variables, here *low*, *medium* and *high*. The weights of the connections between the neurons of both layers describe the function course of the membership functions. Thus the weight represents the μ-value of that output value U_k which is plotted over the corresponding neuron on the U-axis. The dashed line shows the resulting course if a rule strength has the highest value (1) for an adjective. If the μ_i have the maximal values 0.67, 0.33 and 0.5 for the three membership functions *low*, *medium* and *high*, then the solid function course results.

2.3.3.4 Defuzzification

Standard methods of defuzzification, e.g. *center of gravity*, can be easily implemented in neural networks. The defuzzification is computed by the following formula:

$$U_{\text{out}}^{\text{COG}} = \frac{\sum_{i=1}^{n} \mu_i U_i}{\sum_{i=1}^{n} \mu_i}.$$

This is done in two steps: first, the sum $\sum_{i=1}^{n} \mu_i U_i$ and $1/\sum_{i=1}^{n} \mu_i$ are computed, then the product of both terms $\sum_{i=1}^{n} \mu_i U_i \cdot 1/\sum_{i=1}^{n} \mu_i$ is built. Thus the network shown in Figure 2.20 results. In the left neuron of the hidden layer the sum $\sum_{i=1}^{n} \mu_i U_i$ and in the right one the inverse of the sum $1/\sum_{i=1}^{n} \mu_i$ is calculated. The last term is computed with the help of the $1/x$-function as activation function. In the output layer the product is built. Normally the input function computes the weighted sum of the inputs but in this case the product is computed instead of the sum. Since the weights are all fixed, no learning process is necessary.

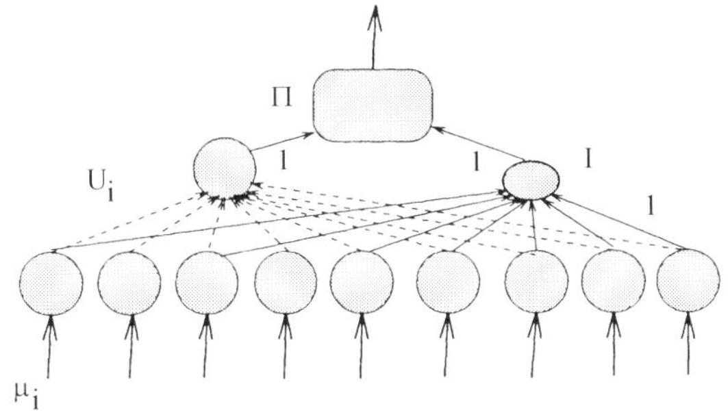

Figure 2.20. Neural network for defuzzification.

A further method of defuzzification is the BADD-method which was introduced by Filev and Yager [26]. BADD means *BAsic Defuzzification Distributions*. The value of the defuzzified output is computed by:

$$U_{\text{out}}^{\text{BADD}} = \frac{\sum_{i=1}^{n} \mu_i^{\alpha} U_i}{\sum_{i=1}^{n} \mu_i^{\alpha}}.$$

This method was generalized by Halgamuge in [25] who chooses for every μ_i another α_i and calls the method *Customizable BADD (CBADD)*:

$$U_{\text{out}}^{\text{CBADD}} = \frac{\sum_{i=1}^{n} \mu_i^{\alpha_i} U_i}{\sum_{i=1}^{n} \mu_i^{\alpha_i} U_i}, \quad \text{with } \alpha_i \in \mathscr{R}$$

2.3.3.5 Methods for the Extraction of Rules

For the extraction of the rules here the two algorithms of Wang and Mendel [27] and of Halgamuge [25] respectively are introduced.

In the first algorithm the structure of the rules is fixed. In this structure every rule depends on every input in the premise. During the training process the parameters of the membership functions in the premises are optimized by the backpropagation algorithm. This method has the advantage that it learns rapidly, but it has the disadvantage that the resulting system of rules is purely transparent if the number of rules resp. inputs is high.

The FuNe-I system of Halgamuge constructs only rules with one or two inputs in the premises. The rules are constructed in three steps as will be explained in the following [28].

In the first step the fuzzy sets of the inputs have to be defined. Then for each input parameter x_i and for each of the three fuzzy sets *low*, *middle* and *high* the values

$$K(i) = \min\{\mu_{\text{fset}}^{(i)}(x_i), \max_{j:j\neq i}\{\max(\mu_{\text{low}}^{(j)}(x_j), \mu_{\text{middle}}^{(j)}(x_j), \mu_{\text{high}}^{(j)}(x_j))\}\}$$

$$D_{\text{fset}}^{(i)} = \max\{\mu_{\text{fset}}^{(i)}(x_i), \min_{j:j\neq i}\{\max(\mu_{\text{low}}^{(j)}(x_j), \mu_{\text{middle}}^{(j)}(x_j), \mu_{\text{high}}^{(j)}(x_j))\}\}$$

are computed. The index *fset* symbolizes one of the three fuzzy sets *low*, *middle* or *high*. Thus K represents a conjunctive and D a disjunctive rule unit. Each of these rule units is connected with each output. The values of the weights are chosen randomly. Then the construction of the training network is finished.

In the second step the weights between the rule units and the outputs are trained. A weight is described by $W(K^{(i)}_{\text{fset}}, v_j)$ which means that it is on the connection between the rule unit $K^{(i)}_{\text{fset}}$ and the output v_j. Now the three weights $W(K^{(i)}_{\text{low}}, v_j)$, $W(K^{(i)}_{\text{middle}}, v_j)$ and $W(K^{(i)}_{\text{high}}, v_j)$ are taken into consideration. If one of these three weights is very different from the others then this fact is interpreted in a way that this rule unit has a high influence on the output v_j, otherwise none of the three rule units is important for the output v_j. If for example the weight $W(K^{(i)}_{\text{f,set}}, v_j)$ is very different from the others then the premise x is fset is included in the rule system. One speciality of the FuNe-I-system is that such a weight can be positive or negative. If it is negative the premise is negative. After that all combinations of two premises are built up additionally the *simple*-rules with one premise are added.

In the third step the rule weights are optimized. If the value of a weight is small then the rule is insignificant and is eliminated from the rule-system.

2.3.4 Equivalent Neural Networks

In this section the functional equivalence between a radial basis function network (see Section 2.2.2.2) and a Takagi-Sugeno-Kang fuzzy system will be proved. If such an equivalence can be shown then a method (e.g. learning procedures) which is useful for one system can be profitably applied to the other system.

2.3.4.1 TSK Fuzzy System

This system consists of rules of the following type:

$$\text{Rule } i: \quad \text{if} \quad x_1 \quad \text{is} \quad A_i \quad \text{and} \quad x_2 \quad \text{is} \quad B_i \quad \text{then} \quad f_i = a_i x_1 + b_i x_2 + c_i.$$

The inference diagram is shown in Figure 2.21. The w_i can either be computed with the help of the t-norm or t-conorm. In the following the t-norm

$$w_i = \min\{\mu_{A_i}(x_1), \mu_{B_i}(x_2)\}$$

will be used. With this $f(\vec{x})$ can either be computed by

$$f(\vec{x}) = \sum_{i=1}^{R} w_i f_i \tag{2.33}$$

or by

$$f(\vec{x}) = \frac{\sum_{i=1}^{R} w_i f_i}{\sum_{i=1}^{R} w_i} \tag{2.34}$$

The corresponding fuzzy system is shown in Figure 2.22.

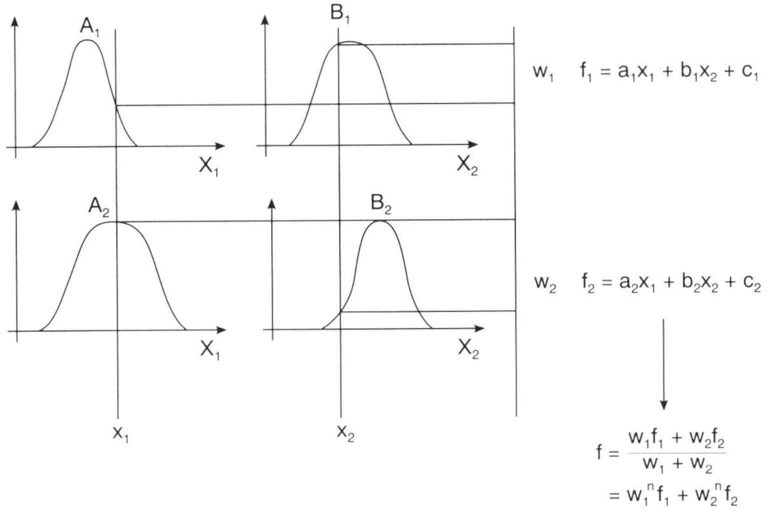

Figure 2.21. Inference diagram.

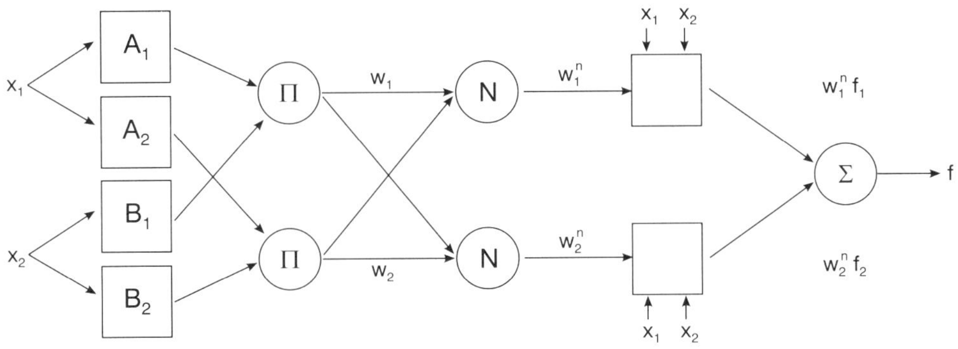

Figure 2.22. TSK system.

2.3.4.2 Functional Equivalence

Equations (2.25), (2.26), (2.27) and (2.28) show the functional equivalence of both systems if the following conditions are fulfilled.

1. The number of units of the receptive fields has to be equal to the number of fuzzy rules.
2. The outputs of the fuzzy rules have to be constant, i.e. $f_i = c_i$, thus $a_i = b_i = 0$.
3. In each rule the membership functions have to be Gauss-functions of the same variance.
4. For the calculation of the w_i the *t*-norm is used.
5. Either equations (2.25) and (2.27) or (2.26) and (2.28) are used.

Under these conditions the membership functions are

$$\mu_{A_1}(x_1) = \exp\left(-\frac{(x_1 - c_{A_1})^2}{\sigma_1^2}\right)$$

$$\mu_{A_1}(x_2) = \exp\left(-\frac{(x_2 - c_{B_1})^2}{\sigma_1^2}\right)$$

and the weights are

$$w_1(\vec{x}) = \exp\left(-\frac{||\vec{x} - \vec{c}_1||^2}{\sigma_1^2}\right) = R_1(\vec{x}),$$

with $\vec{c}_1 = (c_{A_1}, c_{B_1})$. The same can be deduced for $\mu_{A_2}(x_1)$, $\mu_{B_2}(x_2)$ and for $w_2(\vec{x})$.

If those conditions are not fulfilled then the RBF networks are a special case of the TSK systems.

2.4 Applications

2.4.1 Autonomous Driving of a Vehicle with Collision Avoidance

In this paragraph the algorithm of a controller of a vehicle which is able to avoid autonomously collisions with an obstacle will be presented. Therefore a single ultrasonic sensor is mounted on a model truck (see Figure 2.23) [25]. Its signals are used by the controller to compute the steering position.

During the drive the rotating sensor measures the distances of the obstacles (e.g. a wall, another vehicle, a tree, etc.) in the environment. Because of the limited processing speed of the sensor and of the microcontroller the sensor measures only in seven positions (at the angles of $-60°$, $-40°$, $-20°$, $0°$, $+20°$, $+40°$ and $+60°$ (see Figure 2.24). These seven measurements are the inputs of the microcontroller. Its rule-base consists of seven rules, one rule for each position. They are of the following form:

Rule i : IF distance IS small / middle / large THEN turn to direction $\quad i = 1; \ldots; 7$

There are five membership functions for the direction to turn to: lg = left great, ls = left small, s = straight, rs = right small and rl = right large. After the inference process the results of the seven rules are defuzzified and the servo steers to the newly computed direction. Figure 2.24 shows the sensor readings in the six different situations. In the three situations pictured on the left a steering to the left side and in the other situations a steering to the right side results as the line indicates.

In Figure 2.25 it is shown how the truck drives in order to avoid hitting the obstacles. The performance of the sensor has been improved by using corrugated cardboard because it better reflects the ultrasonic signals.

Figure 2.23. Truck.

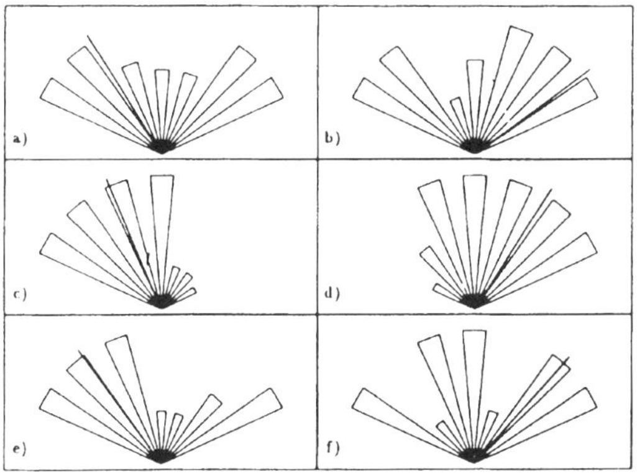

Figure 2.24. Sensor measurements.

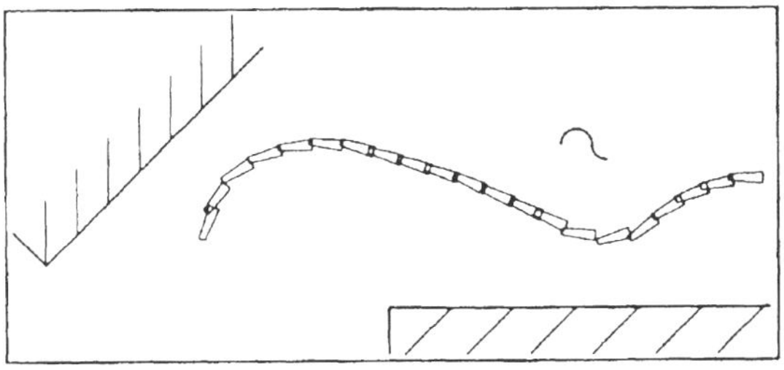

Figure 2.25. Driving of the autonomous truck.

2.4.2 Detection of Smouldering Fire

As presented previously, neural networks can either be used for function approximation or for classification. In this section a classification network will be introduced in order to detect a smouldering fire in an incinerating plant. Here, a lorry dumps the rubbish and the rubbish is stored in a pit (see Figure 2.26). During this storage the rubbish easily starts to smoulder. This is very dangerous because a fire may result. Until now this was supervised by an employee of the incinerating plant. But an additional sensor system should be installed. Therefore four sensors were assembled to a system: an electronic nose. From the sensor signals the following information can be deduced:

- the voltages of the sensor;
- the derived signal;
- the offset-corrected signals.

With the help of the nonlinear mapping of Sammon, sequential backward selection, linear or nonlinear principal component analysis, the significant features can be extracted. In order to test the different classification algorithms the following seven experiments have been used: the burning of a cable, the smouldering fire of coal, the smoke of a cigarette, the open fire, the smouldering fire of clothes, the evaporation of propanol and the normal situation. A neural network has been trained which is able to classify these experiments in the following four situations: normal situation (class 1), little smoke (class 2), smouldering fire (class 3) and open fire (class 4). In Table 2.1 the classification rates are given.

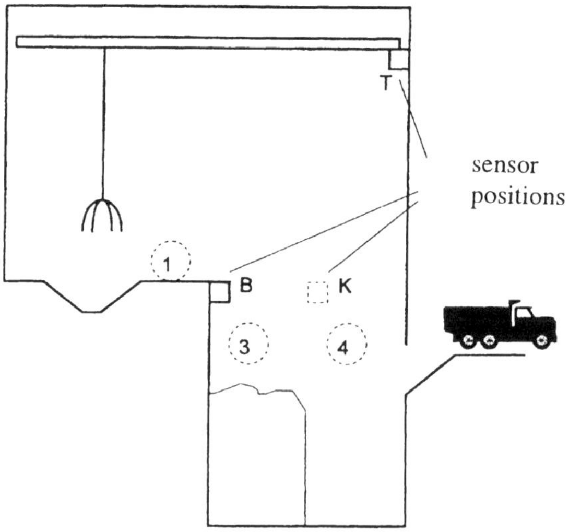

Figure 2.26. Incinerating plant.

Table 2.1. Classification rates.

	class 1	class 2	class 3	class 4
class 1	0.67	0.19	0.07	0.23
class 2	0.15	0.65	0.08	0.14
class 3	0.05	0.05	0.76	0.05
class 4	0.13	0.11	0.09	0.58

2.4.3 Real Time Road Surface Identification

Within a long-term research project (*Sonderforschungsbereich 241*) a system for the identification of the road surface was developed in collaboration with the institute for Vehicle Technique at Darmstadt University of Technology. Since the friction coefficient μ varies between 0.05 on ice and 1.2 on well-textured dry asphalt the identification of the road surface is very important for anti-lock bracking systems (ABS). Therefore a Hall sensor is integrated in the tyre (see Figure 2.27) [2]. The output of the sensor is a voltage which is a measure for the deformation of the tyre.

During the preprocessing the output signal is firstly filtered by a low-pass filter. After that the significant features are extracted from the sensor signal. In the third step these features are used as inputs for the neuro-fuzzy system which identifies the road surface.

Starting from the filtered sensor signals shown in Figure 2.28 the features are extracted. First the offset voltage of the sensor is computed. Then the beginning and the end of the contact patch is detected. Additionally the maximum and the minimum of the filtered signal are detected, the gradients between maximum and minimum and the area beneath the curve are computed.

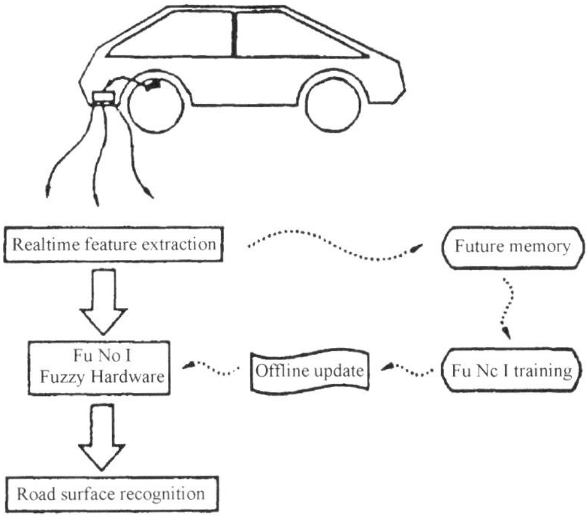

Figure 2.27. Road surface identification system.

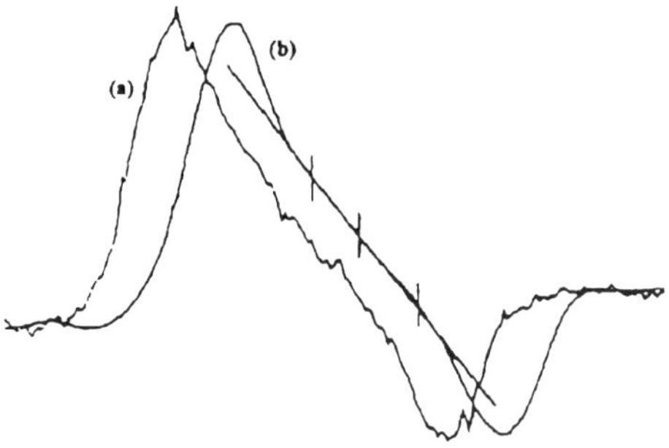

Figure 2.28. Sensor signals of the Hall sensors.

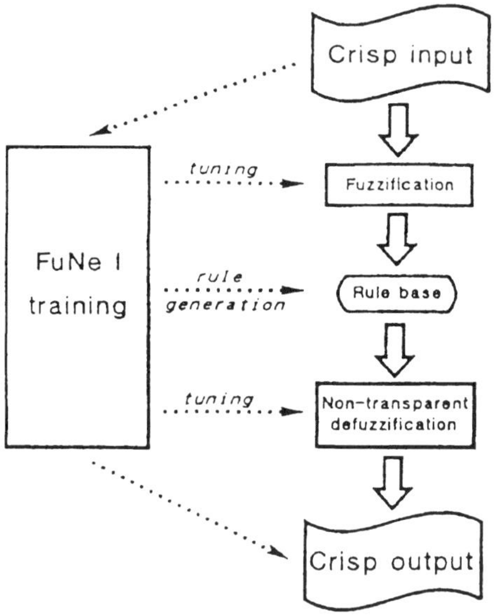

Figure 2.29. Neuro-fuzzy-system.

These features are used by the neuro-fuzzy system (see Figure 2.29) to identify the road surface as dry, wet, icy or sandy. In the first phase the rules have to be extracted from a given data set. After this training process, in the recall phase the network can be used for classification. Here the features are fuzzified in the first step in a block where the membership functions are build (see Section 2.3.3.2). In the next step the results of the rules are computed from the membership values. Then in the third step the values are defuzzified (see Section 2.3.3.4).

In contrast to a pure neural network this system has the advantage of being very transparent because the rules can easily be interpreted. With the help of this system the road surface can be correctly identified with a rate of 99%.

2.5 References

[1] Zadeh, L. A., *Information and Control*, **8** (1965), 338–353.
[2] Glesner, M., Halgamuge, S. K., Runkler, T. A., Genther, H., Hollstein, T., Herpel, H.-J., König, A., and Pöchmüller, W., *VDE Mikroelektronik, Fachzeitschrift für Hersteller und Anwender*, **8**(2): (March/April 1994), 88–93.
[3] Dubois, D. and Prade, H., *Possibility Theory*; New York: Plenum Press, 1988.
[4] Zimmermann, H. J., *Fuzzy Set Theory and Its Applications*; Boston: Kluwer Academic Publishers, 1985.
[5] Pedrycz, W., *Fuzzy Control and Fuzzy Systems*, 2nd edition; New York: John Wiley, 1993.
[6] Kruse, R., Gebhardt, J., and Klawonn, F., *Foundations of Fuzzy Systems*; New York: John Wiley, 1994.
[7] Lee, C. C., *IEEE Transactions on Systems, Man and Cybernetics* **20**(2) (March/April 1990), 404–435.
[8] Preuß H.-P., *Automatisierungstechnische Praxis*, **34**(4) (1992), 176–184, 239–246.
[9] Bibel, W., Hölldobler, S., and Schaub, T., *Wissensrepräsentation und Inferenz*; Brauschweig: Vieweg Verlag, 1993.
[10] Runkler, T. A. and Glesner, M. in: *Proc. of European Congress on Fuzzy and Intelligent Technologies*; Aachen: September 1994, pp. 1206–1210.
[11] Cholewa, W., *Fuzzy Sets and Systems* **17**(3), (1985), 249–258.
[12] Yager, R. R., *International Journal of Man-Machine Studies* **11** (1979), 189–200.
[13] Thole U., Zimmermann H.-J., and Zysno P., *Fuzzy Sets and Systems* **2**(2) (1979), 167–180.
[14] Runkler, T. A., Halgamuge, S. K., Kothe, R., and Glesner, M., *Elektronik, Fachzeitschrift für industrielle Anwender und Entwickler* **15** (July 1994), 92–97.
[15] Jain, A. K. and Mao, J., *IEEE Computer* (March 1996) 31–44.
[16] Rosenblatt, F., *Psychological Review* (65) (1958), 368–408.
[17] McCulloch, W. S. and Pitts, W., *Bull. Mathematical Biophysics* **5** (1943), 115–133.
[18] Zell, A., *Simulation Neuronaler Netze*; Bonn: Addison-Wesley, 1994.
[19] Hush, D. R. Horn, B. G., *IEEE Signal Processing Magazine* (January 1993), 8–39.
[20] Minsky, M. and Papert, S., *Perceptrons: An Introduction to Computational Geometry*; Cambridge, MA: MIT Press, 1969.
[21] Jang, J. S. R. and Sun, C. T., *IEEE Transactions on Neural Networks* **4**(1) (1993), 156–158.
[22] Kohonen, T. *Self-Organization and Associative Memory*. Springer Series in Information Sciences, Springer Verlag, Heidelberg, 1984.
[23] Mamdani, E. H. and Assilian, S., *International Journal of Man-Machine Studies*, **7**(1), (1975), 1–13.
[24] Takagi, T. and Sugeno, M., In: *Proc. IFAC Symp. on Fuzzy Information, Knowledge Representation and Decision Analysis*, July 1983, pp. 55–60.
[25] Halgamuge, S., *Advanced Methods for Fusion of Fuzzy Systems and Neural Networks in Intelligent Data Processing*. Ph.D thesis, Darmstadt University of Technology, VDI-Verlag, 1995.
[26] Filev, D. P. and Yager, R. R., *International Journal of Intelligent Systems* **6** (1991), 687–697
[27] Wang, L. X. and Mendel, J. M., *IEEE Transactions on Neural Networks*, **3**(5) (1992), 807–814.
[28] Nauck, D., Klawonn, F., and Kruse, R. *Neuronale Netze und Fuzzy-Systeme*, published by Bibel, W., Hahn, von W. and Kruse, R., Computational Intelligence Series, Vieweg, 1996.

3 Strategies for Mimicking Olfaction:

The Next Generation of Electronic Noses?

Timothy C. Pearce, School of Engineering, Faculty of Science & Engineering, University of Derby, Kedleston Road, Derby DE22 1GB (UK)

Julian W. Gardner, Department of Engineering, University of Warwick, Coventry CV4 7AL (UK)

Wolfgang Göpel, Center of Interface Analysis and Sensors, Institute of Physical and Theoretical Chemistry, University of Tübingen, Auf der Morgenstelle 8, D-72076 Tübingen (Germany)

Contents

3.1 General Introduction

3.1.1 Nature of Olfaction

Of all the human sensory systems, olfaction is the least understood in terms of the primary receptor mechanism and biological transduction. Our current ability to create artificial sensory systems for sight (e.g. cameras), hearing (e.g. microphones) and touch (e.g. pressure sensors) far exceeds that for smell (e.g. an electronic nose). There are a number of reasons why our understanding of the biological olfaction system is poor. First of all, the nature of the olfactory system is much more complex than any of the other senses, including that of gustation (i.e. taste) which has the four main descriptors of hot/cold/sweet/sour and the more recent addition of metallic. Secondly, the nature of the sensory input is by no means simple. Light, sound and pressure are basic physical measurands that can be defined in simple terms and referenced against international standards. In contrast, odors are usually a mixture of hundreds, or even thousands, of chemical compounds which are carried by an air stream containing many other chemical compounds. To make matters worse, not only can the *relative* concentrations of the chemical constituents vary with time but also the compounds themselves may be chemically unstable! Consequently, it is difficult to define a set of reference odors that describe the olfactory input. The functional relationship between the structure of odorant molecules and their olfactory impact is also, in general, poorly understood. This is mainly due to a high degree of signal processing that takes place in the olfactory system so that compounds of similar chemical structure can give completely different olfactory responses. Finally, there is the additional constraint that not all individuals respond to olfactory signals in the same manner. These differences can be genetic, for instance there are an estimated 1000 olfactory receptor genes that influence our perception of odor, as well as historical effects due to our adaptive response to historical olfactory events. Despite our lack of understanding of the human olfactory system, there has been and is a considerable amount of work that has been done to analyze and classify odors in a scientific reproducible manner. Indeed, there are many practical applications that require a quantitive objective recording of odors, such as in quality and process control of various foodstuffs, perfumes, and petrochemicals.

3.1.2 Possible Strategies for Odor Sensing

A number of strategies have been followed in our attempts to mimic the human sense of olfaction and thus measure smells. One strategy to adopt is the use of sensory analysis, an approach which is widely used in the food industries. Here, the characterization of odors is performed by an organoleptic panel consisting of six or more well-trained persons. The procedure includes a careful cross-check of individual odor perceptions to achieve some degree of objectivity. An agreement between panels of a factor of 3 on signal intensity would be regarded as acceptable.

A second strategy is to use existing analytical instrumentation, an approach often favored by engineers and chemists. On the one hand, *chemical or biochemical sensors* (usually cheap and simple) may be used to measure a limited number of specific compounds, while on the other hand *instruments from analytical chemistry* such as gas chromatographs and mass spectrometers may be used to determine a larger number of chemical constituents [1]. Unfortunately, these approaches are often unsatisfactory because of an inability to characterize an odor in terms of its olfactory impact. For example, individual chemical sensors cannot characterize complex odors such as the headspace of a roasted coffee bean. The same holds for even more sophisticated instruments, such as gas chromatographs, because even though they can record 1000 or so different individual peaks for a coffee sample, they cannot correlate unequivocally these peaks with their sensory impact. Quality or process control in the coffee industry does, however, demand unequivocal correlation between the measurable parameters (e.g. peaks) and perceived aromas.

A more recent strategy is to design *chemical sensor systems*, also termed electronic noses, and this appears to be a more promising development in our attempt to solve this problem. The approach makes use of established knowledge in the fields of materials science and sensor technology to design, in a systematic manner, new sensors with improved chemical sensitivities by synthesizing atomic, molecular, supramolecular, and biomimetic structures with high stability and reproducibility. It also makes use of recent developments in the field of microelectronics to measure the responses of arrays of smell sensors, and computer science to analyze complex data using pattern recognition techniques. A principal advantage of an electronic nose is that it can be used not only for odor characterization, but also for the quantitative determination of the concentrations of individual molecules in a complex environment. Consequently, the instrumentation may be used to monitor the concentrations of toxic molecules which cannot be sensed by the human nose (such as concentrations of CO) or to determine gross parameters (such as amounts of combustible gases, organic solvent molecules, or toxic compounds).

3.1.3 Future Development of Electronic Nose Technology

Although the concept of an electronic nose has been inspired by our attempt to imitate the signal processing in biological noses, in practice the basic elements of signal transduction, signal processing, and identification of chemical patterns currently differ from those believed to be present in the biological system. In both cases, however, a *controlled signal transduction across interfaces* plays the key role. This signal transduction is primarily triggered by odor molecules. It involves the transport of electrons, ions, or molecules, which induce structural modifications with certain time-dependent patterns at the various stages of the complex information cascade.

Progress in the current development of electronic noses on the one hand and progress in our understanding of biochemical, molecular, biological, physiological, and behavioral details of odor sensation on the other now make it possible to outline possible scenarios and characterize milestones for the development of "bioelectronic noses" as one area of future bioelectronics [2–4]. The main task is to bridge the gap between the technical and biological understanding of odor detection by electronically recording "chemical images" of any odor environment with high resolution in both space and time (similar to the existing electronically recorded "optical images" with their excellent resolution in space and time for selected frequencies of electromagnetic radiation [5]). In order to realize this we must achieve a basic understanding of the structural and functional aspects of biological chemosensory systems including their general and subjective learning processes; monitor parameters which relate more closely to the olfactory descriptors; design hybrid systems by complementing or exchanging certain components between the technical and biological sensory systems; understand similarities and differences between both signal processing architectures (e.g. artificial neural nets in computer science and biological neural networks in nerve systems); and finally optimize the hybrid systems.

3.1.4 Outline

This chapter describes the state-of-the-art in our understanding of olfaction and electronic noses and consists of three major sections. In Section 3.2, we first describe the terminology used to describe flavors and odors, followed by an overview of both the sensory methods and analytical methods used to measure them. Next, in Section 3.3, we examine in some detail the mammalian olfactory system. In particular, we discuss the architecture and morphology of the biological system from the olfactory receptor cells through to cortex, and the signal transduction and processing that take place at each stage. We also compare and contrast our current knowledge of the biological system with the design of electronic nose instrumentation. Finally, in Section 3.4 we focus upon the sensor technologies currently employed within electronic noses, and present some of the future sensor technologies that are demanded in order to significantly improve their performance. The chapter concludes with our views on the future of electronic nose technology.

3.2 Flavor Properties and Analysis

3.2.1 Properties of Flavor

Before the advent of instrumental techniques in flavor analysis, the flavor of foodstuffs was attributed to a small group of key-impact compounds. More recently, the use of gas chromatographic (GC) and spectrometric techniques demonstrates that in fact most flavors comprise hundreds of different chemical compounds which exist in a complex mixture. Indeed, further advances in such techniques have made it possible to isolate well over 6000 active flavor compounds. Although the flavor of some foods can be reproduced through the synthesis of a small number of key-impact chemical compounds, this is not generally possible because the vast majority of naturally occurring flavors have a complex make-up. Aishima (1991) expressed the complex nature of flavor in terms of its sensory components through the following equation [6]

$$\begin{aligned}\text{Flavor} = {} & a_1(\text{aroma}) + a_2(\text{taste}) + a_3(\text{texture}) + a_4(\text{color}) + a_5(\text{sound}) \\ & + a_6(\text{temperature})\end{aligned} \tag{3.1}$$

where the sensory components are given in terms of the coefficients a_i and summed in equation (3.1) to form the flavor percept. This model has severe limitations in providing an overall picture of the quality of flavor. In particular, the values of a_i are found to vary for different flavors and also for different individuals. Furthermore, the model ignores the many sensory confusions that occur when flavor is perceived.

Flavor can also be modeled by its m chemical components as given by equation (3.2). Here the concentrations, X_i, of chemical species, i, are found in varying amounts, b_i, and summed to provide the flavor mixture:

$$\text{Flavor} = b_1X_1 + b_2X_2 + \ldots + b_iX_i + \ldots + b_mX_m \quad (3.2)$$

Again, this model does not account for the additive, synergistic, antagonistic and compensative effects that can occur between chemical flavor components. Thus, the high dimensionality and interdependency of flavor components (both sensory and chemically based) have prevented the development of a comprehensive flavor classification scheme. Instead, industries use their own heuristic flavor models based upon practical experience of, say, a particular foodstuff, beverage or cosmetic product.

3.2.2 Flavor Analysis

To date, two disparate approaches have been taken to analyze flavors, namely, the use of either trained sensory panels or instrument-based techniques. Sensory analysis provides mainly qualitative flavor information as the chemical senses are relatively poor at quantifying chemical stimuli. Interestingly, one reason for this poor specificity is that for a particular compound the chemical senses work over a restricted dynamic range of stimuli as compared to the other senses – only three orders of magnitude as compared to eight for vision [7]. In contrast, instrument-based flavor analysis provides accurate quantiative information on the chemical species which make up the flavor.

The evolution of the techniques used in flavor analysis is shown in Figure 3.1. In the 1900s, scientists were restricted to the use of sensory analysis and basic organic chemistry for the purification and structure study of flavor compounds. Due to the limited tools available to the analyst at that time it seemed appropriate to characterize flavor as comprising only a few key-impact flavors. This view was shown to be over-simplistic with the introduction of gas chromatography (GC), infra-red spectrometry (IR), mass spectrometry (MS), and later, in the mid 1950s, nuclear magnetic spectrometry (NMR). Combined GC-MS systems were introduced in the following decade and this permitted rapid on-line sampling of flavor compounds. Such analysis has suggested a highly complex mixture of flavor-active compounds as discussed previously (Section 3.2.5).

During the late 1960s, the emerging complexity of the flavor analysis problem resulted in the utilization of computer-based chemometric techniques to analysis data. Chemometric techniques may be conveniently categorized into eleven subdivisions, all of which are geared to the analysis of experimental data and experimental design. One of these subdivisions, pattern recognition (PARC), which often borrows techniques from multivariate analysis (MVA), has been extensively used for interpreting GC or liquid chromatographic (LC) data. The use of chemometrics in flavor analysis to correlate instrumental data with sensory profiles is considered in more detail later (Section 3.2.3).

More recently, the 1980s have witnessed a large expansion in the development of generic "model" or electronic nose systems, employing an array of differentially

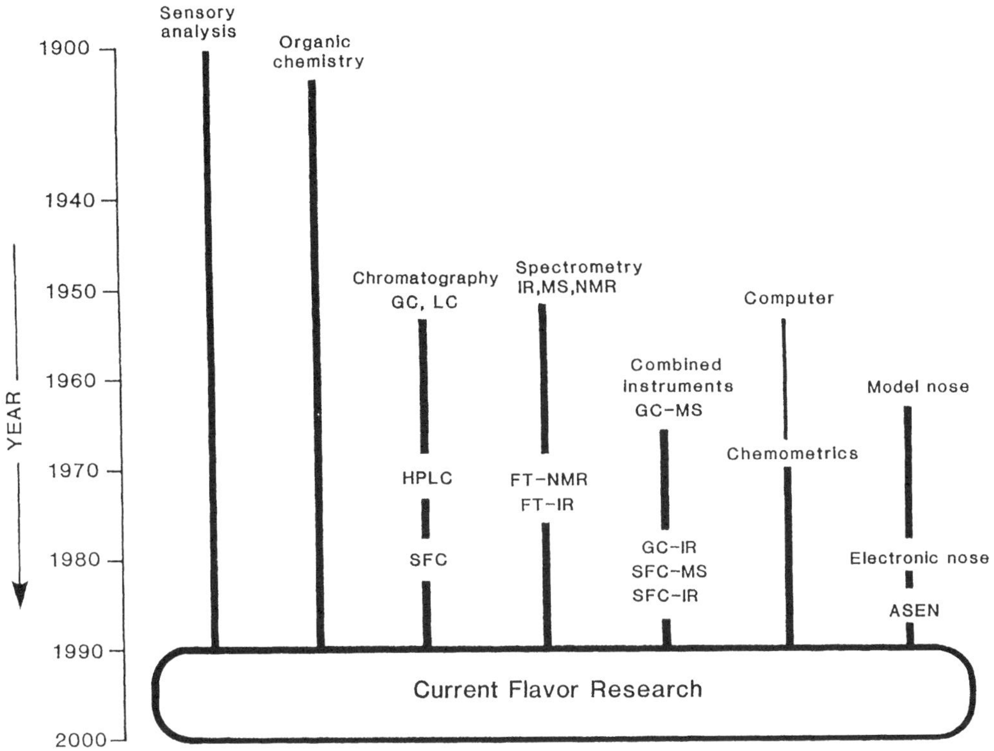

Figure 3.1 Evolution of methods employed to analyze flavors (adapted from Aishima, 1991 [6]).

responding chemical sensors combined with PARC techniques. This has been made possible through the availability of low-cost, localized computing power, increased variety and improvements in chemical sensor technologies and a resurgence of interest in classification and learning systems (including PARC).

A seminal paper by Persaud and Dodd (1982) demonstrated the power of this approach for chemical and, more particularly, flavor analysis [8]. Since this time, the technique has been applied to the flavor discrimination of many simple and complex flavors. Examples of these include coffees, whiskies, lagers and beers, fish, meat, perfumes and grains [9]. It has become clear that the notion of a universal artificial olfactory system, suitable for all types of flavor analysis problem, is still some way in the future. Electronic nose systems need to be specialized to the problem domain through the selection of appropriate sensor types and technologies. Furthermore, the type of odor delivery system. PARC techniques and preprocessing algorithms must be considered carefully in view of the specific application. These design criteria have given rise to the concept of the application-specific electronic nose (ASEN) that is customized to the problem domain [10].

3.2.3 Instrument-based Flavor Analysis

Instrumentation has become an essential tool in flavor analysis since the development of GC and GC-MS systems. Using such systems it is possible to separate out many of the numerous chemical components from a complex mixture of flavor volatiles. This process greatly simplifies the task of the analyst, as each chemical component can be analyzed individually as it elutes from the chromatographic column. A GC system is often combined with a flame ionization detector (FID) which records the presence of most organic compounds: as the flow from the column passes over the FID the signal increases; thus a characteristic series of peaks or chromatograms of the analyte is produced over time. Figure 3.2 shows a typical chromatogram taken for the first 20 minutes on a lager beer. The flavor complexity of the lager beer requires a complete run-time of 225 minutes, which results in a total of 159 separately identifiable peaks.

In 1988, Bartik clearly demonstrated the importance of GC-based techniques for flavor analysis by comparing the number of flavour volatiles identified in apples over a period of 70 years [11]. This number rose from only a few known apple aroma volatiles in 1920 to 47 in 1962 with the introduction of the first GC systems. Further improvements in GC-based technology increased this figure to 145 by 1970 and finally to 380 in 1985. To date, hundreds of complex foodstuffs have been analyzed

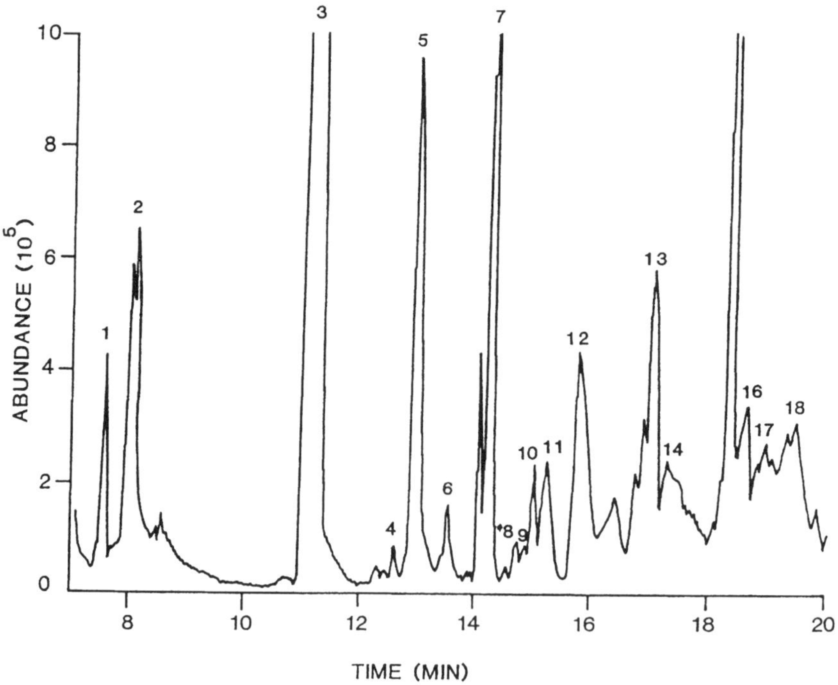

Figure 3.2 Typical chromatogram for the first 20 minutes of a 225-minute GC-run on standard lager beer (courtesy of Bass Breweries, UK).

using this instrumental approach with varying degrees of success. These include wines, sherries, whiskies, beers, perfumes, coffees, water, roasted coconut and cheeses [12–16].

Although GC is an important tool with which to characterize flavor, it measures only one aspect of the quality of flavor. In brief, the flavor analyst makes a running inventory of the compounds discovered in a particular foodstuff. This, however, takes no account of the psychophysical effects this mixture will have during the act of perception or the complex chemical interactions between individual flavor volatiles (Section 3.2.1). Moreover, whilst all odorous compounds are necessarily volatile, the converse does not hold – not all volatiles are odorous. Whilst GC detects volatile chemical species, it provides no indication of whether the detected volatile is an odor [17]. Thus, peaks in the chromatogram shown in Figure 3.2 do not necessarily correspond to actual flavor compounds. The situation is further complicated by the fact that during perception, vastly different sensitivities are displayed to individual flavor-active compounds within a complex mixture. This effect has been summarized well by Engen (1991): "The different flavour components in beer are of very differing importance, and the concentrations will vary very much. Concentration, however, will say nothing about the importance of the components. One component, for instance, may greatly influence beer flavour in concentrations below 1 ppb, while others may be present in several hundreds of ppm without being of any special importance" [18]. Hence, the estimated concentrations of flavor-active compounds, which have been calculated from chromatogram peaks, mean little without some consideration of their olfactory impact during perception. This requirements has been partially addressed by the introduction of the Flavor Unit (FU). This psychophysical measure relates the concentration C of a particular flavor-active component with its detection threshold

$$\mathrm{FU} = \frac{C}{S} \tag{3.3}$$

where S is the detection threshold in 50% of a specified sample population to the particular flavor component. As such, 1 FU corresponds to the concentration level of the flavor component within the mixture at which the average taster can just detect its presence. This measure allows the chemical components of a complex flavor mixture to be classified in terms their contribution to the overall sensation of flavor.

In fact, some flavor-active components in foodstuffs may be present in low enough concentrations to be undetectable by GC (e.g. when $C < 0.01$ ppb). Yet, these components can still be flavor-active in humans. Consequently, hundreds of compounds exist which may affect flavor, yet remain undetected by conventional analytical instrumentation.

3.2.4 Sensory Flavor Analysis and Flavor Terminology

Sensory analysis still provides the key technique in flavor analysis today, because of the superb sensitivity and discrimination of the chemical senses. The lack of qualitative information provided by current analytical instrumentation combined with the fact that the final hedonic judgement on flavor lies with the consumer also favors the human senses.

Sensory-based flavor analysis is carried out by teams of highly skilled tasters that have been screened to ensure that they are not anosmic (complete odor blindness), possess specific hyposmias (lacking sensitivity to specific odorous compounds) or hyperosmias (over-sensitive to specific odorous compounds). Various sensory testing protocols have been developed, such as the triangular taste test and the three-alternative forced choice test (3-AFC), for the detection of taints, off-flavors and also for flavor discrimination. These techniques have been well documented elsewhere by standard associations such as the American Society of Brewing Chemists [19] and the American Society for Testing and Materials [20]. The inherent nature of flavor perception leads to subjectivity is sensory analysis. Employing human subjects as analytical "instruments" can therefore lead to difficulties. As Gregson (1963) elegantly states: "Perception of the world is not a passive process, but an active and selective one. An observer records only those elements of a complex process that he can readily see and associate as meaningful. The rest he eliminates even if they are staring him in the face" [21]. Such effects are particularly acute in flavor perception. As Meilgaard (1991) points out, concept formation of colors, for example, is learnt at an early age whilst individuals are left to form their own ideas about odors and tastes [12]. As a result, semantic misconceptions regarding flavor are common. For example O'Mahoney (1978) found that one in seven university students will typically describe citric acid as "bitter", whereas one in twelve students will find quinine sulfate "sour" [22].

In view of such misconceptions surrounding flavor, a great deal of experimental control is required to conduct a meaningful sensory-based flavor trial. This need has been identified by standards associations for some time, and a vast amount of flavor research has resulted in sets of standards being laid down for various foods and beverages. For example, one of the most sophisticated standards systems has been devised by the ASBC for beer. Known as the International Flavour Terminology System for Beer, this is illustrated by the flavor wheel shown in Figure 3.3. It comprises of 14 classes, 44 first-tier terms and 78 second-tier terms (not shown). It should be noted that of the 122 separately definable elements only 40 or so of these are common and present in most beers. The remaining flavor notes correspond to particular flavor-faults or are characteristics of speciality beers [12]. The terminology system is based upon the following principles [19]:

1. Each separately identifiable flavor characteristic has its own name.
2. No terms are duplicated for the same flavor characteristic.
3. As far as possible, the meaning of each term is illustrated with readily available flavor standards.
4. The system is compatible with the *EBC Thesaurus for the Brewing Industry.*
5. Subjective terms such as good/bad, young/mature, balanced/unbalanced are not included.

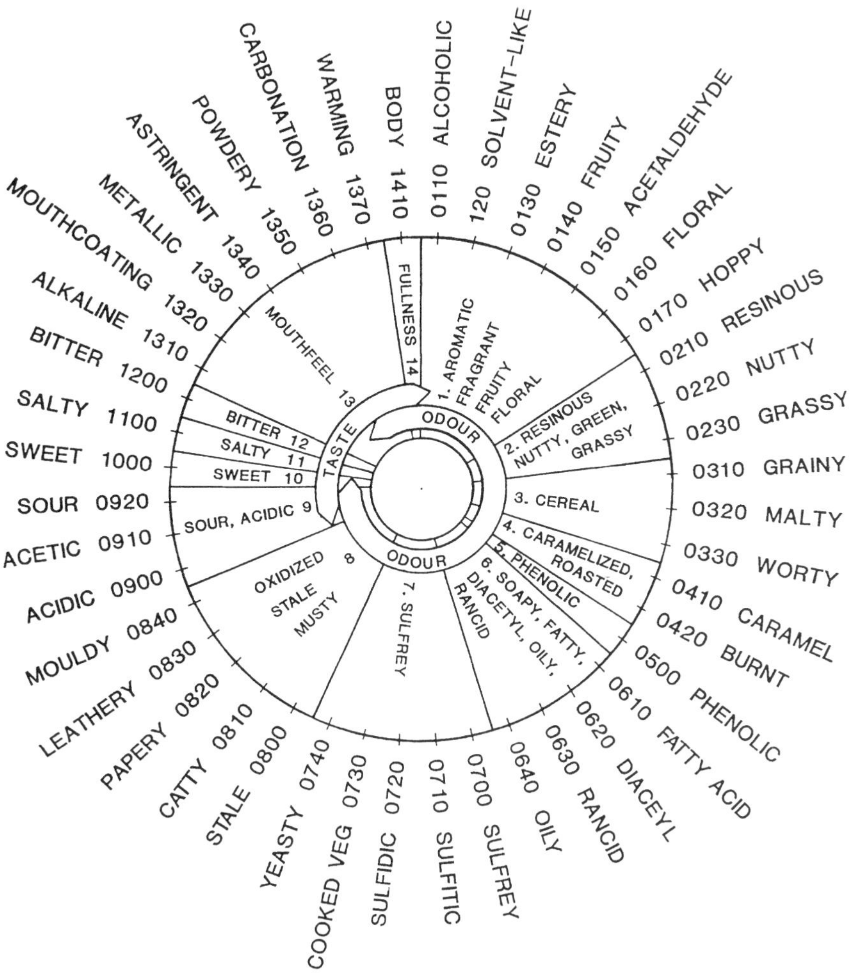

Figure 3.3 Flavor wheel used to illustrate the International Flavour Terminology System for beer [19].

6. Terminology will change with usage and the results of research – every few years a new working group should bring the system up to date.

This terminology system aims to standardize the communication of sensory flavor information used in the brewing industry, and as such is particularly useful in training and testing members of sensory panels. Training of tasters is carried out through the demonstration of individual standard reference compounds within standard beers as outlined by Meilgaard et al. [7]. This time-consuming process acquaints tasters with the individual flavor components present in beer and consequently increases their acuity.

Most sensory-based studies use flavor profile analysis (FPA), whereby the flavor notes under the terminology system present in the beer product are scored by a group of trained panelists. Each flavor note in the terminology system is scored in terms of perceived intensity on a scale of 0 to 10, where a score of 0 corresponds to "not present" and 10 to "very strong".

Whilst the use of terminology systems in sensory analysis are of great help in reducing the subjectivity of panel response, this approach is still subject to a great deal of variation on a test-to-test basis. This is unavoidable due to the nature of flavor perception in humans. For example, a vast range of sensitivities to individual flavor components is displayed by different members of the sensory panel and these thresholds also vary enormously over time (60–80% or more from one test to the next) [12].

Meilgaard also noted that the most sensitive 10% of panelists will have sensory thresholds approximately one quarter of the group mean, and the least sensitive 10% will have approximately five times above it [7]. Furthermore, most healthy people show high sensitivities to particular substances and low sensitivities to others. These factors, combined with at least 15 others (both physiological and psychological in nature) that have been highlighted by Meilgaard (1991), can lead to severe bias in sensory analysis unless great care is taken [6].

3.2.5 Combination of Techniques

After consideration of both sensory and instrumental approaches to flavor analysis, it is clear that neither of these techniques, in isolation, will provide a satisfactory explanation of flavor. Whilst GC-based techniques give objective quantitative information about chemical compounds, the relationship between these and the perceived flavor notes is not straightforward. Sensory analysis, on the other hand, provides reasonably reliable qualitative information regarding flavor, but little quantitative data due to the restricted dynamic range of the chemical senses for individual compounds and also the inherent variability in the sensitivities shown to individual flavor components. As such, there has recently been a trend to unify these disparate approaches to flavor analysis. Such an approach has been applied to the analysis of beer [12], water [23] and wine [24]. Aishima (1991) has also reviewed various studies using this approach on the flavor analysis of orange juice, milk, soy sauce, canned tuna, carrots and cabbage [6]. For example, in the flavor analysis of beer, Meilgaard (1991) has employed a combined analysis approach in order to discover causal relationships between chemical constituents and associated sensory properties.

Aishima (1991) has considered the correlation of instrumental data to sensory properties through the use of chemometric techniques and in particular PARC methods [6]. This approach is summarized in Figure 3.4. In this case the flavor researcher is interested in the cause-effect relationships to be established between a group of chemical compounds, X_i, (as measured via GC or LC analysis) and sensory

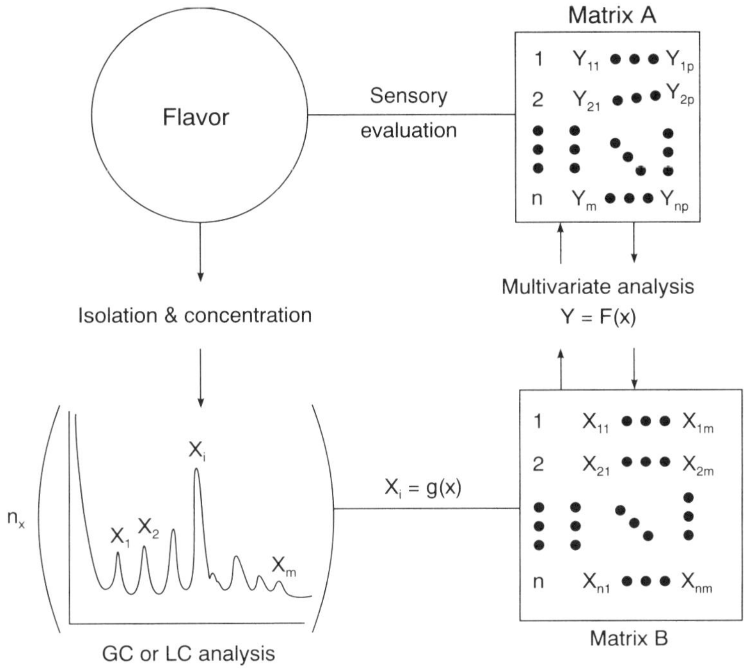

Figure 3.4 Correlating sensory properties of flavor to its instrumental analysis by using pattern recognition (reproduced from Aishima, 1991) [6].

scores, Y_i (such as those resulting from FPA). PARC techniques, such as multiple regression analysis (MRA) and canonical correlation analysis (CCA), have been used to analyze the relationship between the input and output matrices, **X** and **Y**, and in particular to:

1. Discover how the flavor intensity of added substances varies with its concentration.
2. Discover the rules of flavor interaction in mixtures by adding mixtures of compounds.
3. Conduct descriptive (FPA) testing and detailed chemical analysis on numerous beers with added reference compounds.
4. Analyze the data by MVA procedures to discover cause-effect relationships between the chemical compounds and the sensory properties.
5. Develop hypotheses: test compounds having the expected type of flavor; confirm or reject the hypotheses.

Of the 43 most common flavor notes used to describe beer, it has been possible using this combined approach to account for 15 of these [17]. In other words, an estimate of the strength of the flavor note can be calculated from the chemical analysis using the hypotheses developed in step 5 above. It has also been possible to partly explain another 18 flavor terms of the 43. Currently, the remaining 11 cannot be explained, although there may be some promising hypotheses.

In view of this trend to unify the disparate approaches to flavor analysis, there is clearly a need for more intelligent instrumentation. As such, electronic nose systems can potentially be of great benefit to flavor analysts. By virtue of the elements comprising an electronic nose system: the chemical sensor array and PARC engine, this is naturally suited to the task of predicting sensory-like (organoleptic) properties of complex flavors from their chemical constituents. It has recently been possible to demonstrate the application of such a system in providing more reliable organoleptic-like flavor descriptors to be used alongside tasting panels (T. C. Pearce, J. W. Gardner, *Analyst* (1997), in press).

Although GC-MS instruments are now combined with computer systems and capable of classifying flavor samples based upon an on-line database of flavor information, these instruments run at ten times the cost of ordinary GC-MS instruments. As such, they have become prohibitively expensive for wide spread use in flavor monitoring and in addition are not capable of continuous monitoring. It seems likely that such specialized systems will be restricted to flavor research laboratories, whereas an ASEN system could provide an economically viable alternative for **on-line** flavor monitoring.

3.3 Biological Olfaction and its Electronic Analog

3.3.1 Architecture of the Mammalian Olfactory System

We now examine the way in which sensory information is processed in the olfactory system, in order to compare and contrast the biological system and its electronic analog. Each of the layers of the sensory pathway is considered in turn, with special emphasis placed upon key processing principles.

In mammals, the olfactory system consists of three main subsystems–the olfactory mucosa with its associated odor delivery system, the main olfactory bulb (MOB) and olfactory cortex (OC). The elements of the first two subsystems are depicted in Figure 3.5.

Odor is delivered to the olfactory epithelium through a series of nasal passages, or nares. The nasal cavity consists of three parts–the anterior nares through which air is inhaled, the central chamber containing the epithelium and receptors as shown, and the posterior nares through which inhaled air exits into the nasopharynx during normal respiration. Olfaction can also occur retronasally, that is, volatile compounds can diffuse during inhalation from the mouth through the posterior nares to the central chamber. In either case, odorants that have entered the central nasal cavity diffuse through a layer of mucus to bind with the chemically sensitive membranes of the olfactory receptor cells in the epithelium. Individual olfactory cells are then

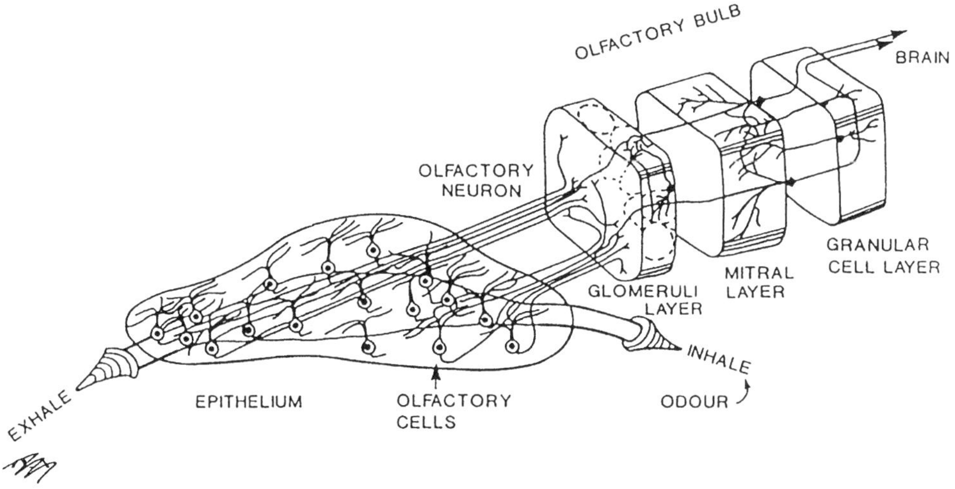

Figure 3.5 Opened view of the olfactory mucosa and exploded view of bulbar layers (from Kauer, 1987) [39].

depolarized differentially, depending on their individual sensitivities and also their position on the epithelium. Consequently, a non-uniform temporal and spatial pattern of receptor cell activity is generated across the epithelium. This pattern of activation is then transmitted (in a broadly topographic manner) to the glomeruli layer of the main olfactory bulb as an array distributed across the olfactory nerve axons. As such, the glomeruli layer represents the first synaptic relay for incoming olfactory information. This layer modulates the original activity pattern to form a smaller, second-order, array of electrical signals in the outputs of the mitral layer. The axons from the mitral layer form the outputs from the olfactory bulb leading to the olfactory cortex, where further processing takes place in the higher-order olfactory structure in the central nervous system (CNS). These structures include the olfactory peduncle, the piriform cortex, the olfactory tubercle and the entorhinal cortex [25].

3.3.2 Morphology of the Biological System

The olfactory system can be divided into a peripheral and a central component [26]. The olfactory epithelium can be considered as peripheral as this lies external to the cranium in the superior region of the nasal cavity in most vertebrates. The olfactory bulb and associated cortex are intracranial and are therefore considered to be the central component.

The olfactory epithelium which covers approximately 5 cm^2 (in humans) of the nasal cavity provides the main area of interaction with volatile chemical stimuli. In humans, the epithelium is host to between 10^7 and 10^8 receptor cells distributed

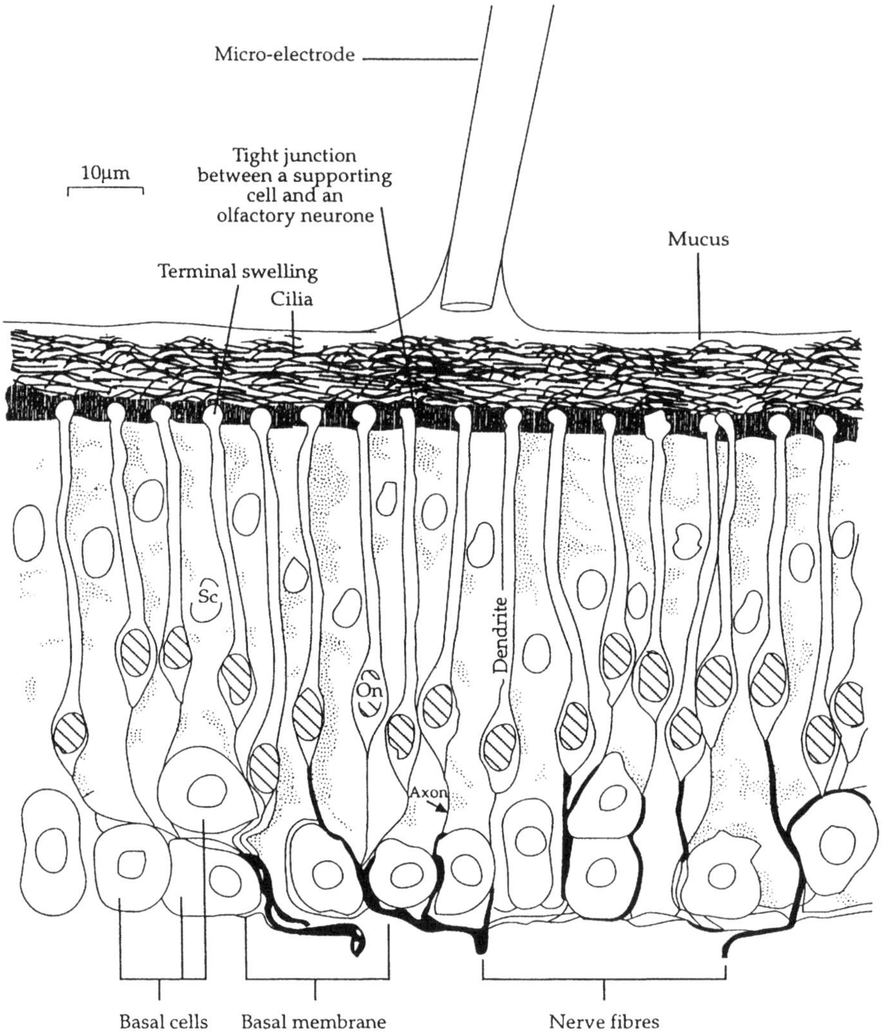

Figure 3.6 The main features of the olfactory epithelium. Olfactory neuron (On), supporting cell (Sc).

across the tissue [27]. A cross-section of the tissue, as depicted in Figure 3.6, clearly shows its morphology. The epithelium is roughly 75 microns thick and is covered by a mucus layer of about 20 microns, which arises mainly from secretions of the Bowmans glands, but also from secretions of supporting cells.

Three types of cell predominate within the epithelial structure – olfactory neurones, supporting cells and basal. Olfactory neurones are unique in the CNS due to their ability to regenerate, since they normally deteriorate after 30–45 days. This form of neurogenesis arises from basal cells deep within the epithelium and are active for the entire life of the individual [28].

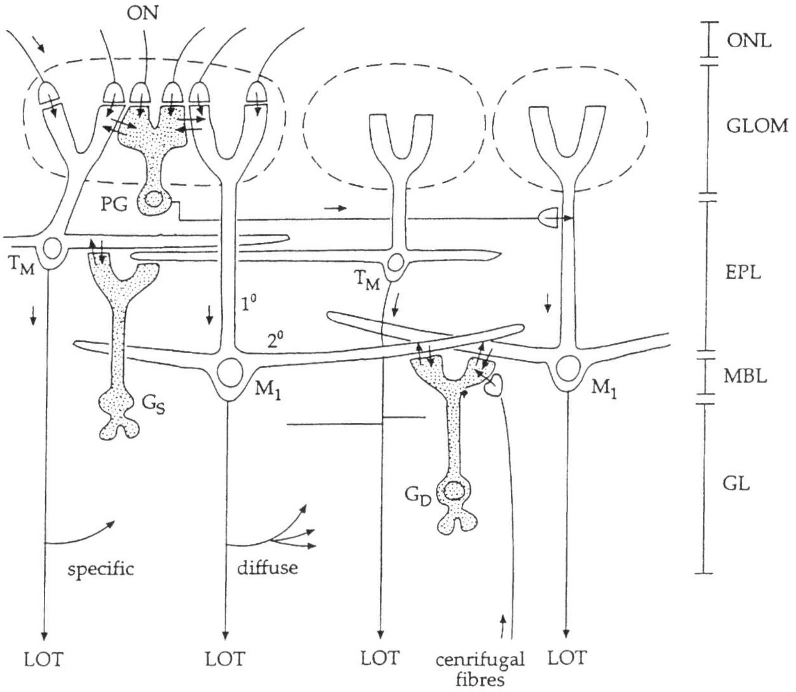

Figure 3.7 Basic circuit for the organization of the olfactory bulb. Olfactory nerves (ON) in the olfactory nerve layer (ONL) which bring convergent input to individual glomeruli (GLOM). Intraglomerular processing is through synapses between dendrites of mitral (M), tufted (T), and periglomerular (PG) cells. Interglomerular processing is through PG axons. Control of output from the olfactory bulb occurs in the external plexiform layer (EPL), by means of synapses between secondary (2°) dendrites of M/T cells and dendritic spines of granule cells (G). Different types of M/T cells connect to different types of granule cells (T_M to G_s, M_1 G_D): middle tufted cells (T_M), type 1 mitral cell (M_1), primary dendrite (1°), lateral olfactory tract (LOT), mitral cell body layer (MBL), granule cell layer (GL) from Shepherd (1991) [30].

The fascicles formed from the bundles of axons from many olfactory neurones ultimately pass through the cribiform plate to connect ultimately with the surface of the olfactory bulb. The olfactory dendrite, on the other hand, extends to the surface of the epithelium to form an olfactory knob, where a variable number (10–60) of sensory cilia arise, extending some way into the mucus layer. Importantly, these cilia have been confirmed to be the site for initial sensory transduction, as odorants bind to receptor proteins to which the cilia are host.

The main olfactory bulb in higher vertebrates is a highly stratified structure with distinct layers, as in clearly shown in Figure 3.7. The most superficial layer of the olfactory bulb is the olfactory nerve layer (ONL). This is represented by the large number of axons arriving through the cribiform plate from the epithelium. The first synaptic contact for these nerve axons within the olfactory bulb is the glomeruli layer, where several hundred olfactory neurones (as many as 10^4 in humans) synapse to form a single glomerular structure. As such, glomeruli are exclusively comprised of nerve terminals and dendrites from olfactory neurones, periglomerular cells, mitral and tufted cells.

Periglomerular cells provide both intraglomerular and interglomerular connectivity through synaptic processes between mitral/tufted and olfactory neurone axons within a glomerulus, and also through axons leading to neighboring glomeruli. The role of glomeruli and periglomerular cells is central to the nature of information processing within the bulb, as will become clear later (Sect. 3.3.3.4).

Mitral cells, located in the mitral cell body layer (MBL), along with tufted cells, located in the external plexiform layer (EPL), make single dendritic remifications with single glomeruli (1°). These cell types constitute the main outputs of the bulb, forming the lateral olfactory tract (LOT), but typical have an additional 4–6 secondary dendrites (2°) that extend laterally within the EPL and MBL [26]. These secondary dendrites form "reciprocal synapses" in the EPL with the apical dendrites of granule cells extending from the granule cell layer (GCL), the relevance of which will be considered in later sections. Centrifugal afferent input is received by the granule cells from several brain regions including primary olfactory cortex, and is thought to modulate olfactory perception depending upon the state of the organism [29].

The main output from the bulb provided by the LOT inervates many subcortical regions of the brain. These regions are collectively known as primary olfactory cortex (POC). The POC is still a relatively unexplored region of the brain, but it is known to mainly be comprised of associative connections between pyramidal cells and the LOT.

3.3.3 Sensory Information Processing

The superb performance of olfaction at molecular discrimination and characterization results, in part, from complex information processing carried out within the olfactory pathway. While olfactory receptor cells have been shown to have a low sensitivity (~ ppm), subsequent processing within the olfactory bulb and cortex enhances the overall sensitivity by at least three orders of magnitude [10]. Collectively these centers are known as early olfaction and combine to provide a key stage of sensory processing. The physiology of the early olfactory pathway consists of a highly stratified structure in order to process sensory information from the nasal mucosa – as is clearly shown in Figure 3.9. These layers carry out a series of neural computations on incoming data, providing a suitable data structure to mediate olfactory perception. We can consider the activity at each layer as a form of *molecular image* and this representation has been verified through the use of functional markers in the neural domain [30]. According to this view, successive layers within the pathway perform a series of complex temporal and spatial computations on the molecular image formed at the epithelium, which refine the odor representation to a point suitable for direct interpretation by the higher cortical regions of the central nervous system (CNS).

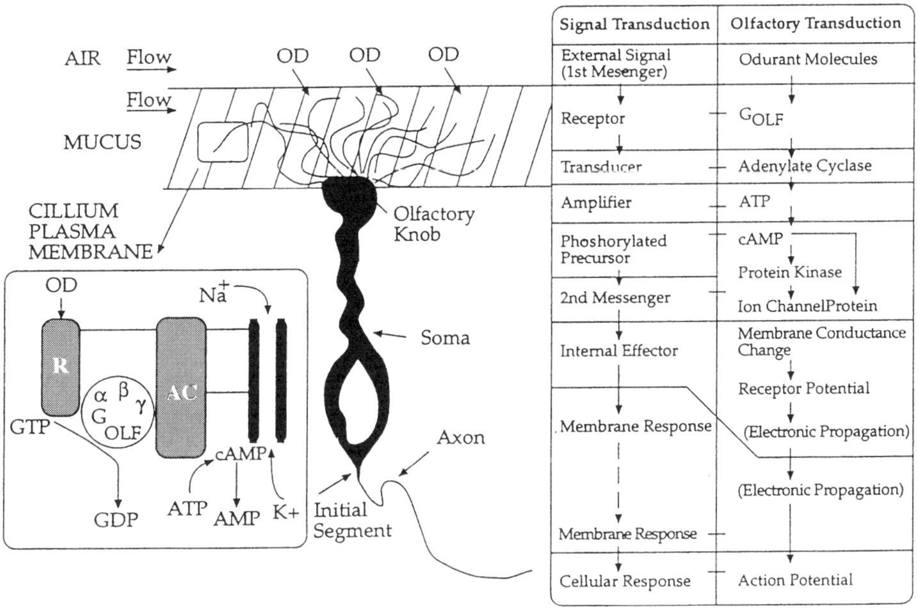

Figure 3.8 Biological model of transduction in the mammalian olfactory system (OD – odorent molecule partitioning into the mucous phase).

3.3.3.1 Odor Discrimination at the Olfactory Epithelium

An important element to understanding the way in which olfactory signals are processed is the relationship between odor stimuli and the receptive surface. In other words, how is the initial molecular image formed at the receptive surface? Clues to the nature of this sensing mechanism have been provided by early studies on the nature of odor transport to the surface of the epithelium and the response of the olfactory receptors themselves. In particular, odor transport is controlled by the physical mechanisms governing odorant delivery to the receptors, such as the aerodynamics of the nasal cavity, the chromatographic effects between the stimulus and the mucosal surface, and diffusion of odorants through the mucous layer overlying the receptor surface [31]. Consequently, molecules with different physical and chemical properties are found to be distributed in different space-time patterns across the olfactory mucosa.

This effect is demonstrated by the distribution of butanol, butyl acetate and octane, as measured at the surface of the epithelium; see Figure 3.10. Such a spatio-temporal distribution of odorants could then allow for an initial sorting of the molecules of different chemicals across the receptor sheet, providing a rudimentary form of odor discrimination [32]. This chromatographic effect of the mucosal layer has since been verified by Mozell (1971), although there is no consensus as to the number of discrete odor qualities or the degree of resolution that may be achieved by this form of physical segregation [33].

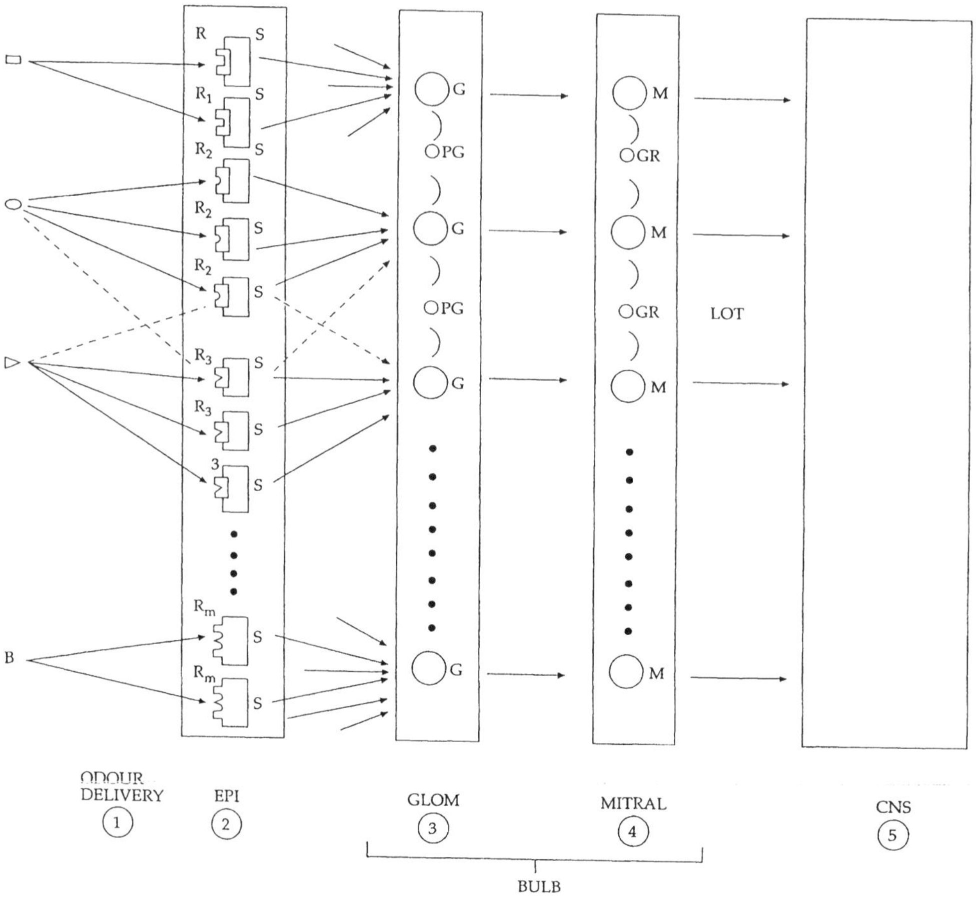

Figure 3.9 The layers of the early olfactory pathway. (1) Odorant delivery to the mucosal surface. Over 10 000 odorant types have been identified. (2) Nasal epithelium host to 2.5×10^7 sensory neurones (S) and in the range of 100–1000 receptor protein types (R). (3) Glomerular layer host to some 1000 glomerular cells (G), leading to marked convergence of 25 000 sensory cells to a single glomerulus. Periglomerular cells (PG), mediate intra- and inter-glomerular inhibition. (4) Mitral/tuffed cells provide the main outputs form the olfactory bulb in the form of the lateral olfactory tract (LOT) [38].

More recently, the discovery of a variety of small soluble acidic proteins known as odorant-binding proteins (OBPs) present in the olfactory mucus have been implicated in perireceptor olfactory events [34]. These belong to a large family of carrier proteins, called licocalins, that are generally involved in the transport of hydrophobic molecules within aqueous media [35]. Thus, OBPs may perform as weakly specific carriers of odor ligands between the mucous surface and the olfactory receptor proteins, to initiate olfactory response. Subsequent to receptor activation, OBPs may also be involved in then removing odors from the receptor sites to allow termination of the olfactory signal [36]. The discovery of more varieties of OBP, possessing weak

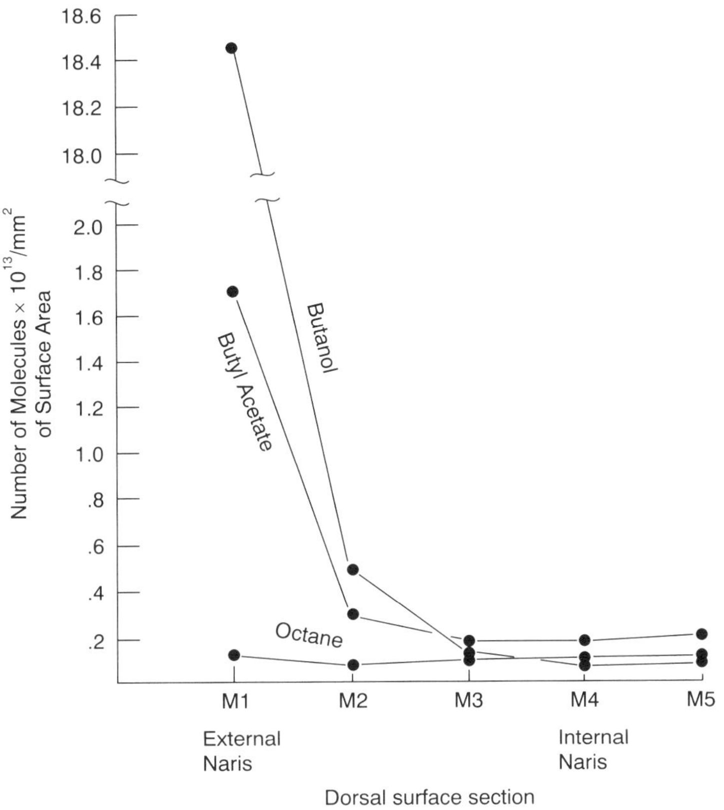

Figure 3.10 Distribution of butanol, butyl acetate and octane at different points along the dorsal surface of the epithelium, resulting from physical segregation of chemical species within the mucous layer (from Hornung and Mozell, 1986) [32].

specificities towards particular classes of odorants has suggested that these proteins could provide a further method of odor discrimination at the olfactory epithelium. This process may reinforce the odor discrimination achieved by physical segregation effects described earlier.

A third form of odor discrimination during the first stage of the olfactory pathway was discovered by early recordings of multi-unit discharges in different regions of the nasal epithelium, and observing differing patterns of activity to various odors [37]. This demonstrated, for the first time, that odor information is also encoded through the differential sensitivity of olfactory neurones across the epithelium. More recently, it has become clear that, in fact, receptor cells can be thought of as "broadly tuned", possessing a wide spectrum of sensitivity to many odorants, with a large degree of overlap of response between different receptor types. Extracellular recordings of single cell responses have made it possible to directly assess the specificity of olfactory receptor cells. For example, an extensive study by Sicard and Holley (1984),

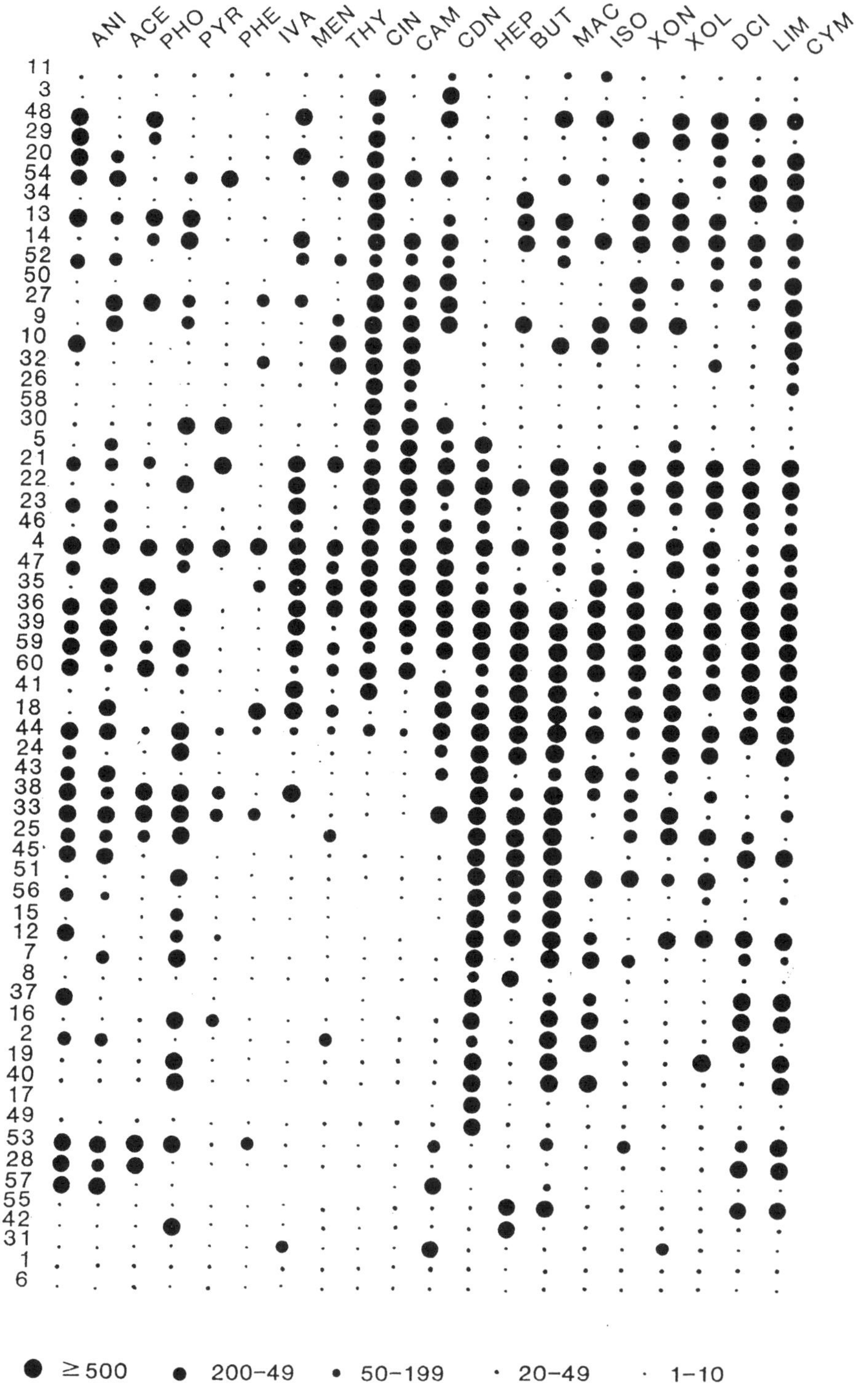
ANI ACE PHO PYR PHE IVA MEN THY CIN CAM CDN HEP BUT MAC ISO XON XOL DCI LIM CYM
11 3 48 29 20 54 34 13 14 52 50 27 9 10 32 26 58 30 5 21 22 23 46 4 47 35 36 39 59 60 41 18 44 24 43 38 33 25 45 51 56 15 12 7 8 37 16 2 19 40 17 49 53 28 57 55 42 31 1 6
≥500
200–49
50–199
20–49
1–10

the results of which are summarized in Figure 3.11, clearly demonstrates the diversity of olfactory receptor response – with each receptor cell displaying a broad but unique profile of sensitivity to the battery of odorants. A considerable amount of overlap between the responses of the cells is also evident, for example, the majority of units responding to camphor (CAM) also respond to cineole (CIN).

In the biological system, overlapping partial specificity at the receptor level of the olfactory pathway clearly provides a high degree of redundancy within the system [38]. This redundancy provides a number of potential benefits. Firstly, a signal of high information content is provided to the first stage of processing, so that the system can suffer massive trauma and still operate effectively. Evidence for this is provided by cases where the removal of large sections of the olfactory bulb have only resulted in restricted perception of odor intensity as opposed to affecting the overall range of odors that can be appreciated. Secondly, Shepherd (1991) has argued that redundancy through overlapping response spectra of the receptors allows the discrimination of odor quality independent of its intensity [30]. Thirdly, the biological system is in a prepared state for a wide range of stimuli, both familiar and unfamiliar, since the non-specific characteristics of the receptors ensure that as broad as possible sampling of the odor space is taken. This provides the olfactory sense with a highly desirable xenobiotic quality that is essential in an environment where there is little correlation between one olfactory stimulus and the next (over time). That is, the next odor cannot be predicted and may not even be familiar [27]. Given such unknown operating constraints the inverse method of biological implementation, as it were, where highly specific receptors are used by the nose seems untenable since the system would need to generate a new receptor class for each newly encountered odor. Such an arrangement would be difficult to envisage in evolutionary terms. It is suggested that the high number of overlapping receptor classes is exploited by the biological olfactory system to implement the processing principles listed above.

As such, both physical segregation mechanisms and differential neuronal sensitivities are employed by the olfactory system for the coding of odor quality and intensity at the epithelium. Physical segregation mechanisms "impose" non-homogeneous delivery of odorant stimuli on to the receptor surface, while "inherent" spatial patterning results from differences in neuronal sensitivity [27]. Interestingly, Kauer has suggested that subpopulations of receptors having similar sensitivity may exploit the physical segregation of incoming stimuli, by optimising their location on the epithelium [39]. In this way, subpopulations of receptor classes selective to particular types of chemical structure may be able to maximize their response through their physical location on the epithelium.

Figure 3.11 Diagrammatic representation of olfactory receptor cell activity following stimulation. The spot size is roughly proportional to spike frequency (spikes/min). Receptor cells are identified by a serial number in the left column. ACE–acetophenone, ANI–anisole, BUT–*n*-butanol, CAM–*dl*-camphor, CDN–cyclodecanone, CIN–*1,8* cineole, CYM–*p*-cymene, DCI–*d*-citronellol, HEP–*n*-heptanol, ISO–isoamylacetate, IVA–isovaleric acid, LIM–d-limonene, MAC–methyl-amylketone, MEN–*l*-menthol, PHE–phenol, PHO–thiophenol, PYR–pyridine, THY–thymol, XOL–cyclohexanol, XON–cyclohexanone (from Sicard and Holley, 1984) [95].

3.3.3.2 Olfactory Receptors and Signal Transduction

The receptor neurones in the epithelium provide the primary processing element of signal transduction in the olfactory pathway. As such, this processing element must provide a number of key functions. Firstly, intensity grading of the stimulus occurs through the selective transduction of the chemical species, resulting in an electrical signal in the olfactory nerve fibers. Secondly, quality coding requires the transcription of all or a significant part of the molecular structure into information modalities–as viewed by Beets (1970) [40]. Furthermore, signal transduction requires the amplification of the primary energy gained by the adsorption of a few odorant molecules to a level sufficient to trigger an electrogenic response of the cell. All of this must be achieved in a relatively small time-frame (a few 100 ms) and without changing the nature of the stimulus.

The last decade has seen rapid progress in our understanding the mechanisms of olfactory transduction which give rise to this impressive process capability. According to Shepherd (1991), from practically no information one decade ago, we now have a "consensus working model" of the main steps in the transduction pathway [41]. Work towards this model began with the verification of the cilia being the initial site of transduction in the olfactory receptor. Early investigations demonstrated how the electro-olfactogram (EOG), a primitive method for measuring the combined electrogenic response of a population of neighboring olfactory neurones, ceased in deciliated receptors and recovered with the regrowth of the cilia over time [42]. Furthermore, recent unequivocal evidence was provided by Firestein et al. using micro-pulse odor stimulation directed towards the cilia and also by computational modeling of receptor neurones themselves [43, 44]. Since modified-cilia and cilia-like microvilli predominate in many other sensory systems (such as the photoreceptor and hair cells), olfactory transduction was shown to be analogous to the other transduction processes, and this signaled the entrance of olfactory transduction into the mainstream of membrane signaling. From here, research has since centered on the mechanisms within the cilia that could be responsible for initial signal transduction. Much of this work has been based upon the finding of Pace et al. (1985) that adenylate cyclase activity was found in olfactory cilia preparations at ten times the normal level observed in any part of the neuroepithelium or the brain [45]. This immediately suggested a whole transduction cascade similar to that uncovered in photoreception, implicating a G protein-mediated cyclo-adenosine monophosphate (cAMP) cascade that was capable of directly gating cationic channels, resulting in a membrane potential and ultimately the depolarization of the cell to produce an action potential. This hypothesis was partially confirmed when Nakamura and Gold (1987) displayed direct gating of the membrane conductance by cyclic nucleotides (such as cAMP and cGMP) [46]. Recently, Firestein et al. (1991) have also shown that not only is there a cAMP activated membrane conductance in olfaction, but that this is also activated directly by odor stimuli [47]. It is now clear that cAMP along with its specific G-protein mediator, G_{olf}, provides a second messenger system in olfactory transduction.

The final component in the olfactory transduction pathway, the long-sought odor receptor, has finally been isolated and cloned in the seminal work by Buch and Axel [48]. The discovery was based upon two predictions – firstly that the receptors may be linked to the G-protein messenger systems (as was found to be the case with light-responsive rhodopsin molecules of the retina) and secondly that these receptors may be members of a large multigene super-family, as suggested by the theoretical arguments of Lancet [49]. This putative olfactory receptor is a member of the family of 7-transmembrane-domain (7TD) proteins that have also been implicated in other G-protein secondary messenger systems. The family shows considerable sequence variation in transmembrane segments 3, 4 and 5, potentially leading to thousands of distinct receptor types of varying specificities, making it one of the largest families in the human genome. Buck and Axel (1991) isolated 18 putative odorant receptor proteins in their work but estimated the number to be between 100 and 300 genes in all. This number of receptor types is far more than many workers expected and thus poses challenging questions regarding the coding of olfactory information in the initial stage of olfactory processing. These 18 clones could be grouped into three subfamilies with higher degrees of sequence homology. It is possible that these subfamilies may correspond to major structurally distinct categories of odorants, and individual members of each subfamily may detect different, but structurally related odorants [50].

Our current understanding of olfaction transduction has been summarized by the physical and biochemical pathways shown in Figure 3.8. The elements of the transducution pathway are also given in a flow chart. Odorants (OD) are generally light hydrophobic polar molecules which are mostly soluble in a lipid membrane. They diffuse rapidly into the mucus flow and reach the neuroepithelial cilia. The interaction of the odorant with a receptor (R) stimulates adenylate cyclase (AC) by interaction with G-proteins (G_{olf}) to produce cAMP which binds with the cation channel and finally promotes neural signalling (adapted from [51] and [52]).

3.3.3.3 The Role of Convergence in the Olfactory System

The first synaptic connection in the olfactory bulb for nerve fibers arriving from the epithelium occurs in the glomerular layer shown in Figure 3.9. The topographical order of this projection between the epithelium and the glomerular layer is fundamental to the way in which information is processed in the olfactory bulb. A key advance in the area of olfaction has been the ability to study the topology of these connections – the projection displays a complex mixture of divergent (from localized mucosal areas to widespread glomerular areas) and convergent (from widespread mucosal areas to localized glomerular areas) connectivity. Overall, strict point-to-point mapping is not observed, suggesting that the topology of connection is determined primarily by the ensemble specificities of receptor neurones, and only secondarily by the spatial constraints of the stimuli [53].

In order to illustrate the topological structure of this connectivity Kauer (1987) has provided a formal framework summarized in Figure 3.12 [39]. Here, receptors with

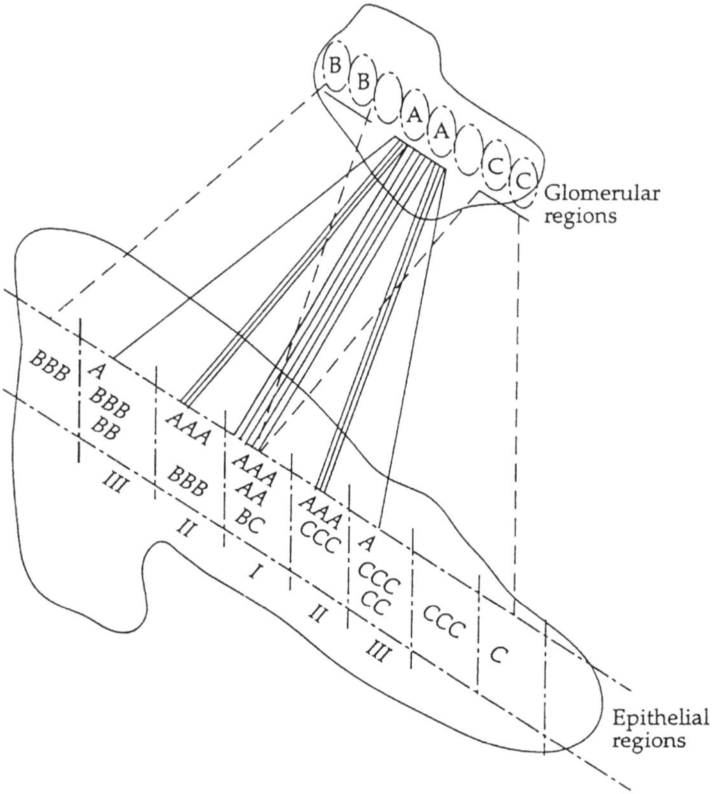

Figure 3.12 Simplified representation of (1) hypothetical distribution of receptor sensitivities across the nasal epithelium to perhaps exploit physical segregation of odors during delivery to the epithelium, (2) convergent and divergent projections to the glomerular layer, and (3) possible different terminal synaptic connections from clustered receptor cells onto mitral/tufted cells versus distributed receptor cells onto periglomerular cells (from Kauer, 1987) [39].

similar spectra of sensitivity are shown to be closely located on the epithelium, as suggested by recent studies (receptor classes A, B and C). Also, Kauer implies in this formalization that convergent connections to a localized group of glomeruli are those coming from receptor cells having similar receptor genes and hence similar spectra of response (formaing receptor class A, for example). This theory has recently been corroborated by monoclonal antibody staining studies [54]. According to such a scheme each glomerulus would act as an "address" for a clone of sensory cells expressing a given olfactory receptor gene and decodes a single molecular determinant or "olfactophore". Divergent connections (localized areas of the epithelium projecting to widespread areas of the glomeruli layer) occur naturally under this model, as can be seen from Figure 3.12. For example, region I of the epithelium holds neurones connecting to all three glomeruli regions (A, B and C).

Marked convergence between the receptor neurones and single glomeruli is a striking feature common to all vertebrate olfactory systems [34]. This suggests that it is a key organizing principle in the olfactory system. In mammals, roughly 25000 olfactory neurones project to a single glomeruli and clearly there is an important combinatorial function for this connectivity – in that a complex mix of convergent and divergent connections can be used to provide varying degrees of specificity in glomeruli response. It now seems likely that such a connection scheme is ideal for providing a more specific, functional representation (in the form of a molecular image) [26].

Averaging of receptor input also provides an important noise reduction method, thus increasing the certainty of measurement. In particular, if there are 25 mitral cells associated with each glomeruli structure, then the average number of olfactory neuron axons terminating on each mitral cell is around 1000. The reduction is noise of this summation of receptor input at the apical dendrite is then $\sqrt{1000}$ or about 30-fold, providing an enormous potential for increased sensitivity at the first stage of processing [55]. This substantial increase in sensitivity in the olfactory bulb as compared to primary olfactory neurones has been demonstrated by Duchamp-Viret et al. (1989), through measuring primary and secondary neuron responses to the same odor stimulus [56].

Convergence may also provide constant response of the olfactory system in view of a continual turnover of primary olfactory neurones. Ontogenic studies of primary neurones have shown that initially these are responsive to nearly all odorants in a test battery and become more selective with muturation. Hence the summed input from a subpopulation of receptors at various stages of development may provide a more constant signal [57].

3.3.3.4 Glomeruli Cells as Functional Units

The most obvious constant in the organization of the olfactory pathway, in both invertebrates and higher vertebrates, is that receptor axons do not terminate in a continuous sheet of synapses in the olfactory bulb, but rather the terminals are segregated into groups of structures known as glomeruli as can be seen in Figure 3.9 and Figure 3.7. This suggests that glomeruli act as important functional units in olfactory information processing that have been retained throughout the evolution of the olfactory system.

The initial molecular image formed at the epithelium is clearly subject to a great deal of conformation during its transmission to the glomeruli layer, as a result of the complex topological structure in the axons connecting the two layers. A key question has been the nature of the resulting spatio-temporal activity at the glomeruli level, and in particular whether this provides a more specific or functional representation of odor quality and intensity. Multi-electrode recordings made at the glomeruli layer first suggested that this layer may act as a kind of molecular map [58]. More recently, it has been possible to use activity mapping methods (such as 2-DG radiography) to reveal such distinct patterns of activity across the glomeruli during odor stimulation;

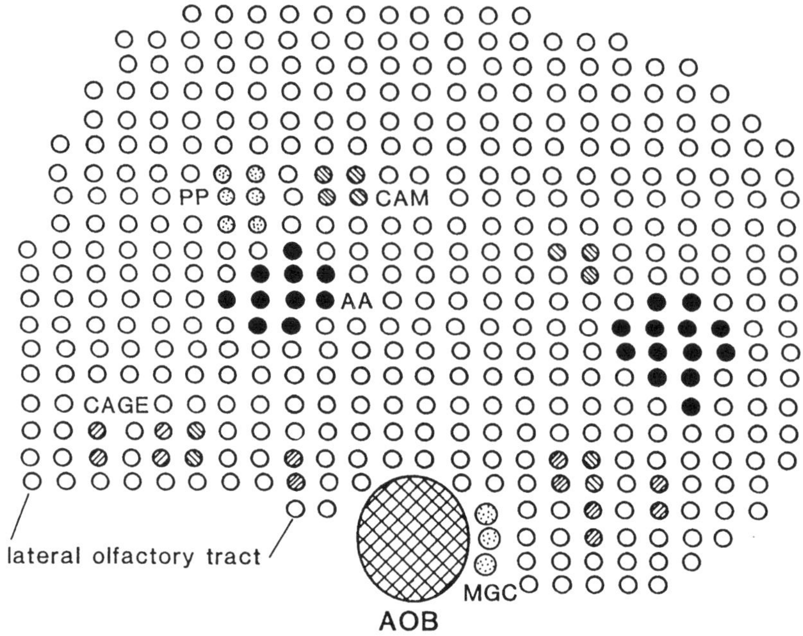

Figure 3.13 Maps of 2-DG activity in the glomerular layer of the rat to five odor conditions: amyl acetate (AA), camphor (CAM), cage air (CAGE), suckling related to the modified glomerular complex (MGC), peppermint (PP). AOB–Accessory olfactory bulb (from Shepherd, 1991) [30].

see Figure 3.13. Even at this early stage in the olfactory pathway, it is possible to discriminate between different odor conditions on the basis of these 2-DG activity maps, showing that the glomeruli layer does indeed provide a more compact and ordered representation of odor quality and intensity [59].

The successes in classification of glomeruli activity have suggested that they may act as functional units, that is, each glomeruli may be largely responsible for particular odor qualities or molecular determinants. It seem likely that each glomeruli has a unique mix of inputs, so that it may be regarded as a complex "labeled line" for that particular mix, thus providing greater specificity [60]. In this way, glomeruli may perform a similar function to first-order cells in vision – their primary role being "grouping" and "discrimination" of complex sensory input, thus providing a coarser (but more compact) representation of odor quality and intensity.

Not only do glomeruli cells play an important role in the organization of sensory input, they also provide a form of dynamic range compression of receptor input. Within the glomeruli layer, periglomerular cells (PG as shown in Figure 3.9 and also Figure 3.7) provide both intra- and inter-glomeruli connectivity through synaptic processes between mitral/tufted and olfactory neuron axons within single glomeruli and also through axons leading to neighboring glomeruli cells. By inducing test pulses in the nerve axons of olfactory neurones, $I(t)$, and measuring the responses of periglomerular and mitral cells, $O_{\mathrm{pm}}(t)$, Freeman (1991) has been able to show that long-lasting logarithmic attenuation of output occurs to subsequent test stimulus that depends upon the output of the bulbar neurones, $O_{\mathrm{b}}(t)$, in the form [61]

$$O_{\mathrm{pm}}(t) = I(t).\exp[-O_{\mathrm{b}}(t).t]. \tag{3.4}$$

This transfer function of the glomeruli layer is in the same form as that which describes light adaption in the retina and reflects a process of logarithmic compression of the input range. It now seems likely that a key function of the glomeruli layer is to scale receptor input as a sort of automatic volume control (AVC), thus allowing odor quality discrimination independent of odorant intensity [26].

3.3.3.5 Mitral Cells Perform Odor Opponency

The next stage of processing is the principal neurones of the olfactory bulb. These are composed of mitral (shown as M in Figure 3.9) and tufted cells. Mitral/tufted (M/T) cells ramifying with glomeruli structures provide the main outputs from the olfactory bulb in the form of the LOT, as shown in Figure 3.9. In mammals, up to 50 M/T cells connect exclusively to each glomeruli, and these act as relay neurones under complex inhibitory control by microcircuits within the bulb [6,32]. The tight electronic coupling of glomeruli and M/T cells would suggest that differential activity maps observed in the glomeruli layer (such as those shown in Figure 3.9) during odor stimulation should be reflected in the activity of cells within the mitral layer. This relationship has recently been verified by the application of activity mapping techniques to these sections of the bulb [62].

Even though mitral cells show relatively broad response spectra to different odors, inhibitory feedback between M/T and granule (shown as G in Figure 3.9) cells give rise to three categories of temporally complex responses (excited–E-type, inhibited–S-type and unaffected by a given odorant). For any one odorant all three response types have been observed across the M/T cells population, although Mori (1994) has shown that the active focus of excitation shifts in response to a series of structurally related molecules [63].

While M/T cells primarily serve to relay sensory information to the higher cortical regions of the CNS, these inhibitory interactions may mediate antagonistic interactions between cells that are preferentially activated by different odor ligands [26]. Such a mechanism may then prevent certain odor conditions being reported simultaneously. These antagonistic interactions between mitral cells of the bulb (mediated by dendro-dentric inhibition by the granule cells) are similar to the "color opponency" mechanism in color perception: where a ganglion cell may be excited by one particular frequency, yet inhibited by another [64]. In olfaction, a multi-dimensional version of such an opponency mechanism (coined "odor opponency") may provide steps toward building the representation of molecular information into a content-addressable form [26].

3.3.3.6 Associative Memory Function of the Piriform Cortex

The fundamental task facing any sensory system must that of object recognition. In terms of smell this requires the recognition and identification of characteristic combinations of molecules in complex blends. Implicated in this process is therefore some form of associative memory where a complex temporal pattern of inputs, comprising sensory data, is transformed into a stable pattern of outputs, representing the perceived odor. How this is achieved in the olfactory pathway can be clarified by considering the physiology of the piriform cortex.

Outputs from the olfactory bulb, in the form of the LOT, enter a number of cortical regions of the CNS including the piriform cortex. The piriform cortex consists of pyramidal cells organized in three distinct layers (Figure 3.14):

1. The superficial plexiform layer is composed of apical dendrites of the pyramidal cells composing layers 2 and 3. These dendrites synapse with axons of the LOT within layer Ia, providing the main connectivity between the olfactory bulb and cortex. Layer Ib consists of an intrinsic association fibers of pyramidal cells.
2. The superficial soma layer is composed of densely packed somata of superficial pyramidal cells.
3. The deep soma layer is composed of the basal dendrites of superficial pyramidal cells and low-to-moderate density of deep pyramidal cells.

Due to the similarities between this cortical structure of the piriform cortex and key memory areas of the brain, such as the entrohinal cortex and hippocampus, it has long been suggested that this region may be responsible for evoking previously stored odor memories. Further evidence for the role of the piriform cortex in odor memory has been provided by electrically stimulating the LOT and showing that in rats such "electrical odors" are perceived and remembered as if they were natural odors [65].

The archetypal associative structure of the piriform cortex has lead to comparisons with existing models of autoassociative memories. These common features are clearly shown in Figure 3.14. Studies of associative memories have demonstrated that these are capable of providing stable and consistent output patterns in responses to previously encountered input patterns [66]. Furthermore, these models are capable of an amount of generalization–an important function of olfactory memory, where odor recognition may have to be achieved given a variety of background odors [67]. These properties, combined with the ability to deal with degraded inputs and provide good pattern separation, make associative memory models very attractive for the task of odor memory and retrieval. We shall later examine how PARC techniques, such as associative memory models, constitute the core of an electronic nose system.

3.3.4 Sensor-based Machine Olfaction

The concept of the electronic nose system emerged from progress in a number of key research areas. The first of these included early studies on finding materials sensitive

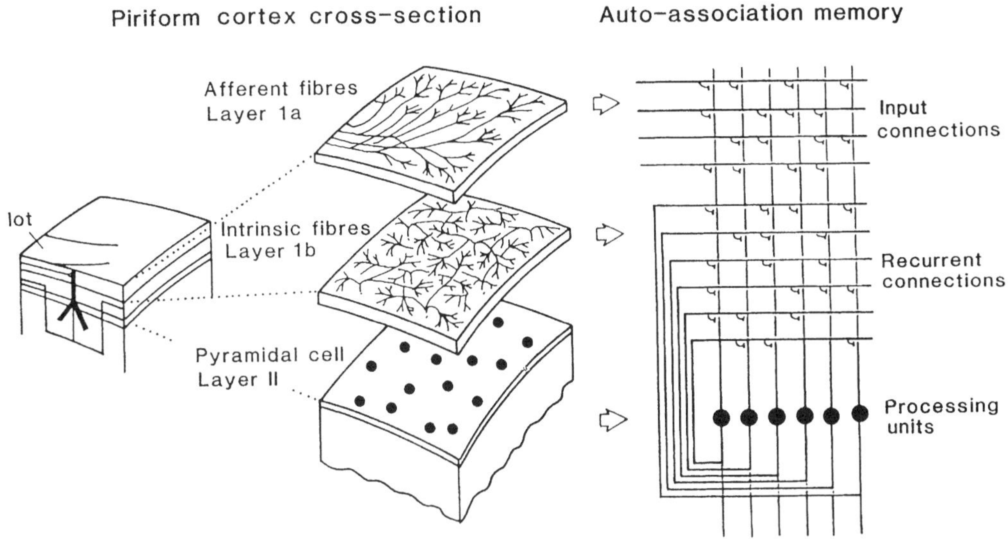

Figure 3.14 Comparison of piriform cortex physiological structure with an autoassociative matrix memory. In layer Ia, axons from the olfactory bulb, in the form of the lateral olfactory tract (LOT) form synapses with the dendrites of piriform cortex pyramidal cells, showing little topographical organization. This connection scheme may be analogous to the input connections to an autoassociative matrix memory. In layer Ib, intrinsic fibers from pyramidal cells synapse with other pyramidal cells in layers II and III, forming auto associative connections. This connection scheme can be compared to the recurrent connections in the autoassociative memory model (from Hasselmo et al. 1990) [104].

to chemical stimuli. One of the earliest of these studies–albeit mechanical rather than electronic in nature–was the measurement of changes in surface tension following adsorption of an odorant [68]. With the advent of semiconductor materials, new opportunities for chemically sensitive electronic sensors became available, and as early as 1953, Brattan and Bardeen performed the first preliminary studies on the possibility of a semiconductor chemical sensor. Another key area of research providing inspiration for an odor sensing system was the modeling of information processing within the biological system, as discussed in the previous section. A seminal work by Deutsch (1967) provided a formal framework for information processing in the olfactory pathway and laid the foundations for a generic "model nose" and in particular the concept of partial specificity of sensor response for odor coding [69]. These factors, combined with a resurgence of interest in learning systems and classification techniques, contributed to the possibility of electronically sensing odors.

The term "electronic nose" was used around the late 1980s and defined by Gardner and Bartlett [10] as: *"An electronic nose is an instrument, which comprises of an*

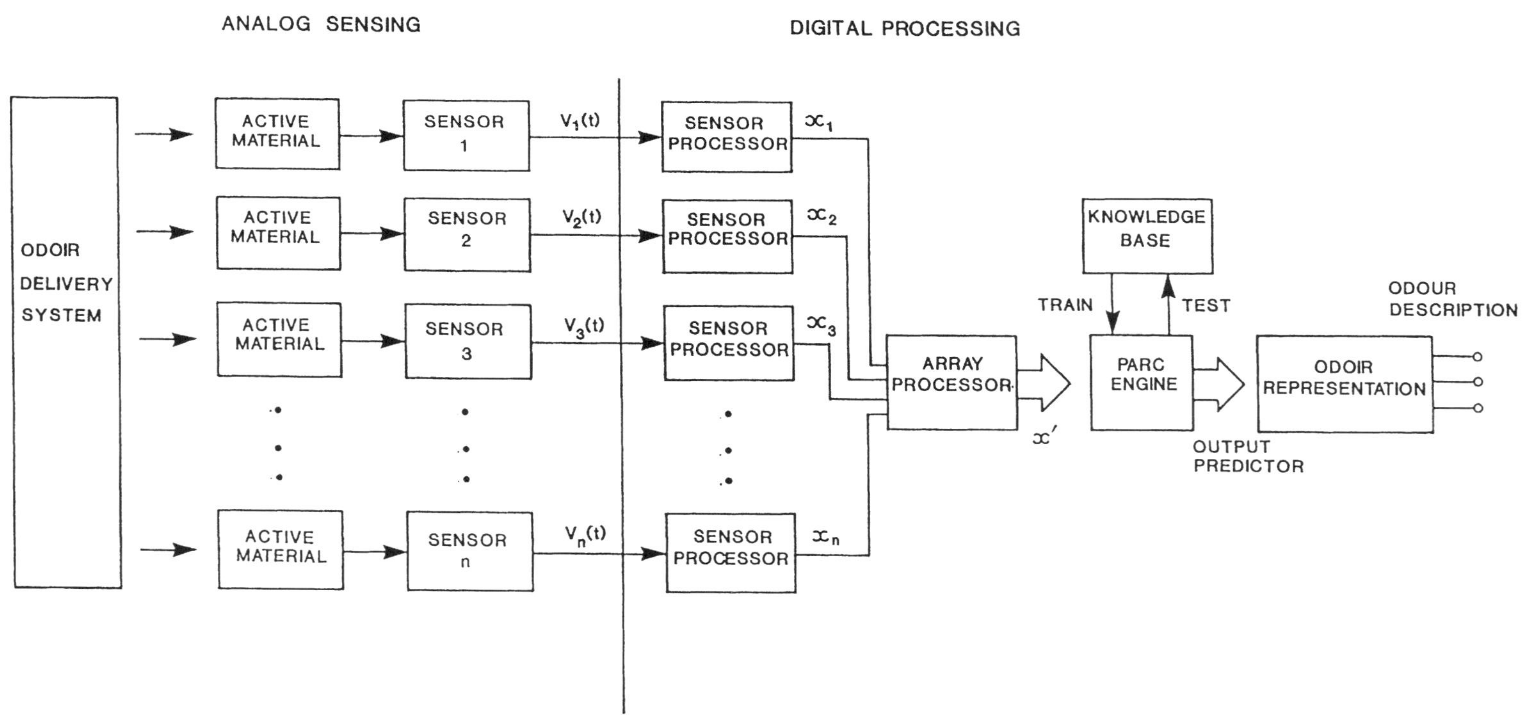

Figure 3.15 Generic architecture of an electronic nose [8].

array of electronic chemical sensors with partial specificity and an appropriate pattern recognition system, capable of recognising simple or complex odours."

Employing this definition, we can construct a generic architecture of an electronic nose system, see Figures 3.15 and 3.16 below. The system comprises an odor delivery or sampling system, an array of chemically sensitive sensors with partial specificity, signal preparation (in the form of sensor and array processors), a pattern recognition engine and some form of odor representation.

An electronic nose system may possess some form of odor delivery to the sensor array in order to elicit a response. Early electronic noses relied on a simple odor delivery system in which the sensor array was located in a closed vessel and odor samples were injected to provide an odorous head-space. Although head-space injection provides a quick method of odor analysis, it lacks reproducibility (due to experimental errors introduced by a human operator) and is also manually intensive. More recently, flow-based delivery systems have been implemented, which allow superior control over the odor stimulus and can also provide automated sampling.

Flow-based odor delivery in the biological system is combined with a complex form of physical segregation effected by the mucosal layer of the epithelium (Section 3.3.3.1). This segregation allows for an initial sorting of molecules (by molecular mass) to different portions of the receptor sheet. In this way receptors with a particular affinity for certain types of odor can maximize their chemoreception through optimizing their physical location on the receptor sheet. How such segregation might be usefully employed within the electronic nose has not yet been fully investigated, although the combination of gas chromatographic-electronic nose (GC-EN) instrument that can be used for measuring complex odors has not yet been sufficiently considered.

An interesting aspect of odors is that an odorant and its source are not necessarily chemically identical. As Gibson (1966) pointed out, "the odour of a source specifies it but it is not chemically identical with it" [70]. This effect is clearly demonstrated in a complex mixture of molecular species, where diffusion in an open environment will take place at differing rates depending upon the molecular mass of each species. This leads to relative concentration differences between each of the components in space and time, causing the quality of the overall stimulus to change. Shurmer and Whitehouse (1993) have considered this aspect of odorants [71]. They agree that, "the very idea of smell being an intrinsic property of a substance, and as such being sensed by a receiver, is false. It has a temporal, spatial, human equation which is demonstrated by anyone's experience". Importantly, they highlight the need to consider such effects for odor measurement, since it may not be possible to devise suitable standards against which to compare complex odors (similar to Standard Temperature and Pressure [STP]). It will not be possible to develop a comprehensive performance definition for electronic nose systems until such odor standards have been developed. In terms of the odor delivery system for an electronic nose, such a diffusion effect can be minimized using an FIA or combined GC-EN system, since odor transport takes place under forced convection of the head-space during the sampling process. Such considerations are of course paramount to the design of a portable odor sensing instrument.

Since the early studies on chemically sensitive materials suitable for use in a electronic nose, we now have a large variety of sensor technologies to choose from. In contrast to the ideal gas sensor, for detecting specific components (which should be highly specific to a single chemical species only), sensors for use in an electronic nose need to possess partial specificity, similar to the "broad tuning" response of olfactory receptors in the nose (as demonstrated by Figure 3.11). In this way the pattern of response across the sensor array provides odor quality information, rather than the response of any one particular sensor.

The benefits of partial specificity of biological olfactory neurones has already been discussed (Section 3.3.3.2). Comparable partial specificity of odor sensors within the electronic nose provides similar benefits of high informational content (providing a degree of fault tolerance) and also xenobiotic capability (an array comprising sufficient numbers of broadly tuned elements is responsive to a whole range of odors not considered at the design stage, or maybe even discovered yet). These benefits result directly from adopting a similar processing architecture to the biological system. However, the use of partially specific sensors for the discrimination of odor quality, independent of intensity, needs further investigation. Further investigations on mimicking this neural mechanism may be of direct benefit to future electronic nose systems [72].

Currently, the most common materials for electronic nose systems are metal oxide semiconductors and polymers. To illustrate typical similarities and differences in Table 3.1 the properties of a typical olfactory neuron are compared with a tin oxide semiconductor, both commercial and thin film, and of the conducting polymer chemoresistors. Recent trends to use nonconducting polymers and other more suitable sensor active materials are discussed below in Chapter 4. The partial specificity of the device can be judged by the molecular receptive range (MRR) given in the table. For all the devices this is wide, although the insensitivity of SnO_2 semiconductor sensors to nitrogen- and sulfur-based organic compounds narrows this range considerably. Polymer devices promise a virtually limitless supply of materials possessing different specificities to different odors. Significantly, functional groups can be built into the polymer or supramolecular compound in order to interact with particular odorants of interest (see Chapter 3.4 also) [10]. This flexibility combined with low operating temperatures and ease of fabrication make such organic compounds particularly attractive for use in an electronic nose.

Importantly, all forms of device, both hardware and wetware, show some form of intensity coding and this is, of course, a primary requirement of any odor sensor. Perhaps not surprisingly, it is clear that the fabricated odor sensors have an inferior operating specification when compared with their biological counterpart, the olfactory neurons. For example, olfactory neurons possess a much faster response time in comparison with odor sensors (20 to 60 s), and this is a limiting time factor in the response of any artificial odor sensing system. Also, olfactory receptors require far less power (*ca.* nW), and although the relatively high power requirements of artificial devices is not currently an issue in existing electronic noses (since today these generally employ less than 50 sensors) it will become a critical factor as array sizes become larger, and portable instruments are also developed. Clearly physical device

Table 3.1 Comparison of the properties of different types of sensor element with those of the biological system (adapted from Barlett and Gardner 1992) [9]

Property	Thick Film (SnO_2)	Thin Film (SnO_2)	Polymeric chemoresistor	Olfactory receptor
Fabrication	Manual	Sputtering Large batch	Electrochemical Growth	Selective expression of cell pool – adaptive
Size	3×1mm	< 1 mm^2	< 1 mm^2	Submicron
Key measurand	Resistance	Resistance	Resistance	Spike frequency
Integrated array	No	Yes	Yes	Yes
Choice of material	Limited	Limited	Wide	Wide
Operating Temperature	250–600 °C	250–600 °C	10–110 °C	37 °C
Power consumption	100 mW	100 mW	$<$ 10 mW	$\sim$ mW
Molecular Receptive range (MRR)	Combustible vapors	Combustible vapors	Wide range	Volatile organics
Response time	$\sim$ 20 s	$\sim$ 20 s	$\sim$ 20 s	$\sim$ 1 ms
Sensitivity	10–1000 ppm	1–100 ppm	$>$ 20 ppm	1 ppm (down to sub-ppb)
Steric sensitivity	None	None	Possible	High
Stability	Poor	Poor	Moderate	Disposable with continuous regeneration
Interferences	Cl_2, water	Cl_2, water	Acidic gases, water	Infection, reactive vapors

size will then also become important and integrated devices will become increasingly necessary.

Ironically, all of these sensing units are destroyed (after gradual deterioration) during operation, and are also susceptible to poisoning by interferences such as particular chemical species and over-exposure. This is a technical aspect of electronic nose design that clearly limits the application of the instrument to repeated odor monitoring. As already discussed, the biological system copes with this degeneration by continually replacing olfactory neurons, whilst having massive stocks of receptors of each type operating at any one time [38]. It seems likely that massive convergence of output from olfactory receptors forming a single class onto a single glomeruli structure may provide a more constant signal in view of this continual neurogenesis. How an electronic nose might benefit from such an arrangement is considered by Pearce (1997) [72].

The key measurand of element response is provided by modulation of spiking frequency for olfactory neurons and device resistance in many odor sensors, as verified by EOG measurement on the olfactory epithelium. Schild and Gardner (1991) have proposed that a frequency-based measurement scheme used in the electronic nose may be another method of compensating for degeneration within odor sensors [73].

Stereochemical properties provide the most sensitive molecular determinant of odor quality, although at least 20 other such determinants or "descriptors" are required for a complete description [38]. For this reason, steric sensitivity in artificial odor sensors is fundamental to the resolving power and MRR of an electronic nose system. From Table 3.1 it is clear that currently only olfactory neurons and polymers (conducting as well as non-conducting [74]) possess this important quality, although the use of modified lipid layers may allow us to design a higher degree of steric sensitivity in the future (see Table 3.2).

That a high number of different molecular determinants are required to describe odor suggests a need for electronic noses employing a hybrid of different sensor technologies. In this way, each sensor transduction mechanism involved could be used to detect different aspects of the same odor molecule. For example, conducting polymers could be used to detect particular functional groups within stimuli; piezoelectric devices could be used to determine molecular mass and provide some steric selectivity through the use of lipid layer or functional sidegroups in polymers; and electrochemical cells with catalytic filaments could be used for determining relative amounts of pyrolised components [75]. Such an approach to an electronic nose gives promising devices results, as shown, e.g., by Winquist et al. [76] using both MISFETs and Taguchi or by Ulmer et al., using OMBs with non-conducting polymers and SnO_2 conductivity sensors [77]. Further possible choices of materials and transducers are discussed in Chapter 3.4.

The principles of operation of artificial sensors currently seem far removed from their biological counterparts. There are two different approaches being used to develop biomolecular devices – the use of natural structures (biological) and the use of completely synthesized (biomimetic) structures (see Table 3.2 [78 and 79]). Biomimetic sensor materials would provide transduction mechanisms that are more closely modeled on their biological counterparts. The use of natural structures would involve cloning olfactory receptors away from the epithelium. As already discussed, the work of Buck and Axel (1991) has provided a suitable method for cloning olfactory receptor genes and it has since been shown that olfactory receptors behave identically away from the epithelium [30,48]. Recent work by Vodyanoy (1996) demonstrates a molecular sensor based upon olfactory transduction [80]. In Section 3.4 we consider the different sensor technologies available for use in an electronic nose in more detail.

In the electronic analog, sensor processing is used to condition input data prior to interpretation by the PARC engine. Briefly, each sensor, i, produces a time-dependent signal (regardless of sensor type), $V_i(t)$, in response to an odor impinging upon the array. In order to simplify the task for later stages of processing, it is often convenient to remove the time dependence of the signal, by isolating the minimum and maximum values attained both before and after odor stimulation. Using these parameters, $V_i^{\min}$ and $V_i^{\max}$ respectively, various preprocessing metrics can be defined such as difference ($x_i = V_i^{\max} - V_i^{\min}$), relative ($x_i = V_i^{\max}/V_i^{\min}$), fractional difference ($x_i = [V_i^{\max} - V_i^{\min}]/V_i^{\min}$) and logarithm ($x_i = \log[V_i^{\max} - V_i^{\min}]$). All of these preprocessing metrics have been applied to various sensor arrays, and specific arguments can be made for their suitability for different transduction mechanisms of particular sensor types or different types of pattern recognition techniques.

The logarithm metric may be used to linearize sensor response for devices that can be modeled using a power law, such as metal oxide semiconductor sensors. Such a linearizing technique may be usefully applied in odor sensing when odor component intensities are of particular importance.

The logarithm metric may also be used when responses to an odor vary considerably across the array, leading to smaller responses being ameliorated during array processing or pattern recognition. In this case the logarithm metric increases the relative contribution of small sensor responses that is presented to later stages of processing and so can aid discrimination. Clearly, the use of the logarithm preprocessing metric increases the overall dynamic range of the system, although this is achieved at the expense of reducing sensitivity to signal size.

For further details about strategies in data preprocessing of sensor elements, signal conditioning, etc., see "Sensors: A Comprehensive Survey", Vol. 2, Chapters 1 and 6 and "Sensor Updates" Vol. 2, Chapter 5.

As reviewed earlier, the biological system performs comparable signal conditioning at the glomeruli layer of the olfactory bulb. The logarithm preprocessing metric may be mimicking the dynamic range compression of sensory input within the biological system described by equation (3.4). Such compression of sensory input, mediated by a logarithmic transfer function within the bulb, similarly increases the dynamic range of the system. Again, this increase in dynamic range is at the cost of increasing the effect of noise within the system, although as has also been considered, the biological system compensates for this by averaging of receptor input. This biomimetic preprocessing mechanism has not yet been fully investigated in electronic nose research.

An array processing function known as array normalization is often used in the electronic nose in order to reduce the effect of variations in sample concentration

$$x_i' = x_i / \sum_{i=1}^{i=n} x_i^2. \tag{3.5}$$

This function sets the length of all response vectors to unity and puts the vectors on the surface of a unit hypersphere [9]. As such, it removes concentration information in the dataset, and therefore promotes the identification of odors independent of their concentration. Interestingly, Shepherd has argued that the analogous process in the biological system (the perception of odor quality independent of intensity) may be achieved by the overlap in the response spectra of both olfactory receptor and glomeruli cells [30].

In analogy with the biological system, marked convergence takes place in the electronic nose after signal conditioning has been performed. As remarked upon earlier, this provides a valuable opportunity for sensitivity enhancement and also selective data reduction. Likewise, there is clearly an opportunity in electronic nose development to provide increased sensitivity through the replication of sensor types and using some form of averaged array processing in order to reduce noise and thereby permit increased signal gain. Such an approach to sensitivity enhancement has not yet been reported although such a biomimetic scheme has been proposed by Pearce (1997) [72].

Pattern recognition techniques for use in electronic nose systems act as a form of content-addressable memory (CAM) by way of function mapping between various input and output vectors, based upon training examples. Such a CAM function is essential to both the artificial and biological systems in their task of odor (object) recognition and discrimination.

The massively parallel architecture of the olfactory bulb and its associated cortex, combined with its impressive ability for odor memory storage and retrieval, clearly suggests that a connectionist or parallel distributed processing (PDP) approach to pattern recognition within the electronic nose may prove beneficial, or at least increase biological plausibility. The first PDP network considered for modeling memory processes occurring in olfaction was the associative memory. These models were highlighted as having excellent qualities for odor memory storage and recall, including reproducible pattern activation, generalization, ability to deal with degraded input as well as good pattern separation. However, the information storage efficiency of associative memories is abysmally low unless sparse coding of the input data is used. This implies a very large number of inputs, with only a small number of these carrying data at any one time. While such conditions may exist for input to the piriform cortex in the biological system, it will certainly not be the case for sensor array data, where partial specificity of sensors within an array ensures a high degree of overlap within the data. Fortunately, PDP mapping networks, such as the MLP do not require particularly sparse coding of the input space to operate efficiently, but still carry out heteroassociative CAM function.

Unfortunately, while the MLP as well as many other competing PDP paradigms bear passing resemblance to some aspects of later stages of processing within the biological olfactory pathway, there currently exist many more differences than similarities. In particular, it was clear from the overview of the biological olfactory pathway that temporal patterning of neuronal activity is just as fundamental as spatial patterning to the way in which sensory information processing is conducted. In fact, oscillation has been considered to be fundamental to processing within the brain for some time [81].

While the precise role of oscillation in olfactory information processing may currently be unclear, connectionist models (such as the MLP) being used to model olfactory processing have only been considered for their static qualities. It is becoming increasingly clear that the initial goal of network stability within connectionist systems may have been a false one, and that far more interesting and useful processing principles will be discovered through consideration of their dynamic properties. Likewise, in sensor-based machine olfaction future consideration of dynamic computational models of the biological olfactory system is likely to provide the key to exploiting information processing that is both biologically more realistic as well as also being invariant to sensor drift and background odor interference (see [79], [82] and references therein).

3.4 Current and Future Sensor Technologies for an Electronic Nose

3.4.1 System Aspects and General Scope

Although general similarities may be seen between natural chemosensory systems and electronic noses, any more detailed view illustrates significant differences. This was outlined for the natural systems in the last section and will be illustrated for technical systems in more detail in the following section. The development of complete sensor systems requires optimizing their ten components which are shown schematically in Figure 3.16.

Today, any progress in the development of electronic and bioelectronic noses is basically determined by progress in the understanding of materials for the individual sensor elements. These consist of chemically sensitive coatings enabling molecular recognition and of transducers enabling signal conversion. The specific recognition site on a certain transducer is the key component of the total sensor system (see schematic component 5 in Figure 3.16).

Typical sensor principles, materials, and transducers utilized to convert chemical or biological information into electronic information are shown schematically in Figures 3.17–3.20. Evidently, a large number of different *materials with different recognition sites* can be utilized today for chemical and biochemical sensor systems (for a survey see Table 3.2). The simpler the structure of these materials, the easier it is to optimize their long-term stability. A clear trend is apparent towards using more complex molecular structures or even biological function units in order to achieve better selectivity and better "odor-relevant" parameters.

Commonly used transducers monitor chemical compositions by monitoring the phenomenological sensor properties listed in Table 3.3. For certain transducers the use of current microsystem technologies will lead to drastic improvements as a result of miniaturization and/or integrated data preprocessing. Examples include interdigital structures (for complex impedance measurements), thermopiles (for temperature measurements), piezoelectric oscillators based upon bulk, surface, or plate waves (for mass measurements), integrated optics components (for optical measurements), multi-electrode arrays (for electrochemical measurements or for electrical connections to axons in nerve systems), or arrays of capacitive sensor elements with integrated transducers as well as data processing by capacitance-frequency converters to be used in integrated "electronic noses" [83, 84].

The present state of the art in electronic noses is determined by the technological problems to design sensitive, selective, and stable sensor elements, to build arrays, and to compose complete electronic nose systems. These problems could be solved more simply by molecular recognition which is based on relatively simple inorganic structures and organic molecules. Details about their functionalities will therefore be discussed first in Section 3.4.2. A variety of commercially available systems make use of such sensor materials (see list in Table 3.4 taken from Chapter 5 in Sensors Update, Vol. 2 with further details of the pattern recognition and chemometrics approach).

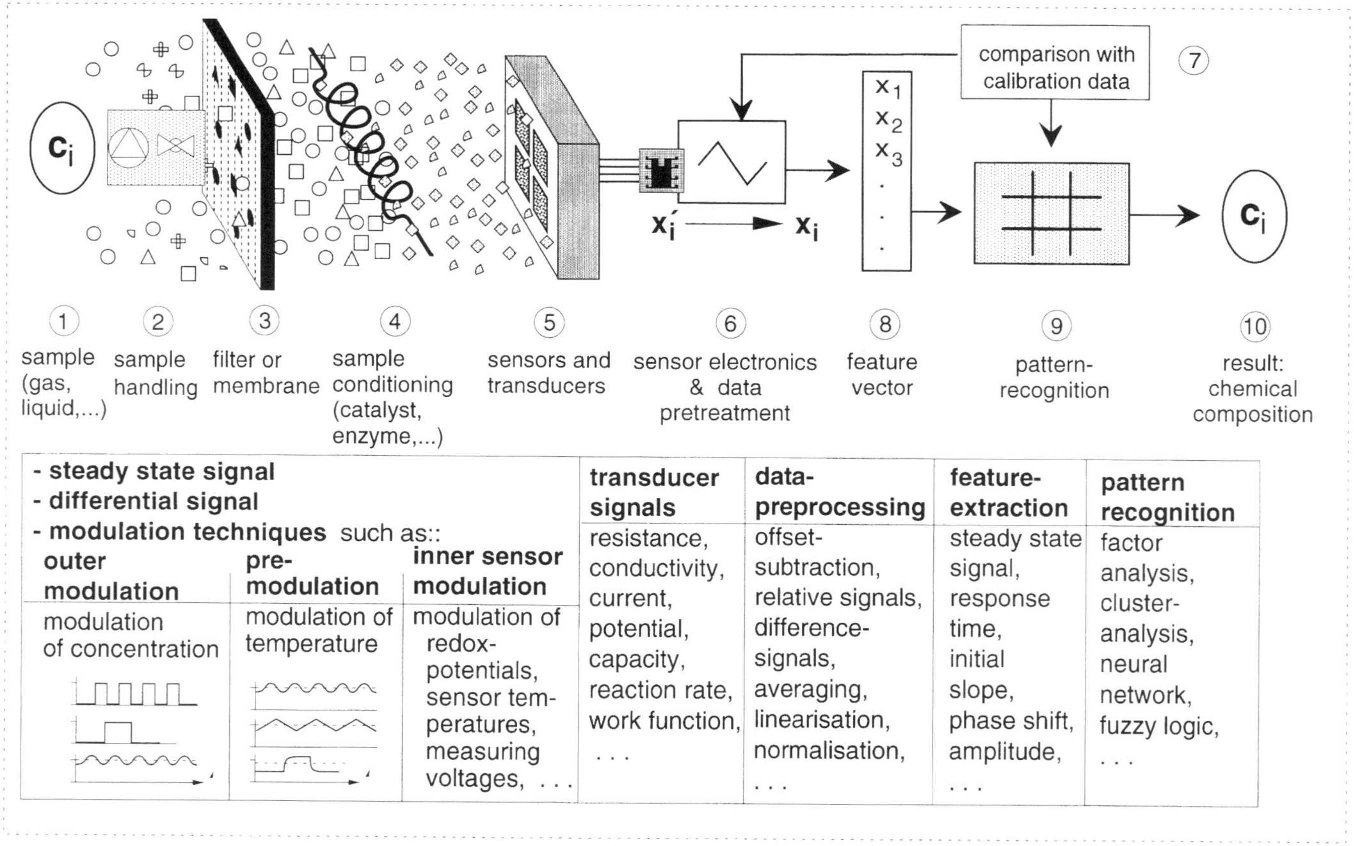

Figure 3.16 Ten components of a complete sensor system. Signal generation is determined by the mode of sample exposure (1–4), by the sensor coating, transducer type and its operation (5), by the data pretreatment (6) and pattern recognition (9). For details, see text and [3,4].

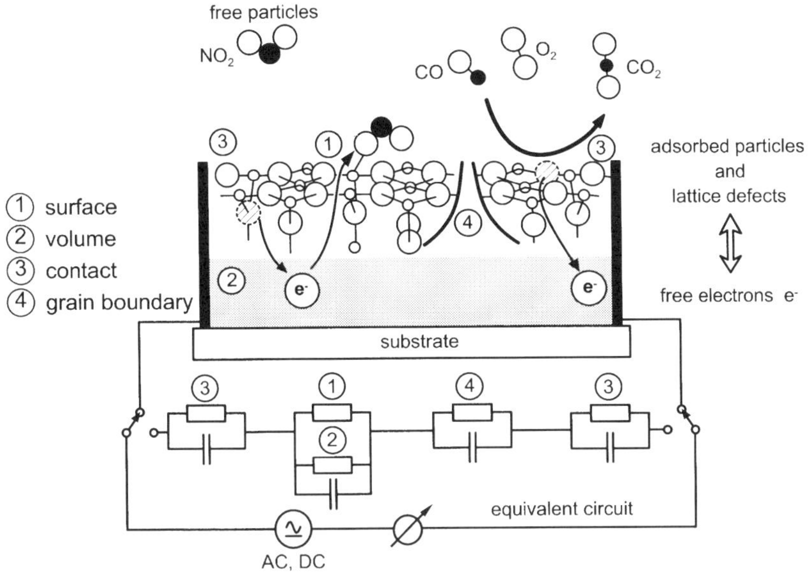

Figure 3.17 Prototype materials for chemical sensing with inorganic materials showing electron, ion, or mixed conduction with reversible changes upon chemisorption (of, e.g, NO_2 at binary oxide surfaces as shown in the upper left part) or upon catalytic reactions (cf. e.g., CO oxidation which involves lattice oxygen and O_2 from the gas phase, as shown in the upper right part). The schematic characterization of different sensor/gas reactions may also be done by simulating capacitive and conductive contributions to the overall impedance with corresponding components of equivalent circuits. The latter describe the frequency-dependent (alternating or direct current, AC or DC) behavior which is influenced by reactions between the gas phase and the oxide. Depending on the choice of oxides and electrode materials, ionic, electronic or mixed conduction may be monitored. For details, see [92].

The most successful electronic noses are based on modular sensor systems, i.e. sensor systems in which different transducers are used whereby each transducer principle is applied to several individual sensor elements. This leads to maximum "orthogonality" in the determination of independent chemical or odor information.

The design of bioelectronic noses which utilize biological function units (including even organelles) is evidently more complex. Usually their stabilities are limited and hence time-dependent signal outputs cause serious problems in calibration procedures. As a result, corresponding systems are not yet commercially available. Significant progress is expected by screening systematically new materials with stabilized biological functions (either biomimetic recognition sites, see Section 3.4.3, or biological systems, see Section 3.4.4–3.4.6) and by screening systematically new materials, which interface the biological function units with inorganic transducer substrates.

(a)

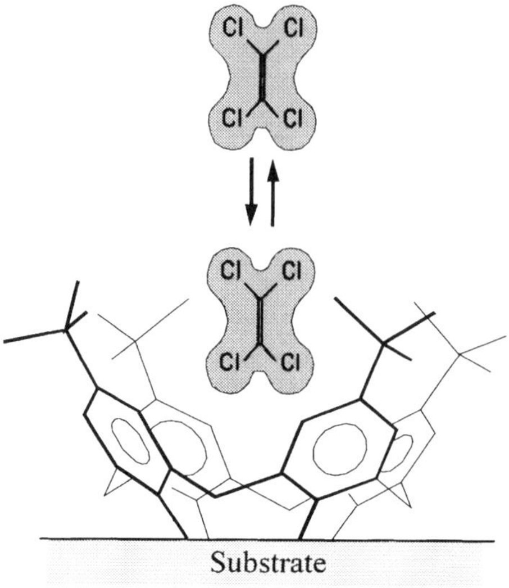

(b)

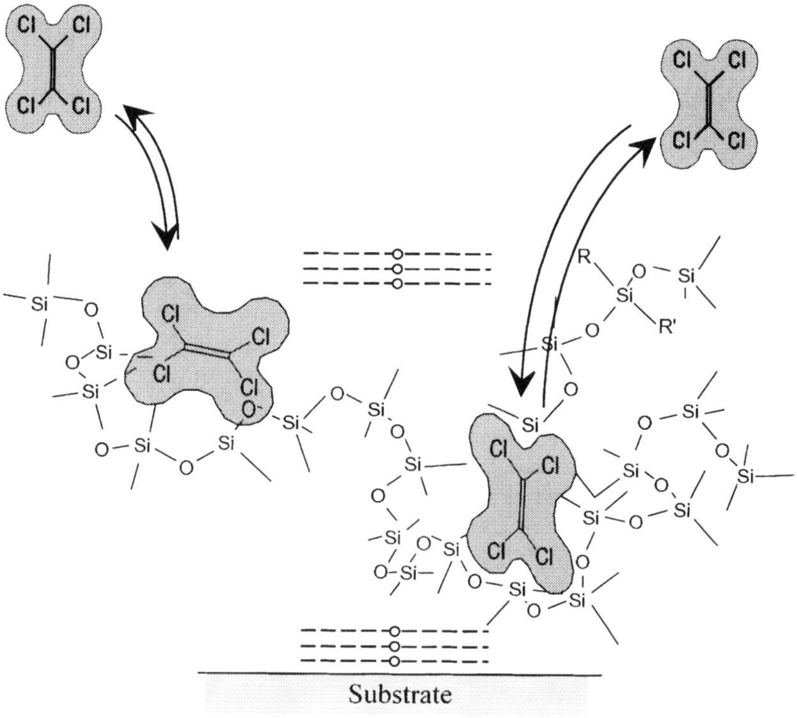

As already mentioned, for all sensor systems the weakest component is the sensor element itself including its transducer. In this context, two contradicting properties have to be realized in any (bio-)electronic nose:

- One interface is with the environment, it contains the recognition site and has to undergo quick and completely reversible bonding of the detectable molecule to the sensor-active coating upon changes in the molecular composition of the environment.
- Another interface links the recognition site with the electronics. This interface has to be completely stable and has to remain chemically inert under all circumstances in order to protect the electronics from the environment.

Hence current sensor research concentrates on three approaches to reduce this bottleneck in the development of reliable systems by:

- Empirical optimization of sensor materials and of transducers in systematic tests under realistic measuring conditions (in air, water, blood etc.).
- Systematic studies of elementary steps of chemical sensing under thermodynamically or kinetically controlled conditions by means of microscopies and spectroscopies (under UHV-, ultraclean-, glovebox-, electrochemical conditions etc.).
- Theoretical calculations of elementary steps of selective "key-lock" interactions (with *ab initio*, semiempirical, or molecular dynamics approaches).

Many researchers are optimistic about developing not only electronic, but also bioelectronic noses realized in the future because of the current progress in an atomistic understanding of interfacial phenomena. This understanding is deduced from spectroscopies and microscopies that already operate down to molecular-scale resolution in the gas and liquid phase. Of particular importance in this context has been the development of scanning probe techniques enabling the analyst to extract signals from atoms or molecules down to the principal limits of space and time resolution. These limits are determined by quantum mechanics and statistical thermodynamics.

Figure 3.18 Prototype materials for chemical sensing with organic materials by key-lock interactions with static and dynamic recognition centers. (a) Interaction of a typical volatile organic compound (VOC, here perchloroethylene, C_2Cl_4) with a simple supramolecular structure (here t-bu-calix(4)arene) with its surface interaction site facing towards the gas phase. Because of the rigid recognition structure, the values of simple changes in energy of this key/lock principle describe the interaction mechanism in this situation. (b) Dissolution of VOC's into the bulk of polymeric thin films (here C_2Cl_4 in polydimethylsiloxane, PDMS). Because of large fluctations in the time-dependent structure of polymers, the values of changes in free energies (including energies as well as entropies) attributed to the induced fit describe the interaction mechanism in this situation [85].

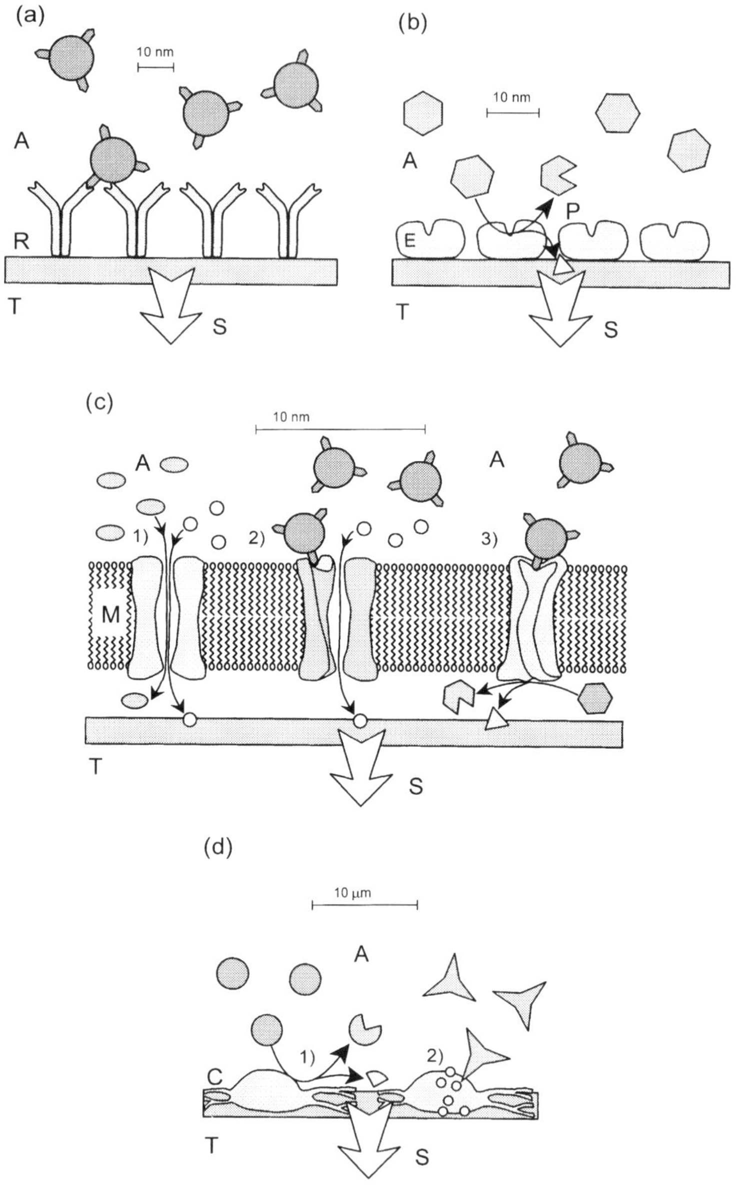

Figure 3.19 Prototype materials and different principles of biosensing via structured interfaces. The analyte is denoted by A, whereas T and the arrow denote the transducer and the recorded signals, respectively. (a) Bioaffinity sensors. The analyte (antibody, antigen) is recognized by immobilized recognition units R (antigen, antibody, DNA). (b) Biocatalytic sensors. The analyte is converted by immobilized enzymes E to products P. (c) Transmembrane sensors. Transport or channel proteins (1) or receptor proteins (2) are incorporated into a membrane M. These structures either transfer the analyte through the membrane (1), bind the analyte and open a channel for another species (2), or activate subsequently a separate enzymatic cascade (3). (d) Cell sensors (note the change in scale). Immobilized living cells C convert (1) or bind the analyte (2). For details see (Göpel and Heiduschka, 1994 [86]; Göpel, 1994 [87] Göpel, 1995 [2].

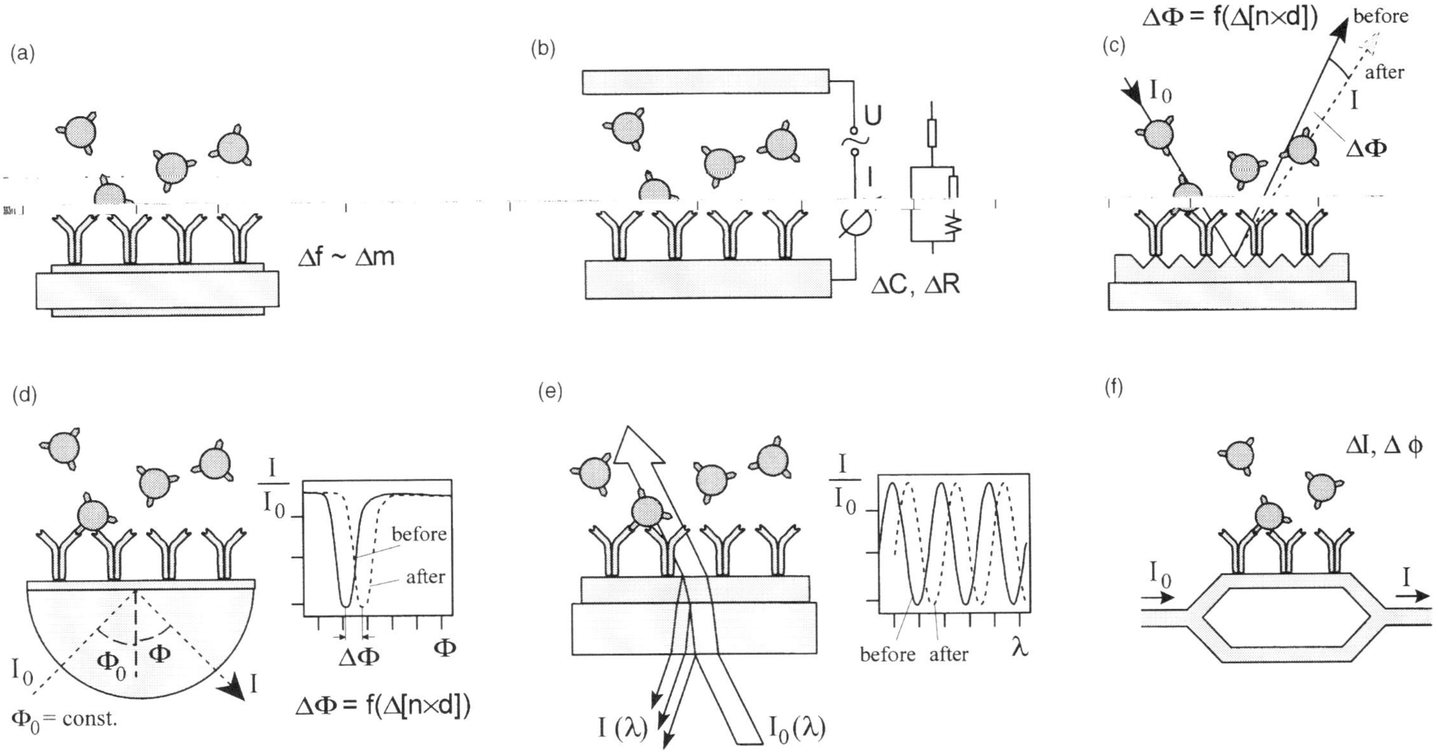

Figure 3.20 Schematic presentation of different types of transducers for monitoring key-lock interactions of chemical or biochemical sensors. For bioaffinity (or immuno-) sensors these transducers make possible direct signal transduction in the absence of any labeling of antigens or antibodies: (a) mass sensitive detection, e.g. by quartz crystal microbalances (mass changes Δm lead to frequency changes Δf). (b) measurements of electrochemical impedance or related electrochemical properties (changes of capacitance ΔC, of resistance ΔR, etc.), (c) grating couplers (intensity I and phase Φ change with changes in optical thickness, i.e. in the refractive index n times film thickness d) (d) surface plasmon resonance (SPR), (e) reflectometric interference (intensity changes for different wavelengths λ), (f) Mach–Zehnder interferometer. For details, see (Rickert et al., 1996 [88, 89]; Göpel, 1996 [3]).

Table 3.2 Typical materials and fields of application used for (bio-)chemical sensors [2, 3, 82]

Metals	Pt, Pd, Ni, Ag, Au, Sb, Rh, Ga, Ln, ...		
Semiconductors	Si, GaAs, InP, ...		
Ionic compounds	electronic conductors	SnO_2, TiO_2, Ta_2O_5, IrO_x, ...	environmental gases
	mixed conductors	TiO_2, WO_3, Ga_2O_3, $SrTiO_3$, $La_{1-x}Sr_xCo_{1-y}Ni_yO_3$, in general: perowskites	exhaust gases
	ionic conductors	ZrO_2, LaF_3, CeO_2, CaF_2, Na_2CO_3 ß-alumina, Nasicon, ...	
Molecular (incl. liquid) crystals	phthalocyanines	PbPc, $LuPc_2$, $(PcAIF)_n$, ...	environmental gases
LB films	phthalocyanines, polydiacetylenes, Cd-arachidate, ...		
Polymers	polyethers, polyurethanes, polysiloxanes, polypyrrols, ... polythiophenes, polyfluorocarbons, polyolefines, PTFE, Nafion, ...		inorganics and organics in air and water
Cage compounds and supramolecular structures	zeolites, calixarenes, cyclodextrines, crownethers, cyclophanes, resorcinarenes, cyclopeptides (valinomycine, ...), ...		
Components of biomolecular systems	synthetic	phospholipids, lipids, FMDV- and HIV-epitopes	substrates, inhibitors,
	natural	glucose-oxidase, lactose-permease, E-coli-cell-membranes, ... in general: enzymes, transport proteins, antibodies, receptors DNA/RNA, membranes, cells, ...	antigens, hormones, drugs, toxins, neuro-transmitters, carbohydrates, ...

Table 3.3 Survey of phenomenological properties monitored in (bio-)chemical sensors (Göpel and Schierbaum, 1991 [90]; Göpel, 1995 [2])

Δm	measurements of mass differences
Q	measurements of reaction heats
σ	measurements of voltage-, temperature- and frequency-dependent conductivities, complex impedances and of photoconductivities
Φ	measurements of work functions
C	measurements of capacitances
$\Delta E, I$	measurements of frequency-dependent electrochemical potential differences and currents
ε	measurements of frequency-dependent optical absorption or reflection (IR-VIS-UV)
$n \cdot d$	measurements of optical thicknesses (product of refractive index and thickness)

Table 3.4 Survey of commercial and laboratory type sensor systems with different sensors, transducers and chemometric approaches to electronic noses. For details and further references see [82]. The following abbreviations denote:

TSMR: Thickness shear mode resonator
MOSFET: Metal oxide field effect transistor
SAW: Surface acoustic wave device
PCA: Principle component analysis
PCR: Principle component regression
CA: Cluster analysis
PARC: Pattern recognition
ANN: Artificial neural network
PMB: Partial model building
PLS: Partial least squares analysis

Address	Trade mark/ Sensing principle	Number of sensors/ Sensing materials	Chemometrics/ Reference
Alpha M.O.S. 3, Av. Didier Daurat 31400 Toulouse France	Conductance	6 metal oxide sensors, temperature and humidity	PCA, PCR, CA, Discriminant function-Analysis
AromaScan Electra House Electra Way Crewe CW1 1WZ.UK	AromaScan Plc Conductance	32 conducting polymers	Fingerprints, PARC PCA, ANN
HKR-Sensorsysteme Gotzinger Str. 56 D-81371 München, FRG	QMB 6 TSMRs	6 Polymers	Fingerprints, PARC, ANN (Homer, 1993)
Nordic Sensor AB Gunnorp Bäckgard S-59050 Vikingstad Sweden	Nordic Electronic Nose MOSFET Conductance	10 MOSFETs, 4 metal oxide sensors, 1 IR-CO_2-sensor	PCA, ANN (Holmberg et al., 1995)
Forschungszentrum Karlsruhe Postfach 3640 D-76021 Karlsruhe, FRG	(a) Conductance at different temperatures (b) SAWs	(a) 39 metal oxides at 2 different temperatures (b) 8 polymers	none (a) Althainz et al., 1995 (b) Rapp et al., 1995)
Microsensor Systems Inc. 62 Corporate Court Bowling Green Kentucky 42103, USA	SAWs	4 polymers, preconcentrators	Chemometrics (Grate et al., 1993)
TU München Institute of Technical Electronics Arcisstr. 21 80333 München, FRG	Pellistors	6 Si/Pt pellistors at different temperatures	PMB, PLS (Gall, 1993)
UMIST Dept. of Electrical Engineering and Electronics PO Box 88 Manchester M60 1 QD UK	Odor Mapper Conductance	20 conducting polymers	Compare Aroma Scan (Neaves et al., 1995)

(*contd.*)

Table 3.4 (*contd.*)

Address	Trade mark/ Sensing principle	Number of sensors/ Sensing materials	Chemometrics/ Reference
IFAK e.V. Magdeburg Steinfeldstr. (IGZ) D-39179 Barleben, FRG	TSMRs	4 Polymers	ANN
SRL (Sensor Research Laboratory) J. W. Gardner University of Warwick Conventry CV4 7AL, UK	Conductance, Calorimetric	Metal oxides Polymers Phthalocyanines Lipids	ANN, PCA; PLS (Gardner et al., 1994)
Neotronics Scientific Ltd, Western House 2 Cambridge Road Stansted Mountfichet Essex CM24 8BZ, UK	Conductance	12 Polymers	ANN (Hodgins, 1995)
Institute of Microtechnology University of Neuchatel A.-L. Berguet 2 2000 Neuchâtel, CH	Microelectrode array	100 individually addressable electrolyte cells	none (Fiaccabrino et al., 1994)
Institute for Chemo- and Biosensorics Stegerwaldstr. 39 D-48565 Steinfurt, FRG	Amperometric measurement	1024 individually addressable electrolyte cells	none (Hermes et al., 1994, Meyer et al., 1994)
Lennarte Electronics and Steinbeis Transfer Center for Interface Analysis and Sensors Auf der Morgenstelle 8 72076 Tübingen, FRG	MOSES II, Hybrid array: TSMRs, Conductance, Calorimetric Electrochemical cells	8 polymers, 8 metal oxides 8 thermoresistor	ANN, PCA, PLS Hierlemann et al., 1995 Mitrovics et al., 1995 Sensors 1997

Table 3.5 Elementary steps of molecular recognition

Weak physisorption
Strong chemisorption
Catalytic or enzymatic reaction
Volume reaction
Intergranular reaction
Three-phase boundary reaction
Incorporation or intercalation into cage structures
Membrane transport

Table 3.6 Survey of different electroanalytical techniques used for chemical and biochemical sensing

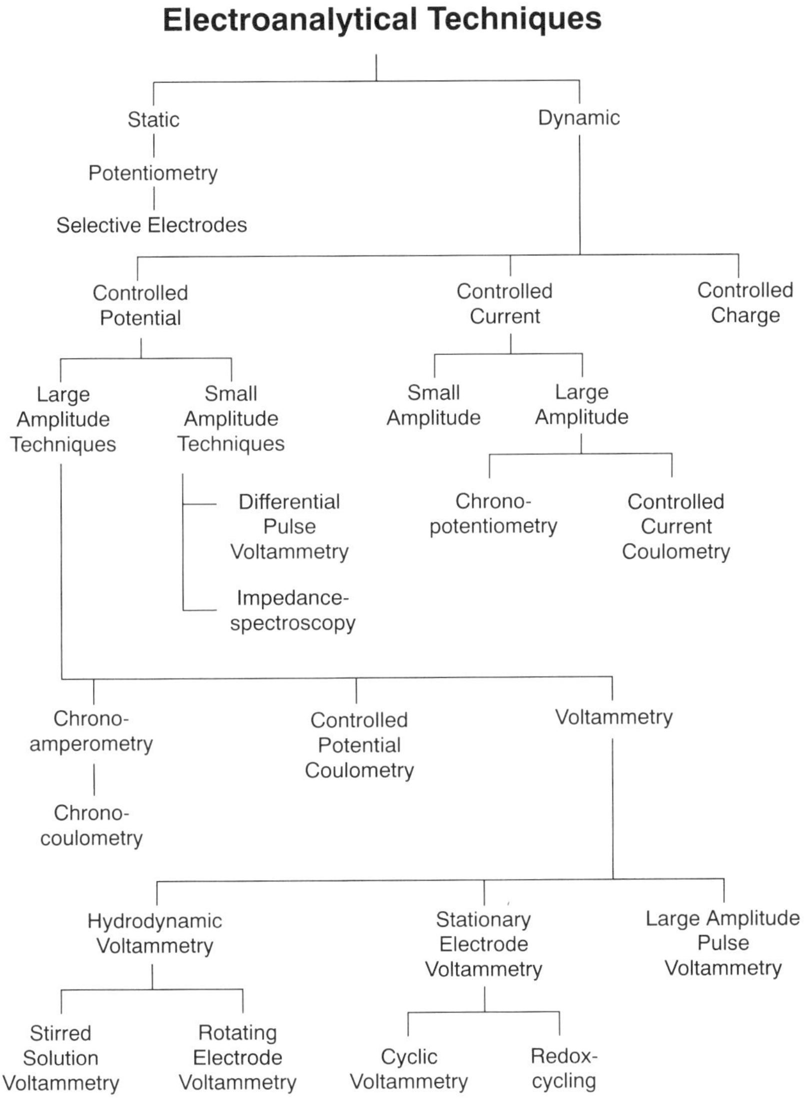

3.4.2 Inorganic Structures and Organic Molecules for Molecular Recognition

3.4.2.1 Recognition by Matching Molecular Orbitals of Electronic Conductors

Electronic conductors are commonly used as chemical sensors to monitor oxidizing and reducing gases such as O_2, H_2, CO, NO_2, or hydrocarbons. The operating temperature determines the dominant interaction step (Table 3.5): with increasing temperature, physisorption, chemisorption, surface defect reactions, catalytic reactions, grain boundary reactions, and bulk reactions can be utilized for chemical sensing, provided that the interaction step leads to reversible changes in the DC resistivity or, more general, by the AC impedance (see schematic Figure 3.17 above). Particularly suitable for electron conducting devices is their operation in the chemisorption or catalysis sensor mode. Depending on the position of the highest occupied (HOMO) or lowest unoccupied molecular orbital (LUMO) of the chemisorption or catalytic complex relative to the Fermi level (of usually n-type conducting oxides such as SnO_2), the donor or acceptor type of interaction is monitored by positive or negative changes in surface conductivities or work functions. Fine-tuning of selectivities, sensitivities and stabilities is performed by optimizing the electrode material and geometry, the crystallinity, chemical composition, doping, operating temperature, and mode of operation (DC, AC, pulsed temperatures or pressure, etc.).

3.4.2.2 Recognition by Matching the Doping of Mixed Conductors

For gas sensing, electron and ion conduction of mixed conductors is usually monitored at elevated temperatures ($T > 700\,\mathrm{K}$) in oxide-based sensors. Metals, like platinum, are often used to improve the selectivity of gas detection by encouraging the catalytic activity of the surface, by adjusting current/voltage behavior across the sensor interface or by enhancing the kinetics of certain interface reactions so that either the surface properties or the subsequent bulk equilibrium of defects determines the sensor response.

3.4.2.3 Recognition by Matching Electrochemical Potentials

This principle is applied for both gas and liquid electrolyte sensors in three basically different modes of operation, i.e., the potentiometric, amperometric, and conductometric mode with further refinements listed in Table 3.6. Typical surface and interface properties to be controlled in such devices are summarized in Figure 3.21.

Gas-sensing electrochemical devices based on solid electrolytes are the best understood sensors in this context. The often cited prototype material is ZrO_2. It is widely used as a chemical sensor to monitor O_2 by utilizing its high temperature ion (O^{2-}) conduction in the bulk phase. In *potentiometric devices*, the cell voltage is determined by the pressure difference between two Pt contacts of ZrO_2 samples (Nernst law). In *amperometric devices*, the limiting current through a diffusion barrier is monitored at a given voltage. Of key importance for both applications is the stability of the three-phase boundary at which oxygen molecules from the gas phase are converted to ions in the solid. Currents and voltages are determined by the difference in the energetic position of the Fermi level, E_F, of both electrodes under various operation conditions.

Externally applying a voltage leads to changed concentrations and chemical reactivities of different electroactive species at the surface, to variations in the kinetics of different electrocatalytic reactions and hence to changed selectivities. As an example the charge transfer rate monitored in any amperometric operation mode at constant voltage is determined by the concentration of those educts in the gas phase which react with surface oxygen. Recent developments of new electrode materials (mostly perowskites) make possible the amperometric detection of CO, NO, and hydrocarbons. The same electrode arrangements are also of practical importance in completely different devices, i.e., in catalysts, fuel cells, or in electro-catalysts based on solid or polymer electrolyte bulk materials.

The huge number of liquid electrochemical sensors can only be mentioned here. The most commonly used ion-selective electrodes (ISE) make use of controlled charge transfer across the liquid/solid interface. Membranes and incorporated selective sites determine their selectivity and cross-sensitivity patterns to detect specific molecules or ions in the gas and liquid state. A variety of different electroanalytical techniques is in use (Table 3.6) to achieve specificity, sensitivity and selectivity of both solid state and liquid state electrochemical sensors. Of particular interest are those electrode arrangements which allow their direct coupling to biological membranes and cells and hence allow us to monitor extracellular potential variations and ion (Na^+, K^+) currents.

3.4.2.4 Recognition by Key-lock Interactions with Static Molecular Cages

A completely different optimization is required for chemical sensors which make use of supramolecular recognition structures with their "key-lock" or "host-guest" principle of recognition via selective surface or bulk structures. Typical molecules to utilize this geometric recognition principle are calixarenes which incorporate small organic molecules. Signal transduction of these sensors is possible by monitoring changes in mass, capacitance, or temperature upon incorporation of organic molecules. The geometry of molecules like calixarenes may be modified by substituting atoms in the ring or by using different numbers of repeat units in the ring. For small rings the molecule/ring interaction energies can be calculated with sufficient accuracy by static force field approaches. Larger rings require the theoretical calculation of dynamically adjusted recognition centers ("induced fits") by molecular dynamics concepts. Such theoretical molecular design concepts also become of increasing interest for developing new pharmaceuticals ("drug design")

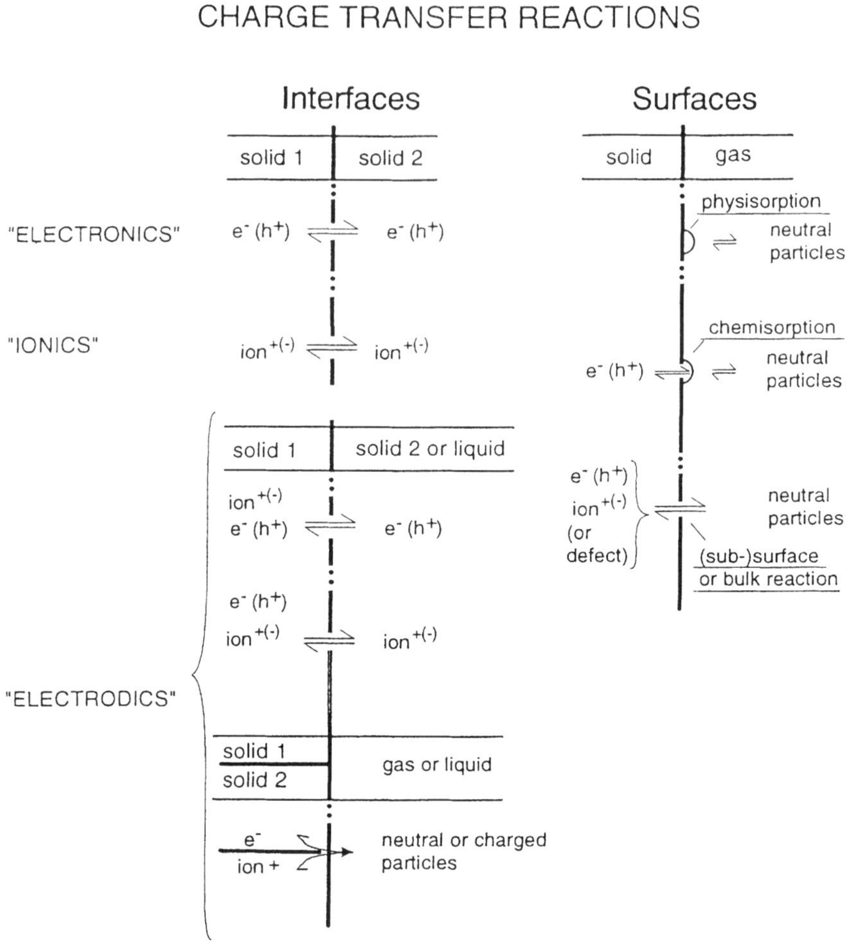

Figure 3.21 Survey of charge transfer reactions at interfaces and surfaces of electronic or ionic devices such as chemical sensors, biosensors, batteries, or fuel cells. Control of these reactions is particularly important for any electrical signal transduction involving biomolecular function units. For details, see the text and Göpel and Reinhardt, 1996 [92].

To achieve a controlled key/lock geometry with easy access of molecules to be recognized, monolayers have been used to advantage in chemical sensors with a covalent coupling of cage compounds via sulfur bridge bonds to Au substrates. As a typical example, a two-dimensional layer of resorcinarenes on Au(111) was utilized as coating of a mass-sensitive transducer to monitor organic volatiles [91].

3.4.2.5 Recognition by Induced Fit with Dynamic Molecular Cages

In contrast to the host/guest interaction treated in the rigid lattice approach to describe the key/lock interaction in the last section, a huge variety of materials are commonly used for chemical sensing in which this interaction is optimized as an *embedding effect* in a flexible lattice ("induced fit" in the liquid state which is theoretically treated by molecular dynamics calculations). Typical materials chosen in this context which usually show certain specificities of their absorption phenomena include:

- liquid crystals of side-chain modified polymers (such as siloxanes) which monitor volatile organic compounds (VOCs) (such as benzene) by the key/lock interaction via changes in the mean capacitance due to the disturbed nematic order;
- molecular liquids with dissolved dye molecules which show color changes upon association of gas molecules because of electronic charge transfer interactions;
- extra- and/or intramolecular cages of supramolecular structures;
- polymers modified by characteristic chemical functional groups or by geometric recognition sites which show mass changes, dielectric constant changes, optical changes, or temperature changes upon host/guest interactions.

The last class of compounds combines two advantages, i.e., first the matrix effect which allows fast bulk diffusion of gas or liquid molecules in the open network of the polymers and hence fast response times, and second the recognition site effect which allows us to utilize higher selectivities of rigid as compared to flexible recognition structures. A huge variety of different polymers shows excellent sensing properties to monitor VOCs in the gas, but also liquid phase. Even enantioselective detection was demonstrated to be possible with individual sensor elements (Figure 3.22). In this context a huge number of materials known as selective coatings in chromatography (mostly polymers) is now being tested as sensor coatings.

3.4.3 Biomimetic Recognition Sites

Biomimetic structures, when compared with biological units, show an enhanced stability in practical sensor applications. Nevertheless, one has to take into account that considering a reasonable investment of synthetic chemistry work a loss of selectivity may have to be accepted for a gain of stability. A typical example was shown earlier; according to the principles of supramolecular chemistry, artificial ion channels were synthesized which, however, were not shown to be selective and which in addition could not be gated by ligands [94]. Nevertheless the synthesis of artificial receptors is advancing, even though their practical use still demands new concepts of integrated signal transduction.

Suitable integrated transduction principles for biomimetic recognition systems make use of covalently bound chromophors or fluorophors which respond by shifts in their spectral properties to a docking of guest molecules [95]. Furthermore, the most

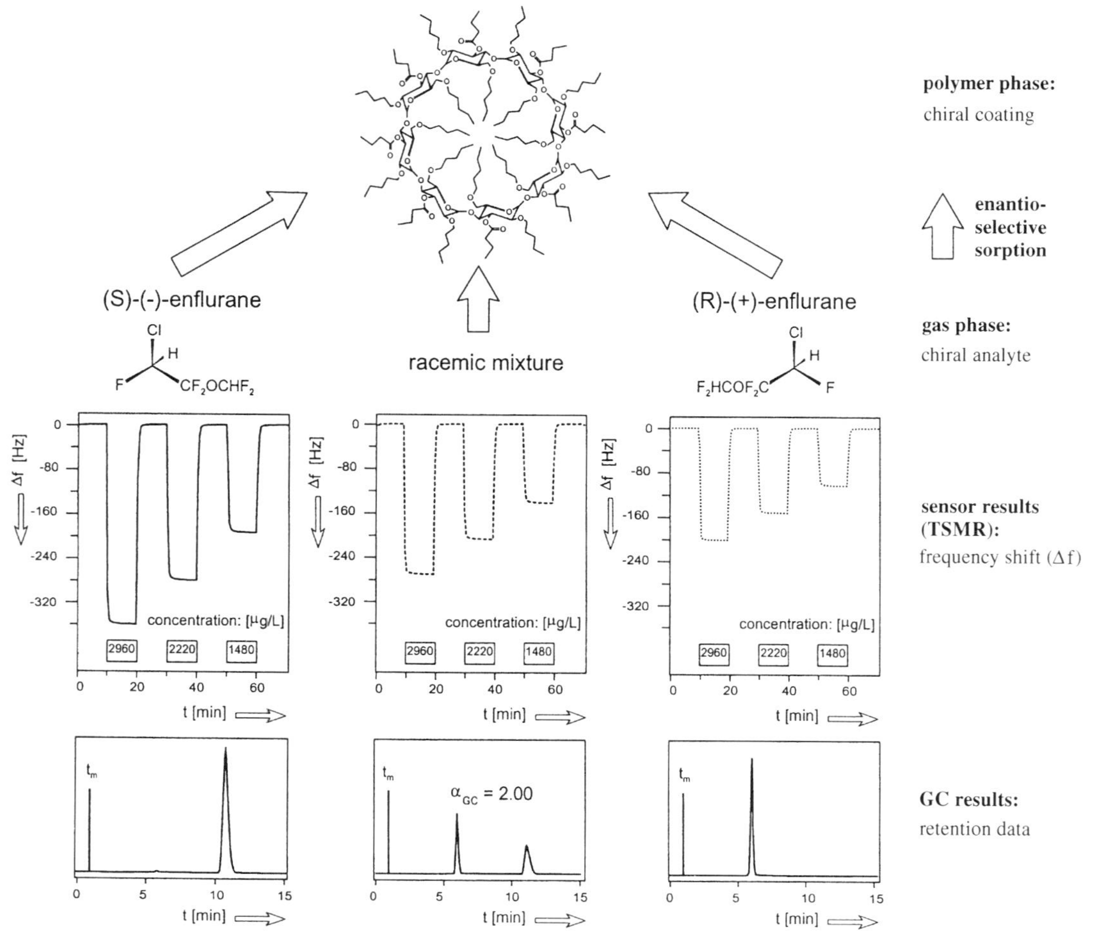

Figure 3.22 Typical example of molecular recognition with polymeric phases. In this case, a chiral γ-cyclodextrine was embedded in a polymeric siloxane phase. This phase was chosen as a sensor coating and as a coating for a gas chromatography set up. Enantioselective detection was possible in both cases of the anesthetic gas enflurane. Since enantiomers are believed to play an essential role in the odor perception, these recent results illustrate new applications and possible trends in the development of sensor arrays. For details see (Bodenhöfer et al.) [74, 93].

common and universal transduction of complementary recognition between guests and supramolecular host molecules in the measurment of a mass change implied in the docking process. This is also true for the recognition of biological antibodies by synthetic peptides resembling the recognition sites (so-called epitopes) of antigens or viruses (Figure 3.23). An array of several piezoelectrical microbalance sensors not only provides the advantage of being able to compensate an interfering background (e.g. water in the gas phase), but also to enhance the selectivity and to integrate a quality evaluation by means of neuronal networks. During the last few years the use of a piezoelectrical sensor and surface acoustic wave systems with artificial affinant surface layers was extended to sensors for aqueous systems.

The most promising trend for practically applicable systems may therefore be to conceive and synthesize better and more versatile biomimetic recognition molecules, and to compensate for their deficit in selectivity as compared to proteins by combining an array of several sensors.

Two different concepts of synthesizing new receptor structures are basically followed today. Synthesis is performed either by a rational design of very specific structures, mostly accompanied by computer assisted molecular modeling and subsequent organic synthesis ("supramolecular concept", see Figures 4.9 and 4.10), or by a statistical approach utilizing synthesizers to prepare a large number of different molecules and to subsequently select the most promising ones ("combinatorial chemistry concept").

3.4.4 Biological Recognition Sites without and with Subsequent Signal Amplification

A variety of biosensors are used to monitor large molecular weight molecules in the environment, in body fluid, etc. Usually four different principles of biosensors are distinguished which were shown schematically above in Figure 3.19.

Sensors of the first category, the bioaffinity-sensors, function like the recognition centers described above (Section 3.4.2.5) by induced fit (dynamical) key-lock interactions in molecular cages. Induced fit occurs here in a natural receptor. A variety of receptor-based interactions may be used for highly selective biomolecular recognition. The controlled preparation of thin films with embedded receptor structures of high molecular weight is a difficult task. One approach is therefore to mimic the biological function by fully synthesizing structures of lower molecular weight which represent a copy of the recognition site of the biological function unit ("epitope"). This "biomimetic" approach was elaborated in more detail above (see, e.g., Figure 3.23). Several different transducers are generally used for such bioaffinity sensors (see Figure 3.20). The signal transduction across the interface determines the sensitivity of these devices and limits their minimum lateral dimensions.

Enzymes as biocatalysts are particularly suitable for biomolecular recognition. The catalytic activity of their active sites may be monitored sensitively by changed electrochemical behavior at the electrode. This, however, requires a "molecular

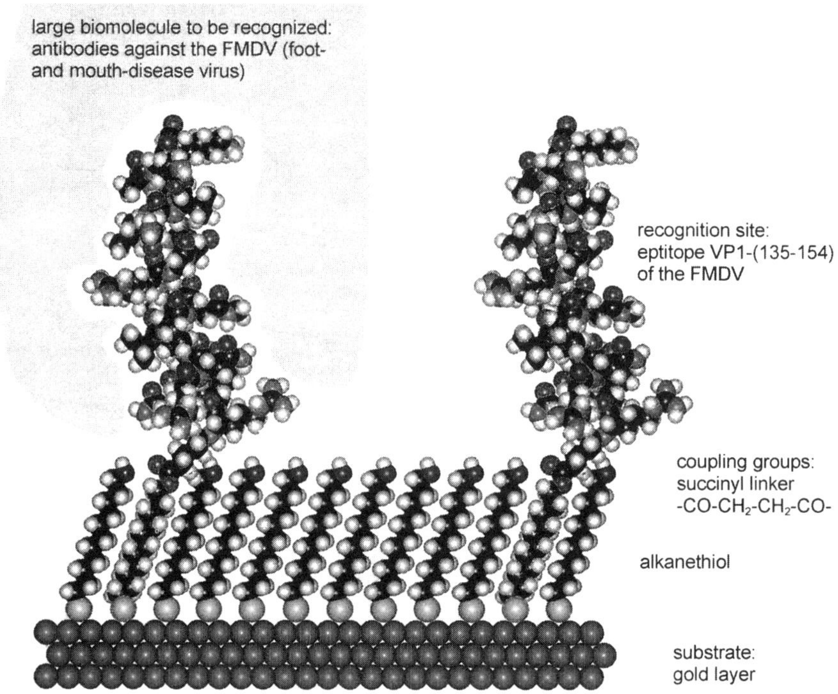

Figure 3.23 Schematic presentation of the covalent coupling of two different alkane thiol units via sulfur atoms to Au(111). The longer component contains an epitope with the amino acid sequence 135–154 of the protein VP1 of the foot-and-mouth disease virus (the so-called antigen recognition site) which selectively interacts with the antibody. For details, see Rickert et al. [96]; Göpel, 1996 [2].

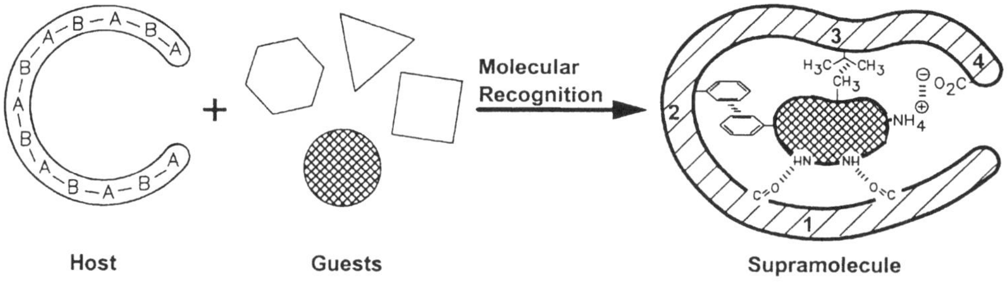

Figure 3.24 Schematic view of molecular recognition and supramolecular bonding with typical local interactions (Dung and Vögtle, 1988 [100].

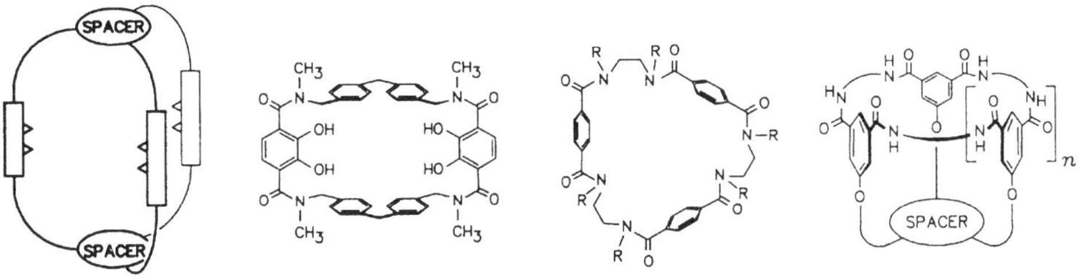

I II III IV

Figure 3.25 Examples of molecular recognition with supramolecular structures: macrobicyclic (**I**), macrocyclic oligoamide (**II**, **III**) and basked shaped (**IV**) receptor molecules (Seel and Vögtle, 1991 [97] and Vögtle et al., 1996 [98]).

wiring" which makes possible an ionic or electronic charge transfer between the active site and the electrode. None of the molecular wiring concepts encountered so far, however, are understood yet on the molecular scale even for the best understood enzyme glucose oxidase with its molecular weight of 186 000 atomic mass units (although glucose sensors currently have a market share of approximately 90% of all biosensors worldwide).

Because of their great importance in the living cell, a variety of biomolecular function units incorporated in membranes has been identified and may in the next step be utilized as components of future "intelligent" biomolecular devices, provided that they can be contacted or interfaced to appropriate transducers. To achieve stability, they must be arranged in natural or synthetic bilayers. Perfect spreading of the modified bilayer and stabilizing the thickness and molecular composition of the inner compartment between the bilayer and the transducer are the main issues of current research. The fourth category includes whole cell sensors as discussed in Section 3.4.5. The long-term stability and reliability of a sensor will, under practical conditions, be determined by the most unstable component. In biosensors utilizing biological systems and principles this is in most cases the biological part. The latter is based on proteins or protein complexes, the functionality of which depend on their native conformation, which is determined predominantly by non-covalent interactions. There are some examples of the immediate use of natural receptor proteins in sensors, because they are easily available today by means of genetic engineering procedures (Table 3.7). However, even these sensors cannot make use of the biological principle of energy provision for signal transduction and active transport. The direct availability of electrical potentials and currents in any man-made sensor is certainly highly superior and much more versatile than the biological counterpart. Therefore, any of the realized mixed sensor-systems using natural ligand gated channel proteins are based on electronic devices for potential or impedance measurements [99]. Correspondingly, the principle of biological signal amplification via cascades of enzyme catalyzed reactions and of the triggering and modulation of signals by induced conformational changes, e.g. by calmodulin, cannot be transferred directly to technical applications.

Table 3.7 Biological and biomimetic systems for chemical signal recognition in biosensors [79]

Chemical signal or analyte	Natural recognition element [main charge carrier]	Biological effect of transduction, amplification	Use of recognition element in sensors direct	biomimetic
Hormone, neurotransmitter, ligand (Fragrance)	Ligand-gated ion channel receptor protein [ions]	Change of membrane polarization or conductivity	Attempts successful	Artificial receptors (with integrated transduction)
Metabolite, ion	Carrier protein Ionophor [ions]	Active or passive transport Selective transport	Attempts Membrane electrode	Artificial channels Synthetic ionophors
Ester, amide	Hydrolase (enzyme) [none]	Secondary chemical signal (H, CO_2, NH_3)	Potentiometric biosensors	Attempts catalytic antibodies
Antigen with artificial determinante	Antibody [none]	Highly selective binding	ELISA, RIA	Affinity sensors with indicator enzymes
Redox active substrate	Oxidoreductase (enzyme) (electrons, mediators)	Direct or mediated electron transfer	Mediated amperometric sensors	Attempts for direct electron transfer

Therefore the general question has to be raised whether it would not be more realistic and convenient to conceive biomimetic systems inspired by the biological model, as discussed in the previous section, instead of directly using the latter. The immediate practical application of biological receptors may be restricted to special investigations in physiology or other fields of academic interest; for real technical applications they still appear to be too unstable.

3.4.5 Cells and Cell Arrays

Until now, the practical applications of biosensors has been almost entirely limited to very few specific examples including in particular glucose and pH measurements. Exciting results may be achieved by expanding the field of biosensors to the adoption of complex biological recognition elements. Using intact cells or tissue slices, "group effects" such as toxicity, mutagenicity, or content of nutrients become accessible to sensor technology as well as the physiological effect of certain molecules [100]. Examples for the direct use of integrated biological systems for biosensors are given in Table 3.8.

Two different approaches for whole cell sensors can be used (Figure 3.26). In the first, up-to-date (bio-)chemical sensors are used to measure metabolic products delivered from cultured cells to the medium. One example of practical importance concerns monitoring pH changes with a light addressable potentiometric sensor (LAP, see Figure 3.27). As a variant the mechanical contact between cells and substrates are measured via conductivity measurements in which the cells act as resistor. In the second approach the direct electrical response of a neural cell or a neural cell network is measured.

The first approach has two main advantages. Established sensors and cell lines can be used. Studies of neuronal cells on the other hand are difficult to perform for several reasons. Neuronal cells cannot be produced in continuous cell lines and they appear more fastidious in their choice of substrate. However, the unique advantage of high specificity through receptor interaction can be made use of in this approach and receptors may be incorporated intentionally for "fine-tuning".

3.4.6 Whole Animals

More than 1 000 000 species of arthropods some of which possess extraordinary sensory abilities present a tremendous natural storehouse of highly optimized chemoreceptors. In order to make these abilities useable for mankind in terms of analytical tools, some interface of chemoreceptive organs of arthropods to microelectronic components of analytical instruments has to be established. One promising possibility is the design of biosensors on the basis of intact chemoreceptors utilizing electrophysiological techniques.

Table 3.8 Direct use of integrated biological systems for biosensors [79]

System	Analyte	Signal	Author
Plant 'tissues'	Phenols, neurotransmitters	O_2-consumption, H_2O_2-production	(He & Rechnitz, 1995; Wijesuriya & Rechnitz, 1993) (Karlsson et al., 1991; 1994; Elwing et al., 1990)
Fish scale	Adrenalin	Absorption change	
Curstaceae olfactory receptor nerves	Amino acids, bases, neurotransmitters	Ion channel current, impedance, capacitance	Rechnitz, several others
Butterfly antenna	Pheromones	Nerve action potential	Several
Moscito olfactory antennae	Amino acids, sweat compounds	Nerve action potential	(Huotari et al., 1996)
Insect chemoreceptor	Fruit ingredients	Nerve potential on ISFET	(Schütz et al., 1996a,b)

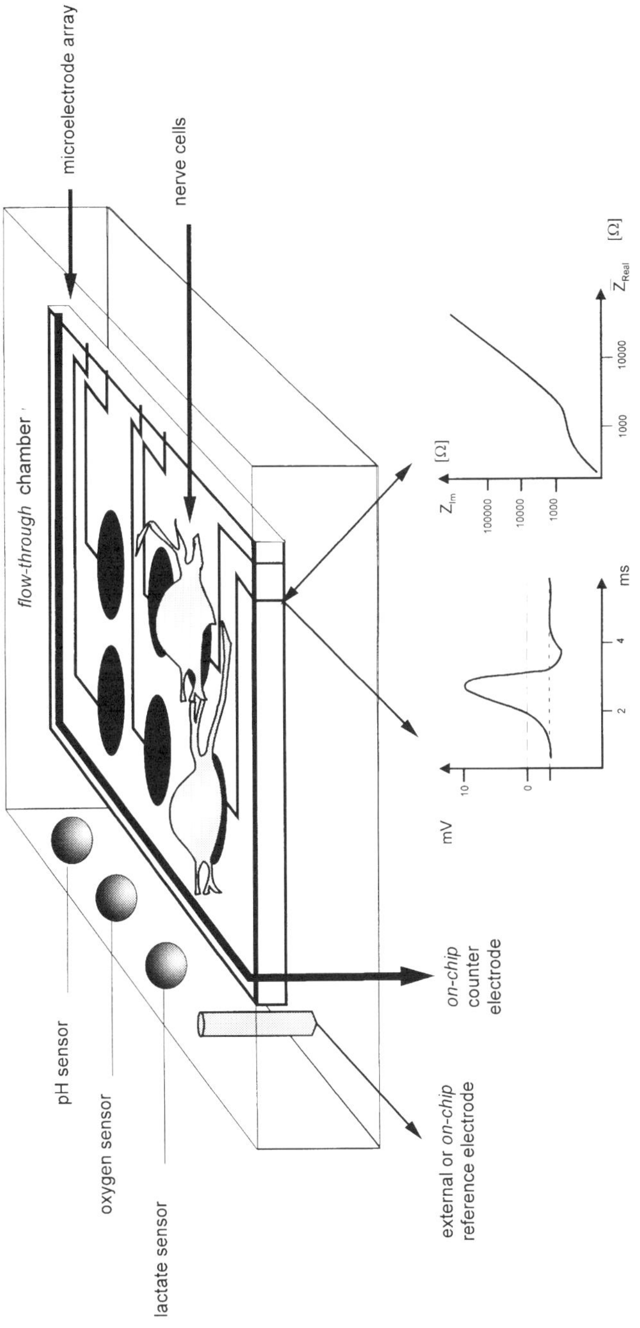

Figure 3.26 Schematic representation of possible set-ups to utilize electrogenic cells as sensor elements [79].

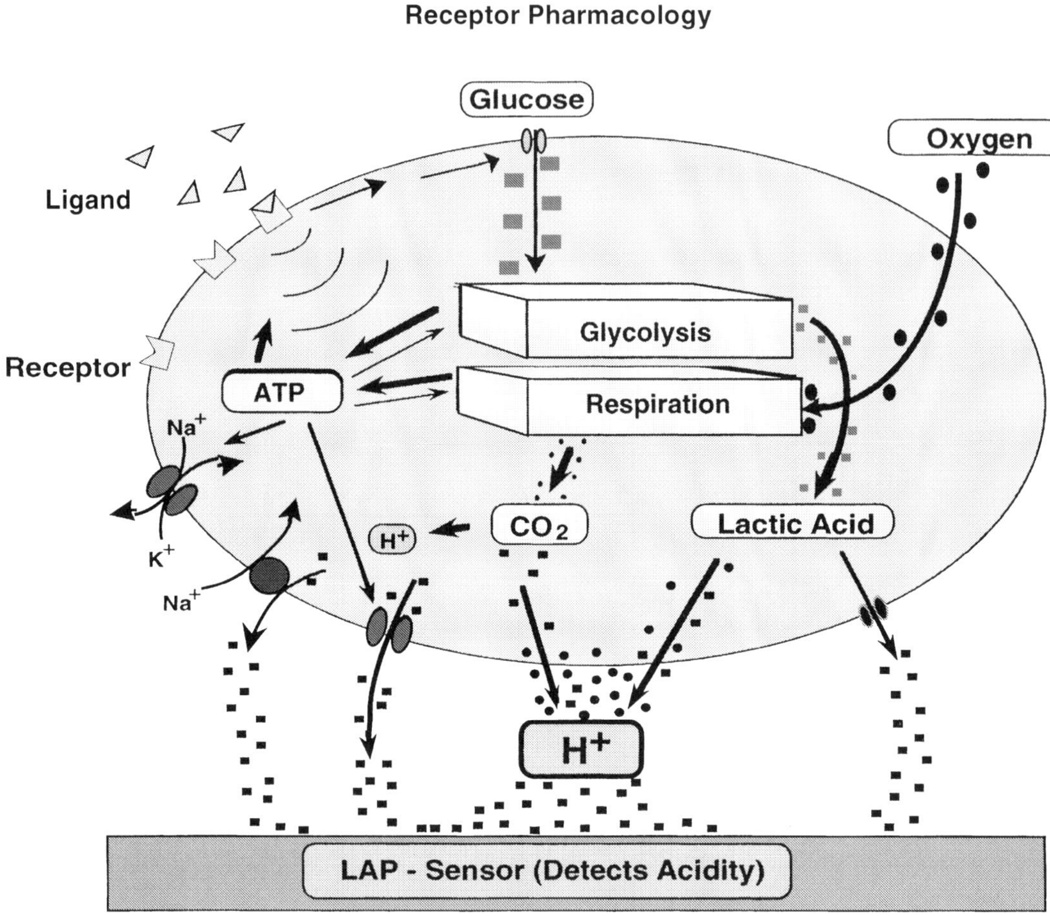

Figure 3.27 Schematic representation of metabolic pathways in which extracellular pH-changes occur (Metzger et al. 1994 [102]; Laxhuber, submitted [103]).

Arthropod olfactory and taste receptor cells are easily accessible for electrophysiological recordings because they occur in small groups, together with usually three auxiliary cells, and innervate a common cuticular apparatus such as a hair, peg, or pore plate [101]. These miniorgans, called sensilla, can also be equipped with receptor cells for other stimulus qualities like mechanical, temperature, and gustatory stimuli.

In spite of the fact that electrophysiological research on olfaction of arthropods has been performed since the 1950s, the first biosensor on the base of a complete olfactory organ (i.e. a crab antennule), was developed by Belli and Rechnitz (1986) in order to circumvent the difficulties associated with isolating, stabilizing, and

Table 3.9 Molecular hierarchy of (bio-)electronic noses

Inorganic structures and organic molecules
Biomimetic recognition sites
Biological recognition sites
odorant binding proteins
olfactory receptors
Recognition sites embedded in biological membranes
olfactory receptors in cilia
Recognition sites in membranes with subsequent signal amplification
Whole cells
olfactory cells
Cell arrays
Neural tissue
olfactory mucosa
Neural tissue with subsequent signal processing step
olfactory mucosa and olfactory bulb
Brain
olfactory mucosa, olfactory bulb, and olfactory cortex
Animals
Humans
odor sensation of distinguished test persons

immobilizing of receptor proteins [104]. This was accomplished by utilizing an intact chemosensing structure (the antennule of *Callinectes sapidus*) as a chemical transducer attached to the tip of an inert pick-up electrode. The prototype demonstrated the feasibility of the concept of receptor-based biosensors; however, problems of lifetime and limited detection range prevented the early "receptrode" from being an analytically useful device. After several improvements and new developments (see Buch and Rechnitz 1989 [105]; Coon et al., 1996 [106]; Van Egeraat and Wikswo, 1991 [107] Schütz *et al.* 1996 and 1997 [108, 109]), a biosensor for volatiles released by damaged plants was developed on the basis of insect antenna (Colorado potato beetle, *Leptinotarsa decemlineata*) using the far more rugged EAG technique in which summed dendrite potentials are recorded with two electrodes placed in the hemolymph space of the antenna. Experience of other groups could be used in designing portable EAG devices [110–112].

A decisive step in the direction of miniaturization and electrical stabilization of this biosensor type was the realization of extracellular recording with a direct FET-insect antenna junction (Figure 3.28 [108, 109]). This more compact set-up combines the benefit of miniaturization with the possibility to achieve increased life time because of the high impedance of the FET which needs only very little antenna injury during preparation.

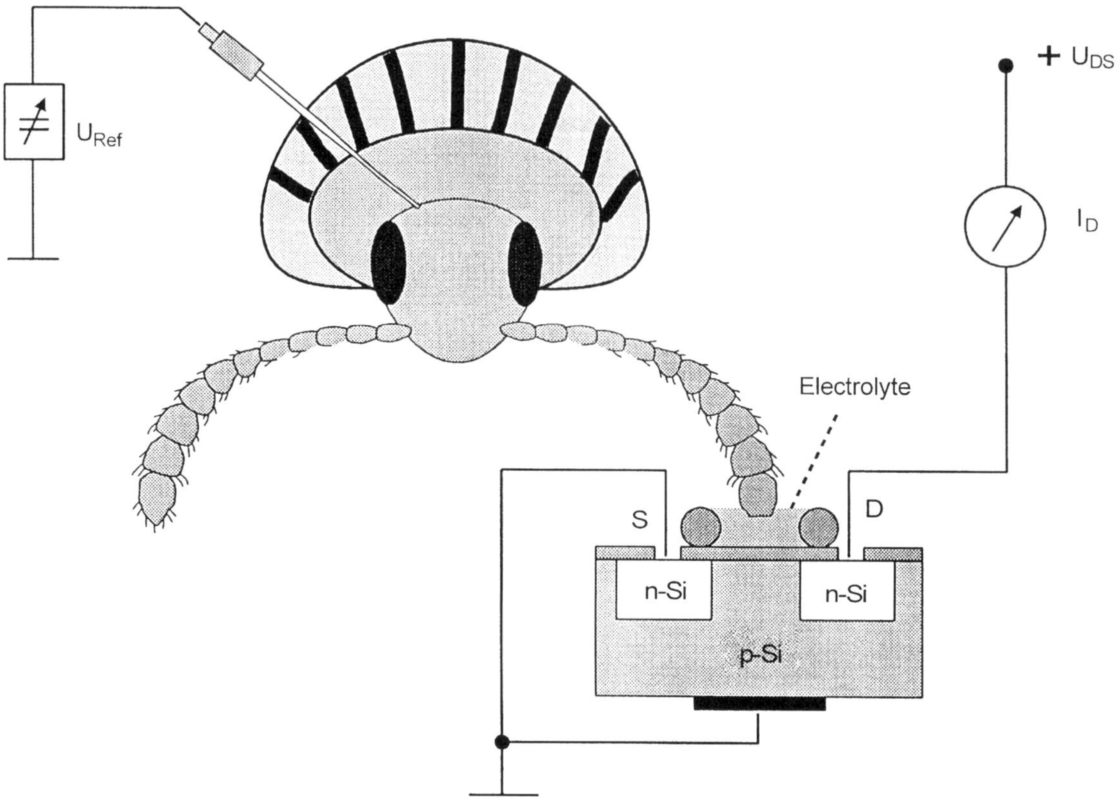

Figure 3.28 Scheme of a junction between colorado potato beetle antenna and field effect transistor. Also shown is a reference electrode (left side). S denotes source, D drain, $U_{Ref(DS)}$ the voltage of the reference electrode (drain source voltage) and I_D the drain current (Schütz et al., 1996; 1997) [108, 109].

3.4.7 Molecular Hierarchy of Electronic Noses

Table 3.9 summarizes the different approches discussed in this chapter in context with the general scope of this article.

3.5 Conclusions and Outlook

The biological olfactory system processes sensory information in a sophisticated manner. Not only is the architecture complex – a massively parallel multilayered recursive neural network – but there are many biochemical and electrochemical processes that occur. Nevertheless our level of understanding of both odors and olfactory transduction has greatly increased over the past decade and should become much clearer over the next few years. Our current understanding of olfaction is outlined in Section 3.2 in which we have used a case study on the flavor of beers to demonstrate the fundamental limitations of conventional analytical techniques, such as gas chromatography.

In Section 3.3 of this chapter we have considered the architecture underpinning existing electronic nose systems, from odor delivery through pattern recognition and finally odor representation. A comparison has been made between this information processing architecture and that comprised by the layers of neural processing making up the biological olfactory pathway. Not surprisingly, clear parallels exist between the two systems that principally derive from the earlier work by Persaud and Dodd [8] which exploits a rudimentary model of information processing within the olfactory system that was proposed by Deutsch as long ago as 1967. It is clear, with hindsight, that while advances in sensing and computing technology have vastly improved the practical performance of current electronic nose systems, such instruments are still limited by the outmoded model of the biological olfactory processing upon which they are based. A number of opportunities for performance enhancement in the practical system have been described. For example,

- SAR studies in olfaction have been unable to locate a single molecular determinant upon which odor quality depends. Rather, it seems that more than 20 molecular determinants are important in defining odor quality. Consequently, there is a clear argument for using a hybrid of sensor technologies within an electronic nose in order to capture different aspects of the odorant under investigation. Future sensor arrays will need to specialized to be the problem domain in question through an appropriate selection of sensor technologies.
- As considered in Section 3.1, broadly tuned overlapping receptor responses provide the ability to discriminate odor quality and quantity (concentration) independently, as a result of subsequent neural processing that emphasizes the differential response of similar receptor classes. Such an approach could be exploited to the advantage of the electronic nose.
- Large numbers of classes of receptors within the biological system provide xenobiotic response. That is, the system takes as broad a sample of the odor space as possible and may respond to odorants even though these may not yet have been experienced. This feature will naturally occur in electronic nose systems that comprise sufficiently large stocks of non-specific sensors.
- In the biological system physical segregation of odorants is mediated by the mucosal layer of the epithelium. This provides an initial sorting mechanism to direct functionally distinct classes of odorants to different portions of the receptor sheet. Such an approach might be usefully employed in a hybrid multi sensor array combined with a GC column. Using such a scheme it would then be possible to direct specific classes of odorants towards the most appropriate sensor technology as odor species elute from the GC column over time.

- Convergence within the biological system has been shown to enhance system sensitivity, through averaging of receptor input, in addition to providing a constant olfactory signal in view of continual receptor degeneration and neurogenesis. An equivalent scheme has been proposed in the electronic analog that demonstrates these important properties.
- Dynamic range compression of sensory input may be mediated in the electronic nose through the use of the logarithmic preprocessing metric. There is evidence for a similar mechanism in the biological system that improves the overall range of odor concentrations that can be perceived.

Finally, pattern recognition techniques applied to electronic nose systems have been compared to the associative models of the piriform cortex in the biological system. Advantages in performance in the electronic analog may be achieved by considering dynamic neural processing, such as computational neuronal models.

We also believe that there is a need for major advances in the sensor technologies that are employed with electronic noses and this topic is discussed in Section 3.4. To this end we have first described some promising candidate technologies based on organic (including biological) rather than inorganic materials. For example, we believe that the recognition of olfactory molecules in sensors may be enhanced through the synthesis and use of new biomimetic recognition sites, perhaps through the use of static and dynamic molecular cage compounds. Biomimetic systems based upon a biological model could lead to far superior and intelligent biomolecular devices which may incorporate whole cells or membranes within a bioelectronic nose. These types of electronic noses are clearly much closer in nature to the biological olfactory system, and combined with more sophisticated signal processing algorithms – like adaptive, non-linear, differential fuzzy neural networks – will represent a new generation of electronic noses.

3.6 Acknowledgements

The authors would like to thank the Department of Trade and Industry, and Ministry of Food and Fisheries, Bass Plc, and Neotronics Technology Ltd in the UK for their financial support of T. C. Pearce during a three year LINK research programme.

A large number of colleagues contributed to this survey and helped to keep the balance between the various disciplines of this interdisciplinary topic. We therefore greatly appreciate contribution by R. Apfelbach, H. Hammerle, G. Jung, A. Zell and Ch. Ziegler of the University of Tübingen (D), H. Hatt, University of Bochum (D), J. Joerges, University of Berlin (D), L. Laxhuber, München (D), H. Breer, University of Hohenheim (D), R. Malaka, University of Karlsruhe (D), D. Schild, University of Göttingen (D), S. Schütz, University of Gießen (D), H.-L. Schmidt, University of München (D), F. Vögtle, University of Bonn (D). Support of a Workshop on Bioelectronic Noses, October 1996, by the Ministry BMBF (Dr. Hache) is also greatly acknowledged.

3.7 References

[1] Göpel, W., Oehme M., in: *Sensors Chemical and Biochemical Sensors*, Göpel, W., Hesse, J. Zemel, J. N. (eds); Weinheim: VCH, (1991), p. 819.

[2] Göpel W., *Biosensors and Bioelectronics*, **10** (1995), 853–883

[3] Göpel, W., *Sensors and Actuators A* **56** (1996), 83–102.

[4] Göpel, W., *Nanotechnology*, Spec. Issue of *Microelectronics Engineering*, **32** (1996), 75–110.

[5] Göpel W., *Biosensors and Bioelectronics*, in press.

[6] Aishima, T., Nakai, S., *Food Reviews International*, **7** (1), (1991), 33–101.

[7] Meilgaard, M. C., Reid, D. S., Wyborski, K. A., *ASBC Journal*, **40** (4) (1982), 119–128.

[8] Persaud, K., Dodd, G.H., *Nature* **299** (5881), (1982), 352–355.

[9] Gardner, J. W., Bartlett, P. N. (eds), *Sensors and Sensory Systems for an Electronic Nose*; Dordrecht: Kluwer Academic Publishers, 1992.

[10] Gardner, J. W., Bartlett, P. N., *Sensors and Actuators, B*, **18–19** (1994), 211–220.

[11] Bartik, K. F., *M.S. Thesis*, Cornell University, Ithaca, New York, 1988.

[12] Meilgaard, M. C., *MBAA Technical Quarterly*, **28** (1991), 132–141.

[13] Jayalekshmy A., Narayanan, C. S., Mathew, A. G., *Jaocs Journal of the American Oil Chemists Society*, **68** (11) (1991), 873–880.

[14] Khiari, D., Brenner, L., Burlinggame, G. A., Suffet, I. H. in: *Water Science and Technology: Off-flavours in Drinking Water and Aquatic Organisms*: Proceedings of the 3rd IAWPRC Symposium held in Los Angeles, USA, 3–8 March 1991, Persson P.-E., Whitfield, F. B., Krasner, S. W. (eds.); Great Britain: IAWPRC, (1991), pp. 97–104.

[15] Marais, J., Van Rooyen, P. C., Du Plessis, C. S., *S. Afr. J. Enol. Vitic.*, **2** (1) (1981), 19–23.

[16] Shimoda, M., Shibamoto, T., *J. Agric. Food Chem.*, **38** (1990), 802–804.

[17] Martin, A. B., *Ph.D. Thesis*, Cornell University, Ithaca, New York, 1989.

[18] Engan, S., *Brauwelt International*, **III** (1991), 217–223.

[19] ASBC Technical Committee and Editorial Committee, *Methods of Analysis of the American Society of Brewing Chemists*; 7th edition, ASBC, St. Paul, Michigan, USA, 1976.

[20] American Society for Testing and Materials, *Standard Practice E1432 for Defining and Calculating Individual and Groups' Sensory Thresholds From Forced-Choice Data Sets of Intermediate Size*; Philadelphia, USA: ASTM, 1991.

[21] Gregson R. A. M., *Food Technol.*, **17** (3), (1963), 44.

[22] O'Mahoney, M. O., *Chemical Senses and Flavour*, **3** (1978), 183–189.

[23] Meng, A.-K., Brenner, L., Suffet, L. H., in: *Water Science and Technology: Off-flavours in Drinking Water and Aquatic Organism*, Proceedings of the 3rd IAWPRC Symposium held in Los Angeles, USA, 3–8 March 1991, Persson, P.-E., Whitfield, F. B., Krasner, S. W. (eds.), pp. 49–56. Published in UK by IAWPRC, Oxford, 1991.

[24] Kwan, W.-O., Kowalski, B. R., *Analytica Chimica Acta*, **122** (1980), 215–222.

[25] McLean, J. H., Shipley, M. T., in: *Science of Olfaction*, Serby, M. J., Chobor, K. L. (eds.); New York: Springer-Verlag, 1992, pp. 126–171.

[26] Shipley, M., Reyes, P. in: *The Human Sense of Smell*, Laing, D. G., Doty, R. L., Breipohl, W. (eds.); Berlin: Springer-Berlag, 1991, pp. 29–60.

[27] Lancet, D., *Nature*, **351** (1991) 275–276.

[28] Morrison, E. E., in: *Science of Olfaction*, Serby, M. J., Chobor, K. L. (eds.); New York: Springer-Verlag, 1992, pp. 126–171.

[29] Schild, D., in: *Encyclopedia of Human Biology*, vol. 5, Dulbecco, R. (ed.); San Diego: Academic Press, 1991, pp. 523–531.

[30] Stepherd, G. M., in: *Olfaction: A Model System for Computational Neuroscience*, Proceedings of a conference held at Wellesley College, May 17–18, 1990, Davis, J., Eichenbaum, H. (eds.), Cambridge, MA: MIT Press, 1991, pp. 3–41.
[31] Moulton, D. G., *Physiol. Rev.*, **56** (1976), 578–593.
[32] Hornung, D. E., Mozell, M. M., in: *Clinical Measurement of Taste and Smell*, Meiselman, H. L., Rivlin, R. S. (eds.); USA: Macmillan, New York, 1986, pp. 19–39.
[33] Mozell, M. M., Spatial and Temporal Patterning in *Handbook of Sensory Physiology*, Beidler L. M. (ed.), pp. 95–131, New York: Verlag, 1971.
[34] Pelosi, P., Baldiccini, N. E., Pisanelli, A. M., *Biochem, J.*, **201** (1992), 245–148.
[35] Pevsner, J., Reed, J. R., Firestein, P. G., Snyder, S. H., *Science*, **241** (1988), 336–339.
[36] Pelosi, P., *Critical Reviews in Biochemistry and Molecular Biology*, **29** (3) (1994), 199–228.
[37] Adrian, E. D., *Br. Med. Bull.*, **6** (1950), 330–331.
[38] Pearce, T. C., *Biosystems*, **41** (1997), 43–67.
[39] Kauer, J. S., in: *Neurobiology of Taste and Smell*, Finger, T. E., Silver, W. L. (eds.) New York: John Wiley, 1987, pp. 205–231.
[40] Beets, M. G. J., *Pharmacological Reviews*, **22**(1) (1970), 1–34.
[41] Shepherd, G. M., *Cell*, **67** (1991), 845–851.
[42] Brohnstein, A. A., Minor, A. V., *Tsitologiya*, **19** (1977), 33–39.
[43] Firestein, S. Shepherd, G. M., Werblin, F. S., *J. Physiol (Lond.)*, **430** (1990), 135–158.
[44] Pongracz, F., Firestein, S., Shepherd, G. M., *Soc. Neurosci. Abstr.*, **15** (1989), 1142.
[45] Pace, U., Hansky, E., Salomon, Y., Lancet D., *Nature*, **316** (1985), 255–258.
[46] Nakamura, T., Gold, G. H., *Nature*, **325** (1987), 442–444.
[47] Firestein, T., Darrow, B., Shepherd, G. M., *Neuron*, **8** (1991), 825–835.
[48] Buck, L., Axel, R. A., *Cell*, **65** (1991), 175–187.
[49] Lancet, D., *Annual Review of Neuroscience*, **9** (1986), 329–355.
[50] Anholt, R. R. H., in: *Science of Olfaction*, Serby, M. J., Chobor, K. L. (eds.); New York: Springer-Verlag, 1992, pp. 51–79.
[51] Burchell, B., *Nature*, **350** (1991), 16–17.
[52] Chastrette, M., Zakarya, D. in: *The Human Sense of Smell*, Doty, R. L., Bereipohl, W. (eds.); Berlin: Springer-Verlag, 1991, pp. 77–92.
[53] Shepherd, G. M., *Microscopy Research and Technique*, **24** (1993), 106–112.
[54] Mori, K., *Microsc. Res.*, **24** (1993), 131–141.
[55] Schild, D., *Biophysical Journal*, **54** (6), (1997), 1001–1011.
[56] Duchamp-Viret, P., Duchamp, A., Vigourous, M., *J. Neurophysiol*, **61** (1989), 1085–1094.
[57] Nickell, W. T., Shipley, M. T., in: *Science of Olfaction*, Serby, M. J., Chobor, K. L. (eds.), New York: Springer-Verlag, 1992, pp. 172–212.
[58] Moulton, D. G. in: *ISOT II*, Hayashi, T. (ed.); Oxford: Pergamon, 1967, pp. 109–116.
[59] Sejnowski, T. J., Kienker, P. K., Shepherd, G. M., *Soc. Neurosci. Abstr.*, **11** (1985) ,970.
[60] Shepherd, G. M., Firestein, S., *Current Biology*, **1** (4), (1991), 204–206.
[61] Freeman, W. J., in: *Olfaction: A Model System for Computational Neursoscience*, Proceedings of a conference held at Wellesley College, May 17–18, 1990, Davis, J., Eichenbaum, H. (eds.), Cambridge, MA: MIT Press, 1991, pp. 225–249.
[62] Kauer, J. S., *Nature*, **331** (1988), 166–168.
[63] Mori, K., in: *Olfaction and Taste XI*, Kurihara, K., Suzuki, N., Ogawa, H. (eds.) Tokyo: Springer-Verlag, 1994, pp 429–432.
[64] Mori, K. Shepherd, G. M., *Sem. Cell Biol.*, **5** (1994), 65.
[65] Otto, T., Eichenbaum, H. in: *Science of Olfaction*, Serby, M. J., Chobor, K. L. (eds.); New York: Springer-Verlag, 1992, pp. 213–244.

[66] Kohonen, T., *Self-Organization and Associative Memory*; Berlin: Springer Verlag, 1984.
[67] McClelland, J. L., Rumelhart, D. E., *Explorations in Parallel Distributed Processing*; Cambridge, MA: MIT Press, 1988.
[68] Tanyolac, N. N., Eaton, J. R., *J. Am. Pharm. Assoc.*, **39** (10), (1950).
[69] Deutsch, S., *Models of the Nervous System*, New York, John Wiley, 1967.
[70] Gibson, J. J., *The Senses Considered as Perceptual Systems*; New York: Houghton Mifflin, 1966.
[71] Shrumer, H. V., Whitehouse, D. J., *Proc. Roy. Soc.*, *A* **443** (1993), 313–331.
[72] Pearce, T. C., *Biosystems*, **41** (1997), 69–90.
[73] Schild, D., Gardner, J. W., in: *Cell to Cell Signals in Plants and Animals: NATO ASI Series H.*, vol. 53, Neuhoff, Friend J. (eds.) Heidelberg: Springer-Verlag, 1991, pp. 96–117.
[74] Bodenhöfer, K., Hierlemann, A., Seemann, J., Gauglitz, G., Koppenhöfer, B., Göpel, W., *Nature*, in press.
[75] Schweizer-Berberich, M., Vaihinger, S., Göpel, W., *Sensors and Actuators*, *B* **18–19** (1994), 282.
[76] Winquist, F., Hörnsten, E. G., Sundgren, H., Lundström, *Meas. Sci. T.*, **4** (1993), 1493–1500.
[77] Ulmer, H., Mitrovics, J., Noetzel, G., Weimar, U., Göpel, W., *Conf. Proc. Eurosensors X*, Leuven (B) (9/1996); *Sensors and Actuators*, in press.
[78] Göpel, W., *Biosens. and Bioelectronics*, **10** (1/2), (1995), 35–59.
[79] Göpel, W., Ziegler, Ch. (coordinating authors), Breer, H., Schild, D. (biological section), Apfelbach, R., Hammerle, H., Hatt, H., Joerges, J., Jung, G., Laxhuber, L., Malaka, R., Schmidt, H.-L., Schütz, S., Vögtle, F., Zell, A. (contributing authors), *Biosensors and Bioelectronics*, in press.
[80] Vodyanoy, V. in: *Molecular Electronics: Biosensors and Biocomputers*, Hong F. T. (ed.); New York: Plenum Press, 1996, pp. 317–328.
[81] Nicolis, J. S., in: *Synergetics of the Brain*, Proceedings of the International Symposium on Synergetics at Schloss Elmau, Bavaria, May 2–7, 1983, vol. 23, Basar, E., Flohr, H., Haken, H., Mandell, A. J. (eds.), Heidelberg: Springer-Verlag, 1983, pp. 330–344.
[82] Hierlemann, A., Schweizer-Berberich, M., Weimar, U., Kraus, G., Pfau, A., Göpel, W. in: *Sensors Update Sensor Technology – Application – Markets*, vol. 2, Baltes, H., Göpel, W., Hesse, J. (eds.), Weinheim: VCH, 1996, p. 119–180.
[83] Göpel, W., Hesse, J., Zemel, J. N. (eds.), *Sensors: A Comprehensive Survey*, and in particular vols. 2 and 3: *Chemical and Biochemical Sensors*; Weinheim: VCH, vols. 1–8, 1991 and 1992.
[84] Göpel, W., Schierbaum, K. D., in: *Sensors: A Comprehensive Survey*, vol. 2: *Chemical and Biochemical Sensors*, Göpel W., Hesse, J., Zemel, J. N. (eds.); Weinheim: VCH, 1991, p. 119–157.
[85] Göpel, W., *Sensors and Actuators*, *B*. **24–25** (1995), 17–32.
[86] Göpel, W., Heiduschka, P., *Biosensors and Bioelectronics*, **9** (1994), III–XIII.
[87] Göpel, W., *Biosensors and Bioelectronics*, **10** (1994), 35–59.
[88] Rickert, J., Weiß, T., Kraas, W., Jung, G., Göpel, W., *Biosensors and Bioelectronics*, **11** (1996), 591–598.
[89] Rickert, J., Heiduschka, P., Beck, W., Jung, G., Göpel, W., *Biosensors and Bioelectronics*, **11** (1996), 757–768.
[90] Göpel, W., Schierbaum, K. D. in: *Sensors: A Comprehensive Survey*, vol. 2: *Chemical and Biochemical Sensors*, Göpel, W., Hesse, J., Zemel, J. N. (eds.); Weinheim: VCH, 1991, p. 1–27.

[91] Schierbaum, K. D., Weiß, T., van Velzen, U. Th., Reinhoudt, B. N., Göpel, W., *Science*, **265** (1994), 1413–1415.
[92] Göpel, W., Reinhardt, G., in: Baltes, H., Göpel, W., Hesse, J. (eds.), *Sensors Update Sensor Technology – Applications – Markets*, vol. 1, Weinheim: VCH, 1996, p. 49–120.
[93] Bodenhöfer, K., Hierlemann, A., Seemann, J., Gauglitz, G., Brendle, H.-G., Schlunk, R., Christian, B., Koppenhöfer, B., Göpel, W., *Analyt. Chem.* **69** (1997), 3058–3068.
[94] Lehn, J. M., *Angew. Chem.* **100** (1988), 91–116.
[95] Inouye, M., Hashimoto, K., Isagawa, K., *J. Am. Chem. Soc.*, **116** (1994), 5517–5518.
[96] Rickert, J., Brecht, A., Göpel, W., *Analyt. Chem.*, **69** (1997), 1441–1448.
[97] Seel, C., Vögtle, F., *Angew. Chem. Int. Ed. Engl.*, **30** (1991), 442–445.
[98] Vögtle, F., Seel, C., Windscheif, P.-M. in: *Comprehensive Supramolecular Chemistry* vol. 2, Atwood, J. L., Eric. J., Davies, D., MacNicol, D. D., Vögtle, F., Lehn, J.-M. (eds.); Oxford: Pergamon Press, pp. 211–265.
[99] Adachi, Y., Sugawara, M., Tanigucki, K., Umezawa, Y., *Anal. Chem.*, **65** (1993) 363–369.
[100] Bousse, L., *Conf. Proc. Eurosensors IX*, **C7** (1995), 356–364.
[101] Kaissling, K.-E. in: *Experimental Cell Biology of Taste and Olfaction, Current Techniques and Protocols*, Spielman, A.I. Brand, J. G. (eds.) Boca Raton: CRC Press, 1995, 361–377.
[102] Metzger, J. W., Kempter, C., Wiesmüller, K. H., Jung, G., *Anal. Biochem.*, **219** (1994), 261–277.
[103] Laxhuber, L. A., in [79].
[104] Dung, B., Vögtle, F., *J. Inclusion Phenom.*, **6** (1988), 429.
[105] Hasselmio, M. M., Bower, J. M., Apparent and Association Fiber in Short-term Purefarm (Olfactory) *J. Neurophysiol*, **19**, 179–190.
[106] Belli, S., Rechnitz, G., *Analyt. Letters*, **19** (1986), 403–416.
[107] Buch, A., Rechnitz, G. A., *Biosensors*, **4** (1989), 215–230.
[108] Coon, R. D., Babb, C. W., Rechnitz, G. A., *Anal. Chem.*, **68** (1996), 1671–1675.
[109] Van Egeraa, J. M., Wikswo, J., in: *Biomagnetism Clinical Aspects*, Proceedings of the 8th International Conference on Biomagnetism, Hoke, M. (ed.); Elsevier Science, 1991, p. 895–899.
[110] Schütz, S., Weibbecker, B., Hummel, H. E. in: *Refereed Abstracts of the Fourth World Congress on Biosensors*, Bangkok, Thailand, 29–31 May 1996: Oxford: Elsevier, 1996, p. 85.
[111] Schütz, S., Weibbecker, B., Hummel, H. E., Schöning, M. J., Riemer A., Kordos, P., Lüth, H., *Naturwissenschaften*, **84** (1997), 86–88.
[112] Baker, T. C., Haynes, K. F., *Physiol. Entomol.*, **14** (1989), 1–12.
[113] Sauer, A. E., Karg, G. Koch, U. T., De Kramer, J. J., Milli, R., *Chem. Senses*, **17** (1992), 543–553.
[114] Van der Pers, J. N. C., Klijnstra, J. W., De Vlieger, J. J. in: *Abstracts of the 12th Congress of the European Chemoreception Research Organization*, Zürich, Switzerland, August 1996, 17.
[115] Sicard, G., Holley, A., *Brain Research*, **292** (1984), 283–296.

4 Electronic Sensing of Tastes

KIYOSHI TOKO, Department of Electronic Device Engineering,
Graduate School of Information Science and Electrical Engineering,
Kyushu University, Fukuoka 812, Japan

CONTENTS

4.1 Introduction

A multichannel taste sensor with global selectivity is composed of several kinds of lipid/polymer membranes for transforming information about taste substances into electric signals, which are input to a computer. The sensor output shows different patterns for chemical substances which have different taste qualities such as saltiness and bitterness, while it shows similar patterns for chemical substances with similar tastes. The sensor responds to the taste itself, as can be understood from the fact that taste interactions such as the suppression effect, which appears between sweet and bitter substances, can be reproduced well. Amino acids can be classified into several groups according to their own tastes based on sensor outputs. The taste of foodstuffs such as beer, coffee, mineral water, milk, sake, soybean paste and vegetables can be discussed quantitatively using the taste sensor, which provides an objective scale for human sensory expression. The taste sensor will open the door to a new stage of food culture.

One of the goals of sensors is to reproduce the five kinds of senses of humans or surpass their abilities. In this case, the required abilities of sensors are the following. The first is high sensitivity, the second is durability except for a disposal type, and the third is high selectivity.

The first two items seem to be fairly reasonable. However, the last is not always true. Certainly, light is received in the case of sense of sight. Sound waves are received in the sense of hearing. Pressure or temperature is received in the sense of touch. In these kinds of senses, only each physical quantity such as light and sound waves must be detected with high sensitivity and selectivity.

For example, let us consider the situation where we measure the body temperature when we are sick. In this case, we set a thermometer to our mouth or under the arm, where it is dark. If the thermometer responds to light, it is quite troublesome; i.e., the thermometer should measure only the temperature. The same situation is also applied to the microphone. It should respond only to the voice, and should not respond to light and other quantities. As understood from these examples, the sensors reproducing the senses of sight, hearing and touch should respond to only one physical quantity.

However, the situation is much different in the senses of taste and smell. For example, over one thousand kinds of chemical substances are included in tea or coffee. Humans do not distinguish each chemical substance. Instead, humans decompose the taste of foodstuffs into five kinds of basic taste qualities, i.e., sour, bitter, salty, sweet and umami. It implies that we need the development of a novel type of sensor for taste and smell.

Taste is constructed from five basic taste qualities as above: sourness produced by hydrogen ions of HCl, acetic acid, citric acid, etc.; saltiness produced mainly by NaCl; bitterness produced by quinine, caffeine and $MgCl_2$; sweetness due to sucrose, glucose, etc.; and umami, which is the Japanese term for 'deliciousness', produced by monosodium glutamate (MSG) contained mainly in seaweeds, disodium inosinate (IMP) in meat and fish and disodium guanylate (GMP) in mushrooms [1–3]. In biological taste reception, taste substances are received by the biological membrane of gustatory cells in taste buds on the tongue (Figure 4.1). Then the information on the taste substances is transduced into an electric signal, which is transmitted along the nerve fiber to the brain, where the taste is perceived. We have developed a multichannel electronic taste-sensing system whose transducer is composed of lipid/polymer membranes [4, 5]. This sensor can detect tastes in a manner similar to human gustatory sensation. The output of the sensor is not the number of specific molecules to exhibit taste but the taste quality and intensity, because different output electric patterns were obtained for different taste groups such as sourness and saltiness. On the other hand, similar patterns were obtained for molecules in the same group such as

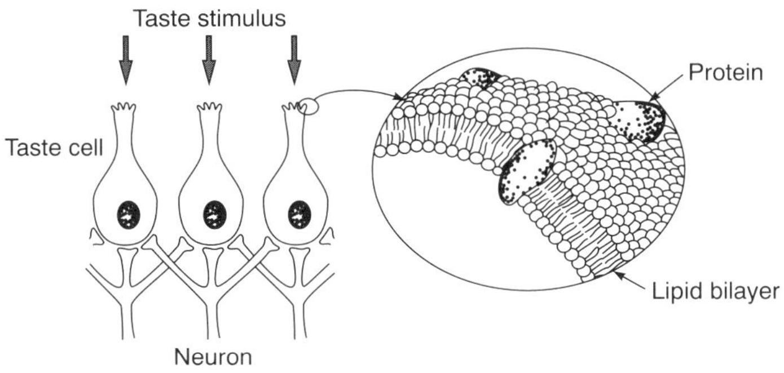

Figure 4.1. Taste reception in biological systems.

MSG, IMP and GMP, which have an umami taste, and NaCl, KCl and KBr for saltiness.

Discrimination of each chemical substance is not important here, but recognition of the taste itself and its quantitative expression must be made. The taste sensor using lipid/polymer membranes has a concept of global selectivity which implies the ability to classify enormous numbers of chemical substances into several groups, as found in the taste reception in biological systems.

We mention here the principle of the taste sensor and show results of measurements of aqueous solution constructed of five basic taste substances and several foodstuffs such as beer, sake, mineral water, coffee, milk and tomatoes. Quantification of the taste is possible using the taste sensor, and hence we can discuss the taste objectively.

4.2 Multichannel Taste Sensor

Previous studies [6–11] showed that lipid membranes had the ability to sense the taste by responding to taste substances; i.e., lipid membranes are useful materials to transform the taste to electric signals. A recently developed multichannel taste sensor, where transducers were composed of lipid membranes immobilized with a polymer, has higher sensitivity and better reproducibility than those of the human sense [12–26]. The reproducibility in short-term measurements, e.g., for one-day use, is very high, as shown later in standard deviations in each measurement. However, slow drift of the response electric potential occurs over several months. Nevertheless, reliable measurement over a long time period becomes possible if a basic standard sample with stable qualities is prepared, because the difference of electric potential pattern between the sample and the basic standard sample is taken as the response electric potential pattern. Of course, the long-term reproducibility also depends on the measured sample. The same membrane can be used continuously for over one year.

4.2.1 Multichannel Electrode

Eight kinds of lipid analogs were used for preparing the membranes. The lipids used are summarized in Table 4.1. According to the object to be measured, and the purpose, we prepared different lipid materials. For example, the mixed, hybrid membranes composed of dioctyl hydrogen phosphate (DOP) and trioctyl methyl ammonium chloride (TOMA) were used for measurements of amino acids [17], astringent substances [19], mineral water [20], milk [21] and sake [23]. Each lipid was mixed in a test tube with polyvinyl chloride (PVC), and a plasticizer (dioctyl phenylphosphonate), which were dissolved in tetrahydrofuran. Successively, the mixture was dried in a glass plate, which was placed on a hot plate controlled at ca. 30°C. The lipid/polymer membrane thus prepared was a transparent, colorless and soft film with ca. 200 μm thickness. The surface structure of a lipid/polymer membrane is illustrated in Figure 4.2.

Lipid/polymer membranes were fitted on a multichannel electrode. The detecting electrode of each channel was made up of Ag wires whose surface was plated with Ag/AgCl, which were embedded in a basal acrylic board of 2 mm thickness, as shown in Figure 4.3. Another acrylic board of 1 cm thickness with eight cone-shaped holes was affixed to this board. The holes were filled with 100 mM KCl solution, and the eight membranes were fitted on the board to cover the holes.

The multichannel electrode was connected to an eight-channel scanner through high-input impedance amplifiers (Figure 4.4). The selected electric signal from the sensor was converted to a digital code by a digital voltmeter and was fed to a computer. Then, a voltage difference between the multichannel electrode and an Ag/AgCl reference electrode was measured.

The commercial taste-sensing system (SA401, Anritsu Corp.) has a mechanism essentially similar to the above. The detecting sensor part is made of seven electrodes of lipid/polymer membranes, and is controlled mechanically by a robot arm, as illustrated in Figure 4.5. Measurements of the tastes of Japanese sake, coffee and milk mentioned below were made using SA401; for coffee, for example, the membranes

Table 4.1. Lipids used in the preparation of membranes. Different sets of lipids were prepared according to the object to be measured and the purpose.

Channel	Lipid
1	Dioctyl hydrogen phosphate (DOP)
2	Cholesterol
3	Oleic acid
4	Decyl alcohol
5	Trioctyl methyl ammonium chloride (TOMA)
6	Oleyl amine
7	Dimethyloctadecylammonium bromide
8	Trimethyloctadecylammonium chloride

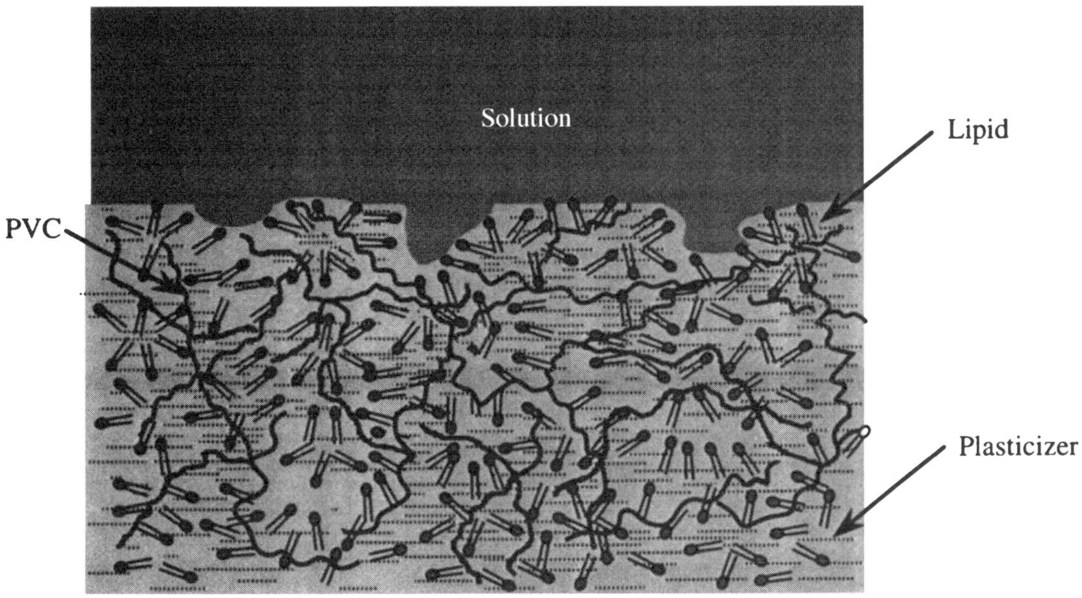

Figure 4.2. Surface structure of a lipid/polymer membrane.

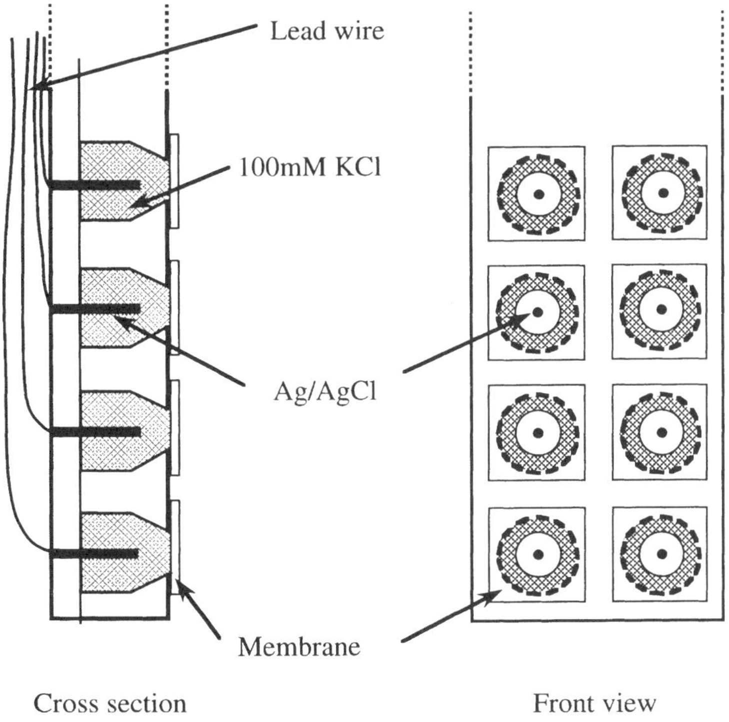

Figure 4.3. Multichannel electrode with eight lipid/polymer membranes.

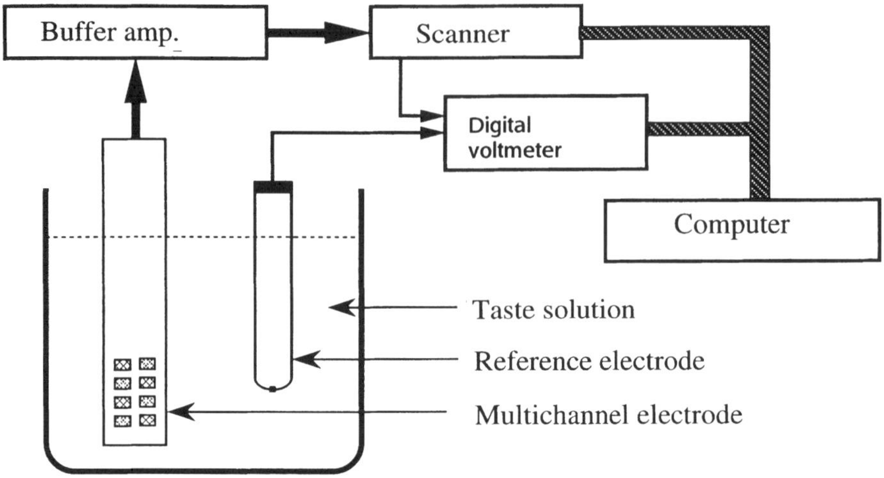

Figure 4.4. Measurement system for laboratory use.

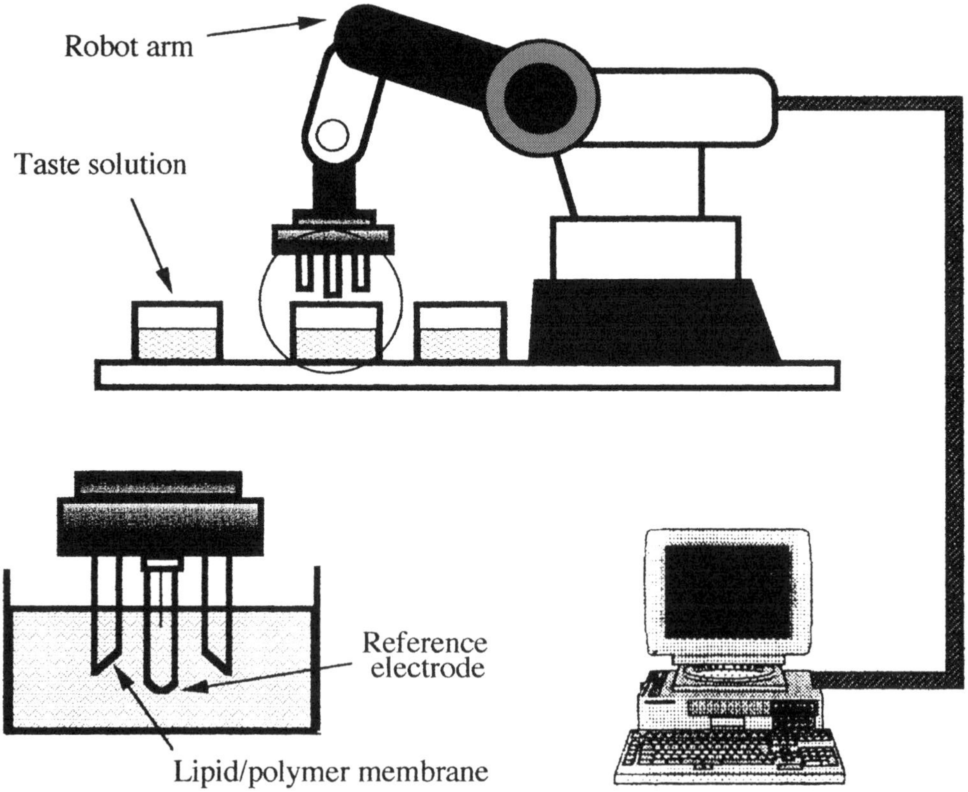

Figure 4.5. A commercially available taste-sensing system (SA401, Anritsu Corp.).

were No. SB00 for ch.1 as a control, No. SB01AAC1 for chs. 2 and 3, No. SB01AAF1 for chs. 4 and 5, and No. SB01AAJ1 for chs. 6 and 7, which were immersed in the standard coffee, Saldvador in this case, for at least a month. The response electric potentials were taken using this coffee as the origin of electric potential pattern. The membranes of chs. 2 and 3 for coffee measurements were negatively charged, those of chs. 4 and 5 positively charged, and those of chs. 6 and 7 were weakly charged negatively.

4.2.2 Global Selectivity to Molecules

Five typical primary taste substances, HCl (sour), NaCl (salty), quinine (bitter), sucrose (sweet) and MSG (umami) were studied.

Figure 4.6 shows the electric potential pattern from eight channels for two kinds of taste qualities of sour and salty, measured by taking the origin to 1 mM KCl [4]. As can be understood from Figure 4.6, the patterns of taste substances showing different taste qualities are much different, and hence each taste can be easily discriminated. The reproducibility was very high, because the standard deviations were smaller than 1%. On the other hand, the taste sensor has similar response patterns to the same group of taste; i.e., as examples of sour substances, HCl, citric acid and acetic acid show similar response patterns. Salty substances, NaCl, KCl and KBr show similar patterns, too. Therefore, we can conclude that this taste sensor can respond to the taste itself.

This fact implies that the taste sensor has global selectivity, as in the taste reception in biological systems. This sensor can be regarded essentially as an intelligent sensor. Of course, the taste sensor can distinguish chemical substances even in the same group of taste; e.g., the patterns for HCl, citric acid and acetic acid are different.

Figure 4.7 shows the mechanism of the potential change in the negatively charged membrane as in ch. 1 in Table 4.1 for quinine showing bitterness [24, 26]. The surface of lipid/polymer membranes made of anionic lipids such as DOP and oleic acid is negatively charged because of dissociation of H^+. The electric double layer is formed near the surface of the membrane in aqueous solution: cations are accumulated near the surface of the negatively charged membrane. The electric potential changes gradually from the negative value to zero in aqueous solution, as shown by the thick solid line in Figure 4.7. Since the right phase in Figure 4.7 is taken as the origin of the electric potential, the measured membrane potential (electric potential difference across the membrane) becomes negative. In the case of positively charged membranes, the membrane potential takes positive values. As shown in Figure 4.7, quinine ions are bound to the hydrophobic part of the lipid, and hence the negative charge due to lipids is somewhat compensated by the positive charge of quinine ions. As a result, the negative surface charge density approaches zero to decrease the magnitude of the electric potential at the surface; thus, the increase in membrane potential is observed. Mechanisms for other taste substances are also understood by taking account of changes in surface charge density (due to H^+ dissociation and binding of chemicals) and surface electric potential.

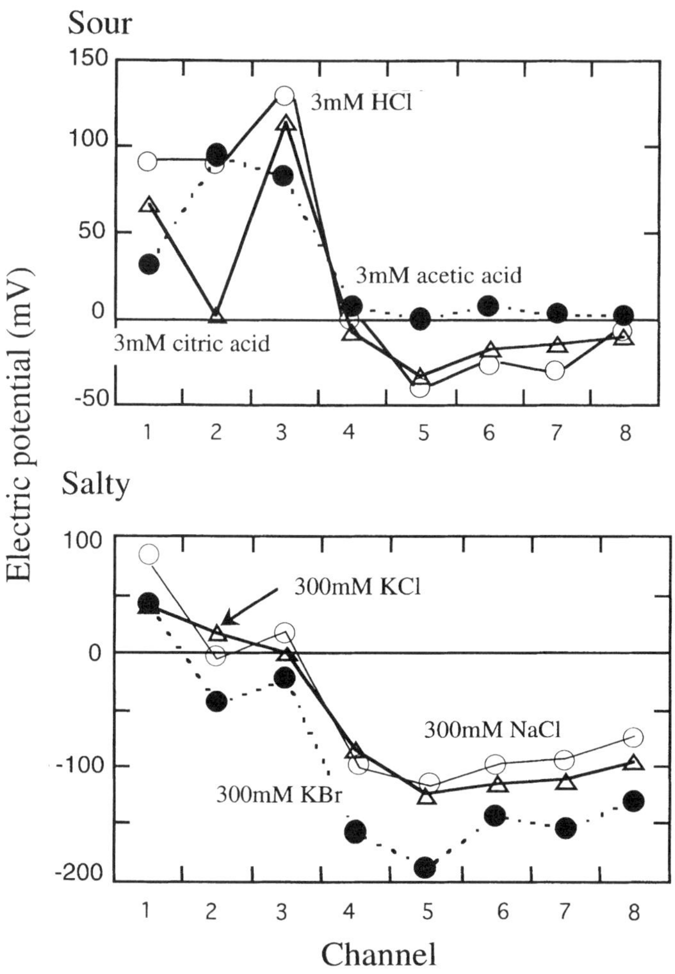

Figure 4.6. Response electric potential patterns for sour and salty substances.

It should be noted that the taste sense is produced by interactions between molecules and biological membranes; i.e., taste is not specific or characteristic of each molecule. Five kinds of taste qualities reflect the difference of the above interactions. For this reason, the situation such as global selectivity arises: the lipid/polymer membrane can reproduce it.

The sensor had detection errors (in units of logarithmic concentration) of 0.73% for saltiness, 0.65% for sourness and 3.4% for bitterness in the aqueous solution containing simultaneously different kinds of chemical substances producing different taste qualities [13]. Humans usually cannot distinguish two tastes with a

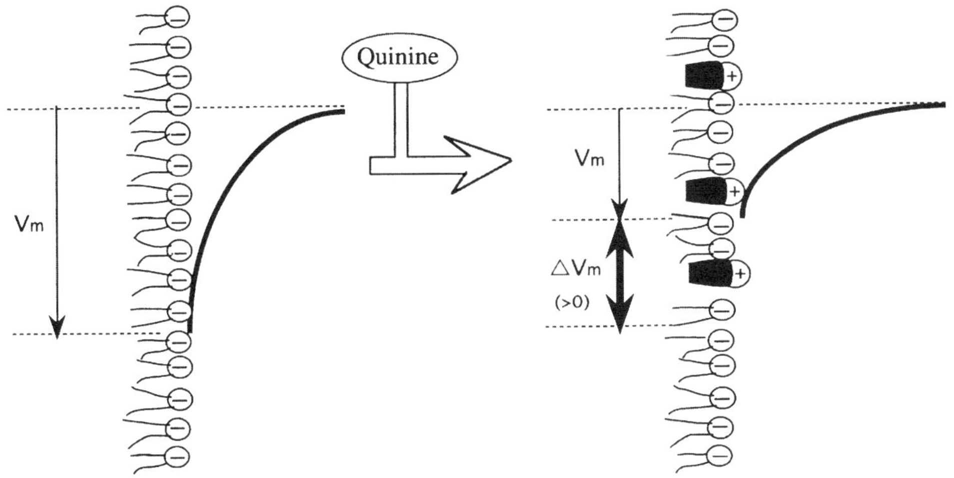

Figure 4.7. Response mechanism of the negatively charged membranes to bitter substances (quinine ·HCl).

concentration difference below 20% [2]. Here, 20% means an error of 7.9% (= log 1.2). Therefore, the ability of detection of the sensor is superior to that of humans.

4.2.3 Discrimination of Amino Acids

The taste of amino acids was studied using the taste sensor [17]. Each amino acid elicits a complicated mixed taste itself; e.g., L-valine produces sweet and bitter tastes at the same time. Thus, there exist detailed data on taste intensity and taste quality of various amino acids by sensory panel tests [27].

Amino acids are generally classified into several groups which correspond to each characteristic taste. Since amino acids show mixed tastes as above, taste interactions such as the suppression effect are automatically included in their taste. The taste sensor should express this situation. The present study is the first trial to study the taste of amino acids using artificial sensing devices.

Figure 4.8(a) shows the response patterns to typical amino acids, each of which elicits a different taste quality in humans. Each channel responded to them in different ways depending on their tastes. L-Tryptophan, which elicits an almost pure bitter taste in humans, increased the potentials of chs. 1, 2 and 3 greatly. Typical examples of standard deviations were 3.16, 7.32, 1.12, 2.51, 2.37, 2.86, 1.31 and 2.33 mV for chs. 1–8, respectively. This tendency was also observed for other amino acids, which mainly exhibit a bitter taste, such as L-phenylalanine and L-isoleucine. L-Valine and L-methionine, which taste mainly bitter and slightly sweet, decreased the potential of ch. 5; the responses of chs. 1 and 2 were small. On the other hand, for L-alanine, glycine and L-threonine which taste mainly sweet, the potentials of chs. 1 and 2

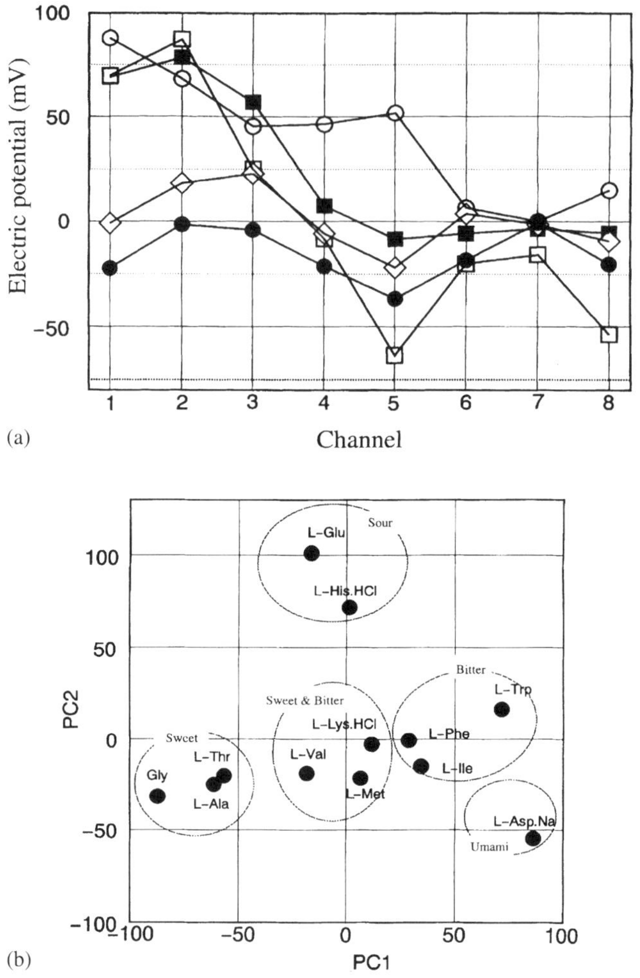

Figure 4.8. Response electric potential patterns for amino acids (a) and principal component analysis (b) [17]. Symbols ○, ■, □, ◇ and ● denote L-glutamic acid (10 mM), L-tryptophan (10 mM), monosodium L-aspartate (100 mM), L-valine (100 mM) and L-alanine (100 mM), respectively.

decreased. L-Glutamic acid and L-histidine monohydrochloride, which taste mainly sour, increased each of the potentials of chs. 1–5 to almost the same degree. Monosodium L-aspartate eliciting mainly umami taste showed a response pattern different from those of the other amino acids.

Figure 4.8(b) shows the data points plotted in the scattering diagram obtained by principal component analysis, which is a kind of multivariate analysis to reduce the dimension of data without losing information. In the present case, eight-dimensional space made from eight channel outputs was reduced to the two-dimensional plane.

The first principal axis (PC1) in Figure 4.8(b) reflects bitterness and sweetness. The second principal axis (PC2) reflects sourness and umami. Amino acids are classified clearly into five groups by the taste sensor.

The response of the sensor to amino acids was compared with the results of panel tests [27] in which intensities of five taste qualities are obtained for each amino acid. Response potentials from the eight membranes were transformed into five basic tastes by multiple linear regression. This expression of five basic tastes reproduced the human taste sensation very well [17].

4.3 Expression of Taste by Basic Taste Qualities

We attempted to make an artificial taste solution, which shows a similar taste to a commercial aqueous drink, by combination of basic taste substances by comparing electric potential patterns of output of the taste sensor [18]. As the basic taste substances, HCl, NaCl, sucrose and quinine were chosen for sourness, saltiness, sweetness and bitterness, respectively.

Four different concentrations were prepared for each of these substances: 1, 3, 10, 30 mM for HCl; 30, 100, 300, 1000 mM for NaCl and sucrose; 0.03, 0.1, 0.3, 1 mM for quinine. The lowest concentrations correspond nearly to the thresholds to be detected by humans. We prepared $4^4 (= 256)$ mixed solutions with different compositions by combination of these four types of basic solutions.

The above-mentioned 256 mixed solutions were measured with the multichannel taste sensor. Therefore, data on the output electric potential pattern were taken for the 256 solutions. While the data on each channel output $V_i (i = 1–8)$ were dispersed discretely in the five-dimensional space constructed from V_i and four different taste qualities, we approximated V_i by a quadratic function of the concentrations. As a result, eight quadratic functions were obtained for V_i.

As test aqueous drinks, two commercial drinks of different brands were chosen. Figure 4.9 shows the electric potential patterns of these two drinks. While they are not so different, the discrimination is easy because of standard deviations of 2 mV. We attempted to fit the above patterns constructed from eight functions to the output electric potential pattern of one of commercial drinks (let us call it 'drink A' for convenience) by minimizing the distance between the output electric potentials for the above mixed solutions (expressed by the quadratic function) and drink A. As a consequence, we obtained one mixed solution whose output pattern was nearest to that of drink A.

The best combination of concentrations for basic taste substances was 2 mM HCl, 50 mM NaCl, 0.2 mM quinine and 100 mM sucrose. As can be seen from Figure 4.9, the pattern for the mixed solution is centainly closer to that of drink A compared to another drink. A sensory evaluation by humans was also made. The above-mentioned two commercial drinks and the mixed solution were tasted, and their tastes were compared to one another. The sensory tests by humans showed that this mixed solution produced almost the same taste as drink A.

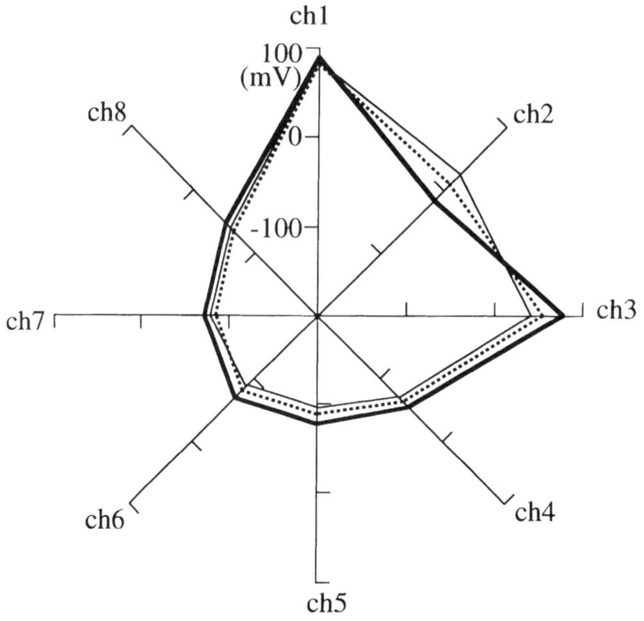

Figure 4.9. Aqueous drinks and mixed solution [18]. - - - - - - -, Drink A; ————, another drink; ———, mixed solution with taste similar to that of drink A. The lipid/polymer membranes used are listed in Table 4.1. (Reprinted with kind permission from Elsevier Sciences S.A.)

As is well known, the τ scale is effective in expressing the taste strength [1]. The concentration of each taste substance can be transformed into the taste strength. The above mixed solution is thus composed of 4.04 sourness, 2.03 saltiness, 5.01 bitterness and 2.24 sweetness in terms of the τ scale. Therefore, drink A has the above taste qualities and intensities.

The taste of every drink can be quantified using the multichannel taste sensor. In other words, the sensor can make quantitative measurements of taste more accurately than the sense of humans can.

4.4 Quantification of Taste of Foods

The present sensor could easily discriminate between some kinds of commercial drinks such as coffee, beer and aqueous ionic drinks, as shown in Figure 4.10 [18]. These three output patterns are definitely different, since the standard deviations were 2 mV at maximum in this experimental condition, where the output for 1 mM KCl was taken as the origin. If the data are accumulated in the computer, any food can be easily discriminated. Furthermore, the taste quality can also be described

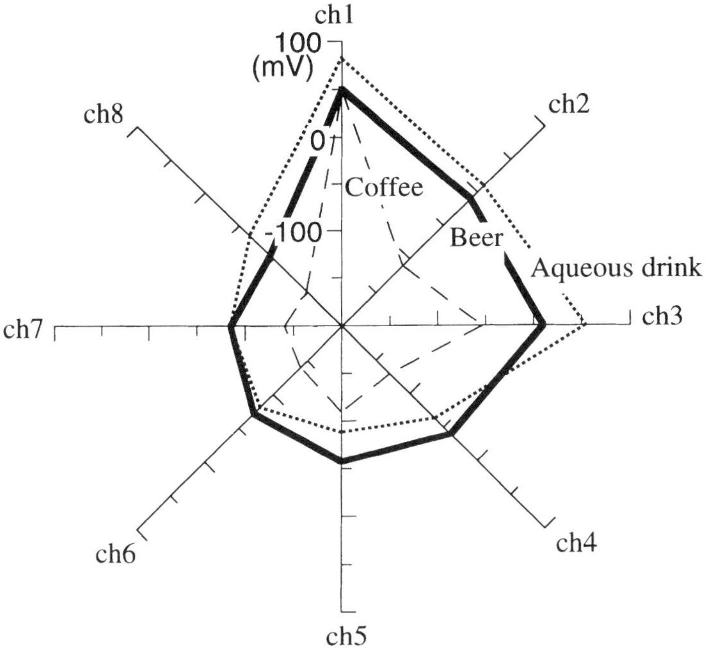

Figure 4.10. Response electric potential patterns for coffee (— — — —), beer (———) and commercial aqueous drink (- - - - - - -) [18]. The lipid/polymer membranes used are listed in Table 4.1.

quantitatively by the method mentioned below. In biological systems, patterns of frequency of nerve excitation may be fed into the brain, and then foods are distinguished and their tastes are recognized [2, 3, 28, 29]. Thus, the quality control of foods becomes possible using the taste sensor, which has a mechanism of information processing similar to biological systems.

4.4.1 Beer

Figure 4.11(a) shows the response patterns to eight different brands of beer [15]. The patterns were measured relative to a certain beer as a standard. Although the difference in electric potential between different brands was a few mV or more, each beer was easily distinguished from the others by these patterns because of the high reproducibility and sensitivity of the sensor represented by 0.2 mV of standard deviations, which are smaller than the above usual conditions by one order. This superior ability is due to the following methods [12]; i.e., the first is the use of a certain beer as an initial reference of the electric potential pattern, although the KCl solution was used as the initial reference in Figures 4.8–4.10. The second is the use of regular repetitive measurements, the period of which was 20 seconds or so.

(a)

(b)

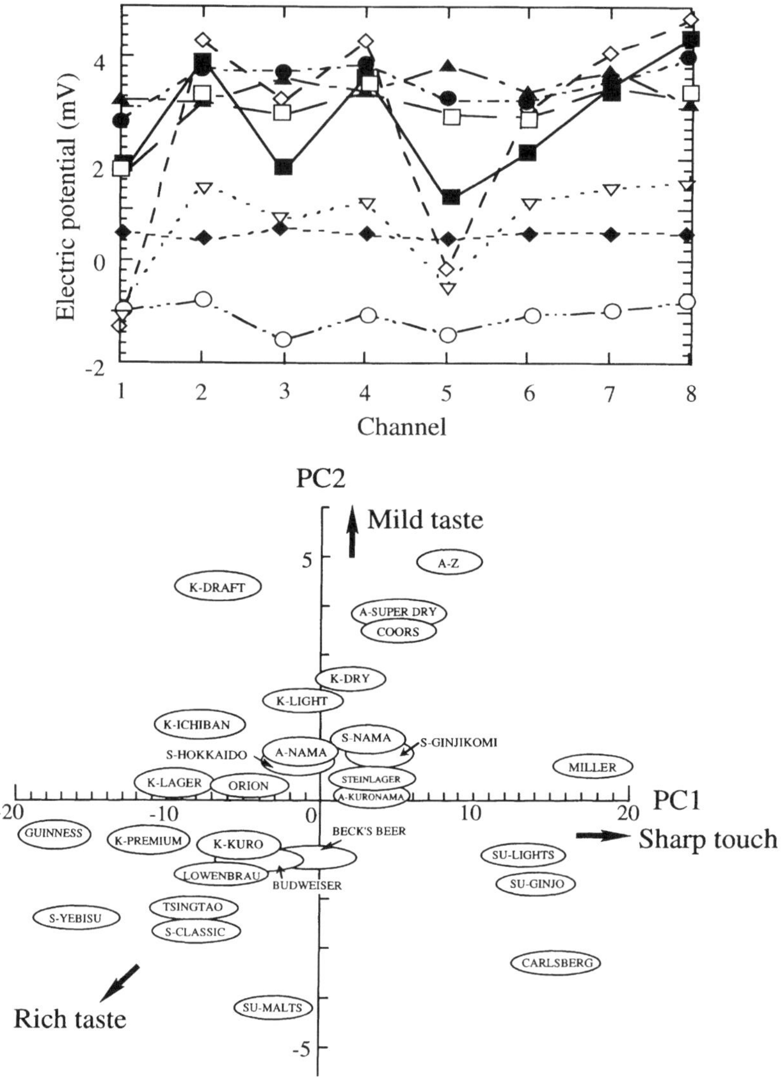

Figure 4.11. Response electric potential patterns for beer (a) [15] and the taste map (b) obtained using principal component analysis. (Reprinted with kind permission from Myu K.K.)

The taste of beer can be expressed by four terms, i.e., 'rich taste', 'soft taste', 'sharp touch' and 'smooth touch' [30]. The rich or soft taste may be mainly related to the concentration of barley, whereas the sharp or smooth touch may arise from the concentrations of alcohol, hops and so on. We tried to express these taste qualities quantitatively by transforming the output pattern of the taste sensor using a principal component analysis. Figure 4.11(b) shows the result. Comparison with the human

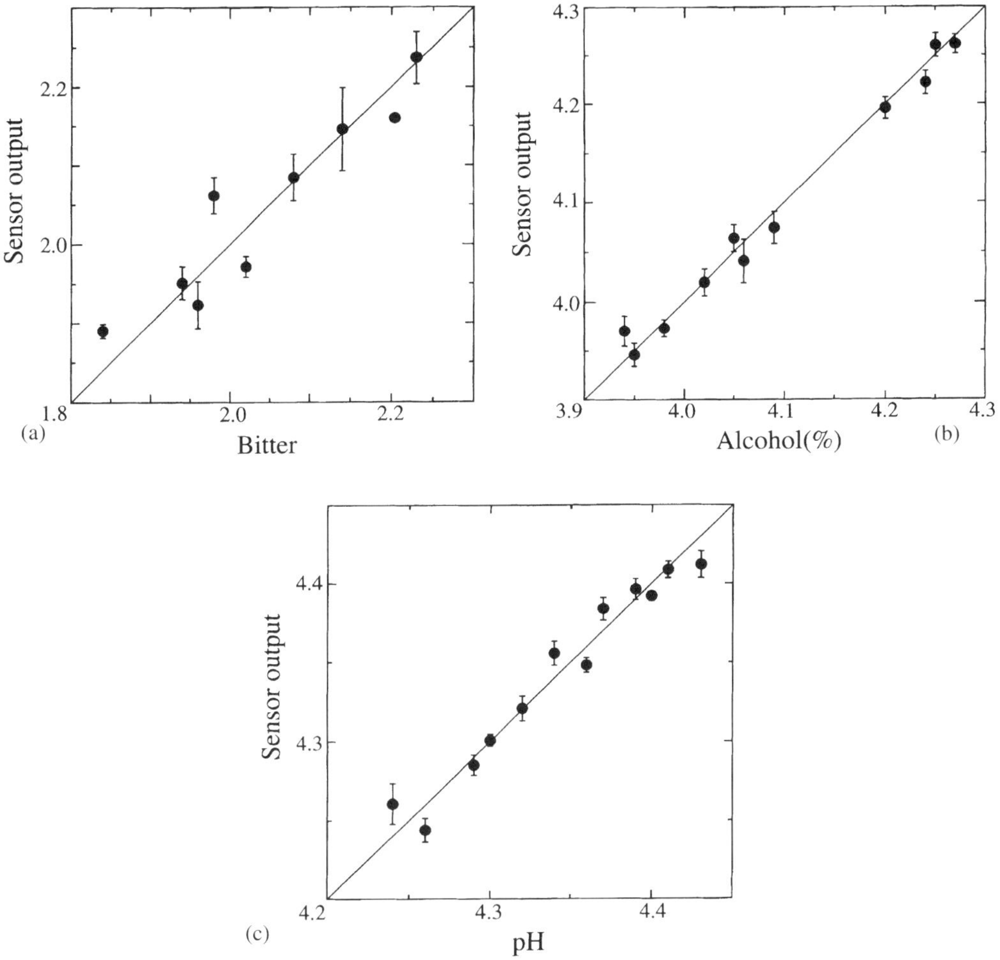

Figure 4.12. Correlations of sensor output with 'bitter' sensation of humans (a), alcohol content (b) and pH of beer (c).

taste sense implied that PC1 corresponds to 'sharp touch', PC2 to 'mild taste' and the lower-left direction to 'rich taste'.

Figure 4.12 shows correlations of sensor output with 'bitter' felt by humans (a), alcohol concentration (b) and pH (c) [31]. A good linear relationship between the resultant sensor output and the human sensory expression (a) and the physico-chemical quantities (b) and (c) of beer can be found in Figure 4.12. High correlations with other sensory expressions such as 'rich taste' and 'sharp touch' were also found, as easily guessed from Figure 4.11(b). These results imply the usefulness of the taste sensor for quantifying human sensory expressions and physico-chemical quantities in beer.

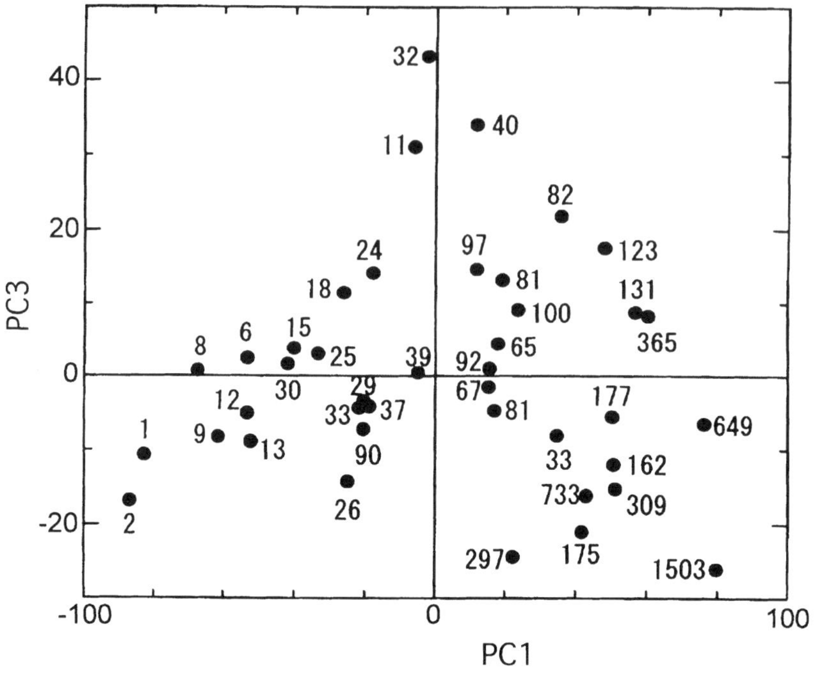

Figure 4.13. Result of principal component analysis applied to response patterns for 41 kinds of commercial mineral water. The figures give the hardness of each sample.

4.4.2 Mineral Water

Recently, interest in drinking water has grown rapidly. Many people are requesting safe and delicious water. The importance of water has also been recognized as a result of interest in environmental pollution issues. However, no convenient measuring system sufficient for quality evaluation of drinking water is available.

Figure 4.13 shows the result of principal component analysis applied to response patterns for 41 kinds of commercial mineral water [32], which were measured in a similar way to that reported before [20]. We can see that the lower-right plane contains mineral water with strong hardness, i.e., hard water, whereas the lower-left plane is occupied by soft water.

The taste of mineral water is quite subtle, and hence it is difficult for humans to discriminate between different brands. However, the taste sensor responded well to mineral water from soft to hard categories. In spite of low concentrations of taste substances in mineral water, the taste sensor can discriminate between different brands because of its high sensitivity to inorganic ions.

Factor loadings of all the channels to PC3 were negative, whereas factor loadings of chs. 1, 2, 6 and 7 to PC1 were positive and those of chs. 4, 5, 8 were negative. Channels 1, 2, 6 and 7 were made of negatively charged membranes and channels 4, 5 and 8 were positively charged; hence PC1 expresses the sum of the effects of cations and anions, while PC3 expresses the simple average of both the effects. The sum and difference of magnitudes of the effects are represented by PC1 and PC3, respectively. If the interaction magnitudes of cations and anions with lipid/polymer membranes are similar to each other, PC3 will become nearly zero. In the samples measured here, PC2 had no physical meaning in the sense of those above.

Sensory evaluations of seven brands of mineral water were also performed. Four terms expressing the taste were selected: 'complicated', 'soft', 'fresh' and 'sharp'. Each term was expressed by five levels, and panelists estimated the level. Principal component analysis was carried out for the result of sensory evaluation. PC1 had a contribution rate higher than 93% in all five trials. The transformation matrix showed that PC1 reflects a 'sharp' taste. This implies that the taste of mineral water evaluated by humans has one-dimensional information. In addition, the order of seven brands of mineral water along the PC1 axis differed from trial to trial. That is, good reproducibility of human sensory evaluation was not obtained. This suggests that the expression of the taste of mineral water by human sensory evaluation is difficult.

4.4.3 Coffee

Figure 4.14 shows the response electric potentials [25] for 10 brands of coffee from different origins, Brazil (Santos No. 2), Guatemala (SHB), Jamaica (Blue Mountain No. 2), Hawaii Kona (Extra Fancy), Kenya (AA), Tanzania (AA), Colombia (Excelso), Indonesia Mandheling (Grade1), Indonesia WIB 1 and Indonesia AP 1, relatively measured from the electric potential for Salvador (CS), which was used

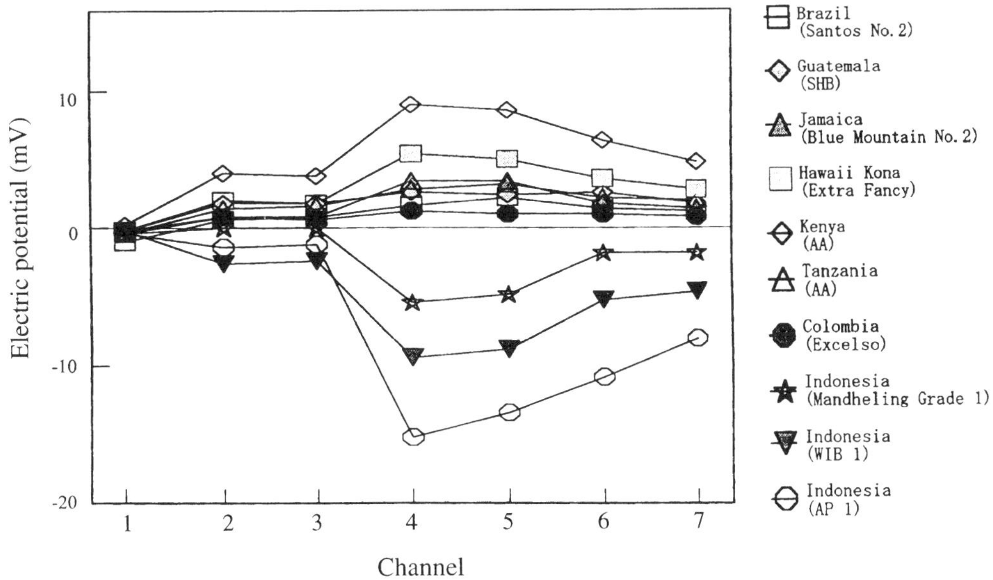

Figure 4.14. Response electric potential patterns for 10 different kinds of coffee [25].

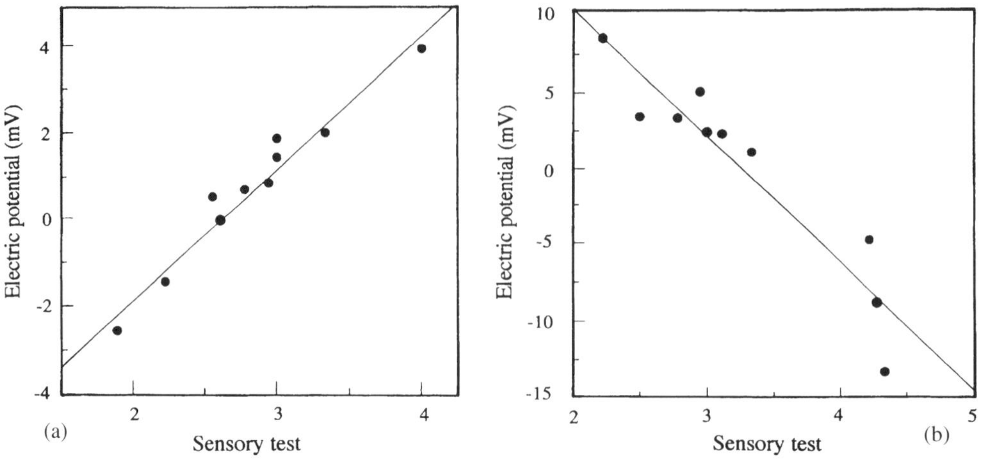

Figure 4.15. Correlations between the response electric potential and the sensory panel test for acidity (a) and bitterness (b).

here as the standard coffee. The measurements were made at 60 °C. We can see that Kenya shows the largest pattern, while AP 1 shows the smallest pattern. Since the standard deviations of the measuring system were 0.1 mV, the origins of the coffees could be easily discriminated.

As is well known, the basic words characteristic of the taste of coffee are 'acidity' and 'bitterness'. We use the term 'acidity' for coffee instead of 'sourness', because 'acidity' is always used to express a basic, desirable sharp and pleasing taste produced by organic acids contained in coffee [33].

Figure 4.15 shows the comparisons between the response electric potential and the sensory panel test for acidity (a) and bitterness (b). The correlation coefficients were 0.98 for acidity using ch. 2 and −0.94 for bitterness using ch. 5. This result implies that chs. 2 and 5 can be utilized to quantify acidity and bitterness, respectively. For example, Kenya has strong acidity while Mandheling shows strong bitterness; this fact can be expressed by using the quantified scale obtained from the sensor output.

The sensory expressions for 'acidity' and 'bitterness' cannot be deduced from chemical components contained in coffee. In fact, caffeine-free coffee tastes bitter, although caffeine can be considered as a main component to produce bitterness in coffee. It implies that the sensory expressions for taste qualities originate in a complex combination of many chemical components. In this situation, a sensor with high selectivity to a chemical substance is not useful. The taste sensor with lipid/polymer membranes can respond to many kinds of chemical substances at the same time; i.e., the sensor has a global selectivity to chemical substances. For this reason, the acidity and bitterness of coffee can be quantified.

The present result is a first step in objective measurement for conventional sensory expressions for coffee production. A desirable blend of several origins of coffee may become possible by using, for example, a method of pattern matching with a database constructed from measurements using the taste sensor.

4.4.4 Milk

The sensor was applied to milk by focusing on the taste variance of milk induced by heat treatment [21]. Some kinds of treatments, such as low-temperature – long-time or high-temperature – short-time, are usually performed on commercial milk for pasteurization or sterillization. It is pointed out that heat treatment considerably affects the taste of milk. Therefore, it is necessary to clarify its effect quantitatively.

The heat-treated samples are listed in Table 4.2. All the samples were prepared from the same original milk. The temperature was gradually increased to the processing temperature, kept at the same temperature during the processing time, and then grudually decreased. Hence, milk with 65 °C, 0 min treatment differs from milk with 80 °C, 0 min treatment. The precondition for stabilizing the electrochemical property of the membranes was achieved by immersing the electrodes in a standard milk sample (140°C 2 sec heat treatment) for a long period of over a month. The output pattern for the standard milk was used as a reference pattern for all measurements over half a year.

Since all the samples listed in Table 4.2 were of the same origin, the difference in taste among the samples was considered to be slight. It is necessary to detect the electric potentials with high sensitivity and reproducibility. Hence, newly separated standard milk was used as a reference sample and also as the rinsing fluid of electrodes at every measurement. Each measuring time was set to 60 sec, and rinsing of electrodes after each measurement was repeated 10 times for about 20 sec.

Three expressions for the taste characteristics of milk, i.e., 'richness', 'cooked flavor' and 'deliciousness', were evaluated by 15 well-trained panelists for the seven samples listed in Table 4.2.

Whey protein denaturation was also measured by an optical method to obtain the whey protein nitrogen index (WPNI), which was estimated from the absorbance at 420 nm of milk treated with saturated NaCl. A high WPNI value means that a small amount of whey proteins was denatured [34].

Figures 4.16(a) and (b) show the relationship between the output from channel 7 (oleyl amine) and richness and that between channel 1 (decyl alcohol) and WPNI, respectivity, where the correlations are −0.885 and 0.953.

Channels 1 and 3 (DOP) showed a correlation as high as 0.95 with the degree of protein denaturation caused by the heat treatment of the original milk. The sensory

Table 4.2. Heat treatment of milk.

Sample no.	Temperature – time
1	65°C – 0 min
2	65°C – 30 min
3	80°C – 0 min
4	80°C – 15 min
5	100°C – 0 min
6	100°C – 15 min
7	100°C – 3 sec

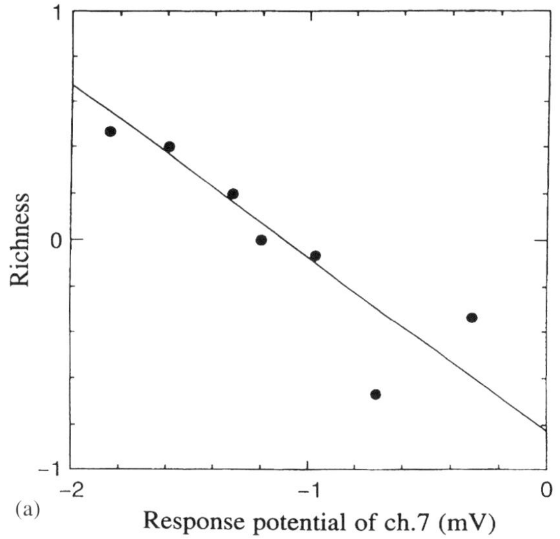

(a)

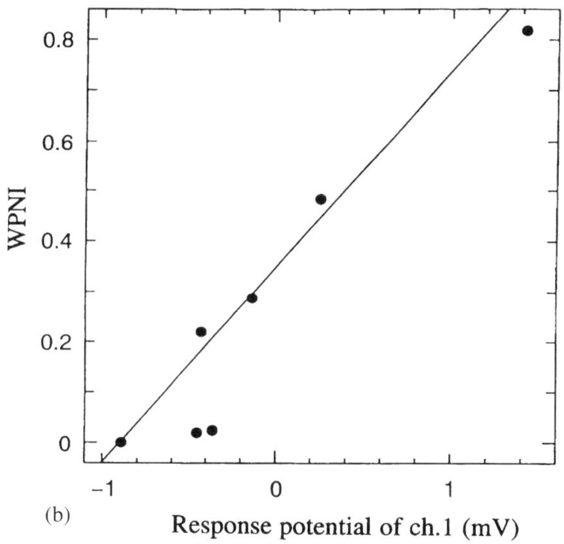

(b)

Figure 4.16. Relationship between the output from ch. 7 and richness (a) and that between ch. 1 and WPNI (b) [21]. Channels 1 and 7 are the membranes made of decyl alcohol and oleyl amine, respectively.

test was also carried out; the sensory evaluation did not show a high correlation for the taste sensor with deliciousness, but that with richness was 0.885.

It is shown that activated sulfhydryl (SH) groups become bare due to denaturation above 76–78 °C [35]. As shown in Figure 4.16(b), the response potential of channel 1 (i.e., weakly negatively charged membrane) decreased with protein denaturation by heat treatment; the potential of channel 7 (i.e., positively charged membrane) decreased at the same time. This result implies that negatively charged groups with the ability to bind to lipid membranes appeared at the surface of denatured proteins. However, the exact biochemical steps are not clear. Utilization of the taste sensor might contribute to the elucidation of the denaturation process.

Discrimination of taste of the samples listed in Table 4.2 by humans is not easy in general, and hence the usefulness of the taste sensor may further increase. In fact, five kinds of milk treated at 100°C for different processing times (0, 1, 5, 15, 30 min) were discriminated using the taste sensor. The sensory evaluations by humans showed that significant discrimination was possible between only two, 0 min and 30 min, treatments.

The detection of the degree of protein denaturation can also be applied to other fields, because the measurement using the taste sensor is very convenient compared with other methods using optical equipment.

4.4.5 Tomatoes

While the sensor was applied to beverages as shown in Figures 4.10 and 4.11, it can also be used for analysis of the taste of gelatiniform or solid foods [16].

For quantification of the taste of tomatoes, the taste sensor was first applied to commercial canned tomato juice, to which four basic taste substances had been added. Data were analyzed by means of principal component analysis. The taste of several brands of tomatoes was expressed in terms of four basic taste qualities by projecting the data obtained from these tomatoes on to the principal axes. As for basic taste substances, we used NaCl for saltiness, citric acid for sourness, MSG for umami and glucose for sweetness. No taste substance for bitterness was added because tomatoes taste only slightly bitter. Changes in response electric potentials by addition of the basic taste substances were measured as a preliminary experiment.

The test foods studied here were five kinds of tomatoes. When eating food, humans first masticate the food with their teeth and then taste it. Therefore, we used a mixer in place of teeth and crushed tomatoes before measuring them. The preconditions were established by keeping the electrode immersed in standard juice, i.e., commercial canned tomato juice without added NaCl, for a long period of time. Standard juice was also used for the reference electric potential pattern, as in the case of beer.

The five kinds of tomatoes were fresh-market tomatoes: *Ryokken*, *Kiss*, *Fukken*, *TVR-2* and a kind of tomato for processing (hereafter named *PT*). *Ryokken* and *Kiss* were cultivated to be sweeter.

Different brands of tomatoes were distinguished by the shapes of the output electric potential patterns, as shown in Figure 4.17 [16]. A typical example of standard deviations for one brand, *TVR-2*, was 0.17, 0.49, 0.37, 0.34, 0.40, 0.27, 0.68 and 0.14 mV for chs. 1–8, respectively.

We can predict the taste of each tomato by projection of the above electric potential patterns in Figure 4.17 using the transformation matrix obtained in the preliminary experiment. High and low scores of the first principal component (PC1) corresponded

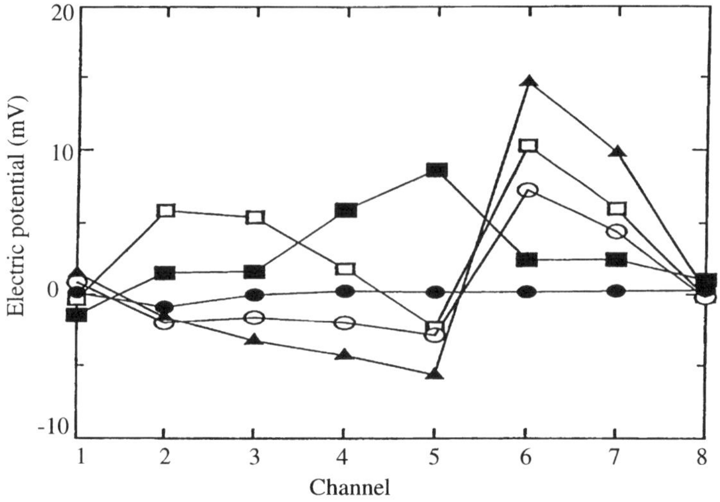

Figure 4.17. Response electric potential patterns for five species of tomatoes [16]. Symbols ○, ■, □, ▲ and ● denote *TVR-2, Fukken, Ryokken, Kiss* and *PT*, respectively. Channels 1–8 are the membranes made of dioctyl hydrogen phosphate (DOP), 9:1, 5:5, 3:7, trioctyl methyl ammonium chloride (TOMA), decyl alcohol, oleic acid and oleyl amine, respectively. The membranes of chs. 2, 3 and 4 are composed of two kinds of lipids, DOP and TOMA, with the above molar ratio.

to 'sourness' and 'umami', respectively. Increasing 'sweetness' decreased the scores of the third principal component (PC3). The result showed that PC1 for *TVR-2* tended to be highly positive and hence *TVR-2* can be considered to taste sour. This result agrees well with the human taste sensation, because *TVR-2* is a relatively old brand of tomato and is generally acknowledged as a sour tomato. For sweet tomatoes such as *Ryokken* or *Kiss*, PC3 tended to be negative; hence these tomatoes are presumed to be sweet, as expected. Since PC1 for *PT* was highly negative compared to other brands of tomatoes, *PT* is presumed to have a relatively strong umami taste. Tomatoes with a umami taste are generally used for processing; hence, this prediction also agrees with human sensation.

In this way, the taste of solid foods was studied by crushing them before the measurements. The preliminary experiment with basic taste qualities enabled us to quantify the taste of foods by projecting the response electric potential patterns on to the plane whose axes are assigned to have the meaning of each taste.

4.5 Quality Control of Foods

4.5.1 Sake

The quality of sake is evaluated by factors such as taste, smell and color. The evaluation of taste is the most difficult among these factors. To date, the evaluation of the taste of sake has been carried out through various kinds of chemical analyses; the relationship between chemical substances and taste has been investigated. However, sensory tests by humans are considered to be more important, although they provide no quantitative evaluation.

Samples for measurements using the taste sensor are as follows [22].

Junmaishu: This is made from a combination of highly refined polished rice (polished up to 70% of its original size), *koji* (fermented rice) and fresh stream water. Sugar and alcohol are not added.

Daiginjoshu: This is made from highly refined polished rice (polished up to 50%) which is slowly fermented at low temperatures. As a result this sake has a rich and fruity flavor.

Honjozoshu: Polished rice (polished up to 70%), *koji*, water and alcohol are blended together using traditional techniques to give this sake a unique color and a rich, smooth aroma. It most resembles ordinary sake among the four kinds of sake.

Alps ginjoshu: This is an original sake made in Nagano Prefecture. It resembles *Daiginjoshu* except for the degree of rice polishing (polished up to 60%) and the alcohol density (low-alcohol sake: diluted with water).

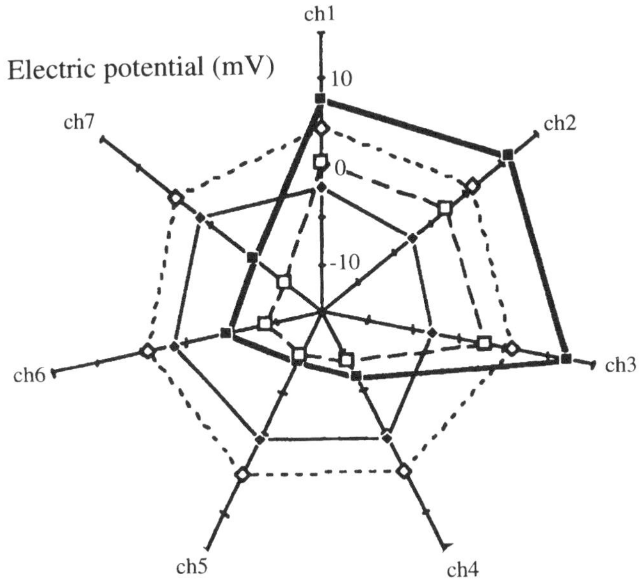

Figure 4.18. Response electric potential patterns for various kinds of sake from one brewery [22]. *Daiginjoshu* (■), *Junmaishu* (□), *Honjozoshu* (◆), and *Alps-ginjoshu* (◇).

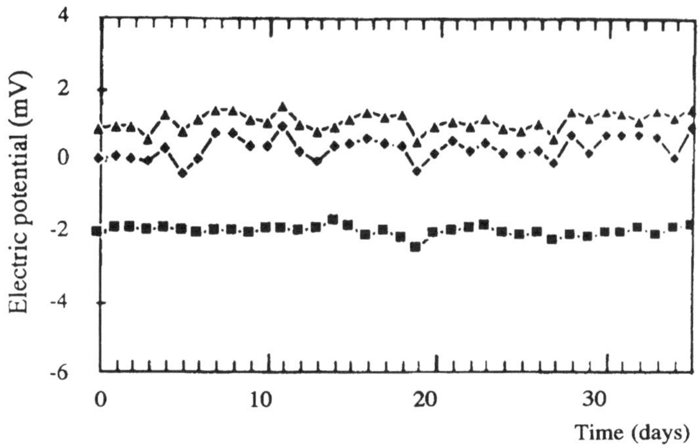

Figure 4.19. Change in response of chs. 1, 4 and 7 to sake with days [22].

Figure 4.18 shows sensor responses for four kinds of sake from one brewery, measured by adopting an ordinary sake from the same brewery as the standard [22]. The same lipid/polymer membranes were used in chs. 1, 2 and 3 (charged negatively), chs. 4 and 5, the same charged positively, and chs. 6 and 7, the membrane with an intermediate positive electric charge.

As found in Figure 4.18, the responses of these three kinds of membranes to the four kinds of sake were different. Distinction of the kind of sake was very easy because of the low standard deviations, e.g., ch. 1 (0.16 mV), ch. 4 (0.31 mV) and ch. 6 (0.25 mV) in the analysis of *Honjozoshu*, repeated 40 times.

When highly alcoholic drinks such as sake are measured, the durability and stability of the lipid/polymer membrane are important. Figure 4.19 shows the changes in response to *Honjozoshu* with repeated use of the same electrode [22]. Every membrane was stable for more than 1 month. The changes in response of three membranes to *Honjozoshu*, whose components are similar to the standard sake, were within 1 mV. Actually, the same electrode has been safely used for more than two years; a similar situation also holds in multichannel electrodes for other foodstuffs.

Figure 4.20 shows multiple regression analysis of the titratable acidity using the two sensor outputs in the fermentation process of sake. The correlation was very high, as represented by 0.99. It is known that the factor that influences the degree of titratable acidity is not only the pH of samples but also the concentrations of organic acids and amino acids. Increasing organic and amino acid concentrations induce greater buffer action resulting in the increase of titratable acidity. On the other hand, the correlation between the response of ch. 1 and a function of pH, organic acid concentration and amino acid concentration was as high as $r = 0.95$. The high correlation between the titratable acidity and the sensor outputs is due to the fact that both the values reflect the pH and the organic acid and amino acid concentrations. It is also related to Figure 4.8, where the taste sensor can discriminate between the taste of amino acids according to their taste qualities.

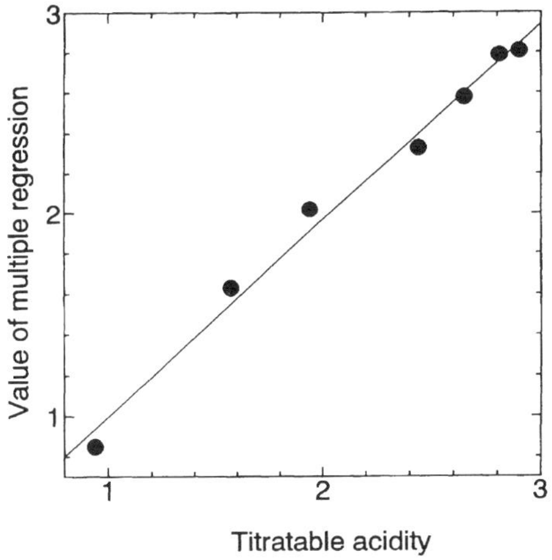

Figure 4.20. Correlation between the response electric potential (chs. 1 and 4) and the titratable acidity of sake mash using multiple regression analysis. During fermentation, the titratable acidity increased almost linearly. The data from the lower to higher values are concerned with 3, 5, 7, 11, 13, 19 and 17 days of the fermentation period in order.

This finding is very important in sake making, because the titratable acidity is one of the most important factors in sake making; in fact, it increases with the fermentation, as shown in Figure 4.20. It is used for the adjustment of taste 'sweetness' and 'thickness', blending of sake and control of fermentation of sake mash 'moromi'. However, automatic analysis has been difficult so far because the analytical method of determining the titratable acidity is actually a titration. The present result shows that ch. 1 of the multichannel taste sensor is promising for use as an automatic acidity meter. The taste of sake will be expressed more quantitatively when measured with the addition of a glucose sensor [23]. It is also shown that the fermentation process of sake can be monitored using the taste sensor [36].

4.5.2 Soybean paste

Miso (soybean paste) is a traditional, fermentative food in Japan. One of the original materials, so-called *koji* (cultivated *kome* or *mugi*), is made by adding seeds of *koji* fungus to steamed rice or barley. As the second step, *koji*, boiled soybean and salt are mixed. The immature mixture is put into a big tank for fermentation. When the fermentation finishes, it turns into a delicious, preservative foodstuff, i.e., *miso*.

The degree of fermentation of *miso*, i.e., to what extent the mixture of *koji*, soybean and salt is ripe, has been mainly judged by the experience of humans. This way of judgement is influenced by the daily physical conditions of humans and the environment. Therefore an objective measure which is not influenced by any environment has been desired for a long time.

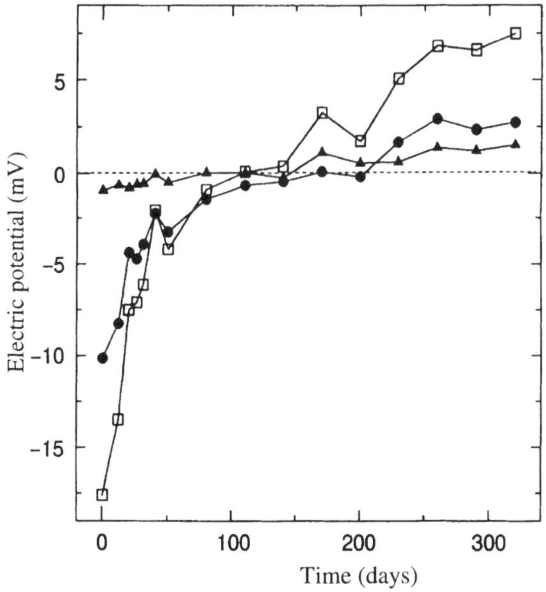

Figure 4.21. Changes in response of chs. 1, 5 and 6 for *miso* with fermentation and storage days [37]. Symbols ●, □ and ▲ denote changes of chs. 1, 5 and 6, respectively.

On the other hand, monitoring of the fermentation of *miso* is made partly using various kinds of chemical analyses, such as titratable acidities. The relationship between the titratable acidities and the fermentation process of *miso* showed a fairly good correlation. Moreover, the content of amino acids in *miso* is known to increase with fermentation. However, it takes time and effort to monitor the fermentation using such chemical analyses. Therefore, these methods are unsuitable for real-time or on-line measurement in *miso* fermentation. A simpler and more useful way to judge the *miso* fermentation is necessary for the assurance of reliable quality and automated production of *miso*.

Figure 4.21 shows the time courses of the responses of chs. 1, 5 and 6, which are made of DOP, TOMA and oleyl amine, respectively, to *miso* with fermentation and storage days [37]. This result indicates that chs. 1 and 5 responded linearly to samples in the fermentation process (0–40 days), while the changes of responses to ripe samples which were shipped (50–320 days) were smaller than those in samples in the fermentation process. On the other hand, ch. 6 showed a little, gradual increase with days. This result suggests that chs. 1 and 5 can be used to monitor the fermentation of *miso*.

As can be found in Figure 4.21, the response potentials of chs. 1, 5 and 8 become positive values beyond about 150 days (about 5 months). This fact is very suggestive because it is natural that *miso* products on sale beyond 5 months should be returned to the factory. Thus, the taste sensor can provide a criterion of reliable quality of *miso* products.

It was also found that the response potentials of chs. 1 and 8 of the taste sensor reflect the amount of amino acids such as L-glutamic acid, L-leucine (including L-isoleucine), L-aspartic acid and L-arginine. Furthermore, the correlation coefficients between the sensor outputs of chs. 1 and 8 and titratable acidity were 0.87 and 0.88, respectively. The titratable acidities are affected not only by the pH of samples but also the concentrations of organic acids and amino acids. Thus, the increasing titratable acidities in the fermentation process are compatible with the increasing amino acid concentrations. These physico-chemical quantities have the possibility of being utilized to monitor the fermentation process. However, control of fermentation of *miso* has not been made by measurements of these quantities so far because of its intricacy. Therefore, analysis using the taste sensor is very feasible because of its convenience as an on-line measurement of *miso* fermentation.

4.6 Summary

The taste sensor showed a similar pattern for chemical substances producing the same taste qualities, while quite different patterns were obtained for substances with different taste qualities. In fact, similar outputs were obtained for, e.g., NaCl, KCl and KBr eliciting saltiness, whereas their patterns differed clearly from those for other basic taste qualities. The taste of amino acids was also classified into several groups in accordance with the human taste sense.

The taste sensor has a new concept of global selectivity. What is important in recognition of taste is not discrimination of minute difference in molecules but to transform molecular information contained in interactions with biological membranes into several kinds of groups, i.e., taste intensities and qualities. This function is of a high level, where intelligent sensing is required. In this sense, the taste sensor is essentially an intelligent sensor to reproduce the taste sense, which is a complex, comprehensive sense of humans.

Several kinds of commercial drinks such as coffee, beer and mineral water were discriminated easily using the taste sensor. The taste sensor has a sensitivity, reproducibility and durability higher than those of humans. Human taste expressions such as ‘bitterness’ and ‘sourness’ were quantified, and also the physico-chemical indices for food production such as alcohol concentration and titratable acidity were shown to be measured by the taste sensor.

Improvements in the membranes and measurement methods have also been made for the purpose of measurements, e.g., by adopting the Langmuir–Blodgett method [38], a monolayer of thiol-containing lipids [39] or impedance measurements [40].

The taste sensor will be applicable for quality control in the food industry and will help the automation of production. The sense of taste is vague and largely depends on subjective factors of human feelings. If we compare the standard index measured by means of the taste sensor with sensory evaluation, we will be able to assess taste objectively. Moreover, the mechanism of information processing of taste in the brain as well as the reception at taste cells will also be clarified by developing a taste sensor which has output similar to that of the biological gustatory system. In fact, a recent study [19] has shown that astringency is a taste expressed by bitterness and sourness, whereas astringency has so far been considered not as a taste sense but a tactile sense. The result using the taste sensor indicates that the reception mechanism of taste substances in biological systems should be reconsidered.

Objective scales of physical quantities such as length and time were proposed over two thousand years ago. We are now familiar with quantitative, objective scales, although there exists subjective feeling as well. As for taste, however, we have had no scale so far. The taste sensor provides a quantitative, objective measure of taste. 'Deliciousness' will be measured simultaneously using other sensors. Excellent odor or gas sensing systems based on various principles have also been developed [41–47]. We are now standing on the threshold of a new age of food culture.

4.7 References

[1] Bartoshuk, L. M., *Physiol. Behav.*, **14** (1975), pp. 643–649.

[2] Pfaffmann, C., in: *Handbook of Physiology*, Field, J., ed., Sec. 1, *Neurophysiology*, vol. 1; Washington DC: American Physiological Society (1959), pp. 507–533.

[3] Kawamura, Y., Kare, M. R. (eds.), *Umami, A Basic Taste*, New York: Marcel Dekker, (1987).

[4] Hayashi, K., Yamanaka, M., Toko, K., Yamafuji, K., *Sens. Actuators*, **B2** (1990), pp. 205–213.

[5] Toko, K., Hayashi, K., Yamanaka, M., Yamafuji, K., *Tech. Digest 9th Sens. Symp.* (1990), pp. 193–196.

[6] Iiyama, S., Toko, K., Yamafuji, K., *Agric. Biol. Chem.*, **50** (1986), pp. 2709–2714.

[7] Toko, K., in: *Characterization of Food-Emerging Methods*, Gaonkar, A. G., ed.; Amsterdam: Elsevier, (1995), pp. 377–401.

[8] Hayashi, K., Yamafuji, K., Toko, K., Ozaki, N., Yoshida, T., Iiyama, S., Nakashima, N., *Sens. Actuators*, **16** (1989), pp. 25–42.

[9] Iiyama, S., Toko, K., Hayashi, K., Yamafuji, K., *Agric. Biol. Chem.*, **53** (1989), pp. 675–681.

[10] Toko, K., Hayashi, K., Iiyama, S., Yamafuji, K., *4th Int. Conf. on Solid-State Sensors and Actuators (Transducers '87), Tokyo, Digest of Technical Papers* (1987), pp. 793–796.

[11] Hayashi, K., Matsuki, Y., Toko, K., Murata, T., Yamafuji, Ke., Yamafuji, Ka., *Sens. Materials*, **1** (1989), pp. 321–334.

[12] Ikezaki, H., Hayashi, K., Yamanaka, M., Tatsukawa, R., Toko, M., Yamafuji, K., *Trans. IEICE Japan*, **J74-C-II** (1991), pp. 434–442 (in Japanese).

[13] Ikezaki, H., Toko, K., Hayashi, K., Toukubo, R., Yamanaka, M., Sato, K., Yamafuji, K., *Tech. Digest 10th Sens. Symp.* (1991), pp. 173–176.

[14] Murata, T., Hayashi, K., Toko, K., Ikezaki, H., Sato, K., Toukubo, R., Yamafuji, K., *Sens. Materials*, **4** (1992), pp. 81–88.

[15] Toko, K., Murata, T., Matsuno, T., Kikkawa, Y., Yamafuji, K., *Sens. Materials*, **4** (1992), pp. 145–151.

[16] Kikkawa, Y., Toko, K., Yamafuji, K., *Sens. Materials*, **5** (1993), pp. 83–90.

[17] Kikkawa, Y., Toko, K., Matsuno, T., Yamafuji, K., *J. J. Appl. Phys.* **32** (1993), pp. 5731–5736.

[18] Toko, K., Matsuno, T., Yamafuji, K., Hayashi, K., Ikezaki, H., Sato, K., Toukubo, R., Kawarai, S., *Biosens. Bioele.*, **9** (1994), pp. 359–364.
[19] Iiyama, S., Toko, K., Matsuno, T., Yamafuji, K., *Chem. Senses*, **19** (1994), pp. 87–96.
[20] Iiyama, S., Yahiro, M., Toko, K., *Sens. Materials*, **7** (1995), pp. 191–201.
[21] Toko, K., Iyota, T., Mizota, Y., Matsuno, T., Yoshioka, T., Doi, T., Iiyama, S., Kato, T., Yamafuji, K., Watanabe, R., *J. J. Appl. Phys.*, **34** (1995), pp. 6287–6291.
[22] Arikawa, Y., Toko, K., Ikezaki, H., Shinha, Y., Ito, T., Oguri, I., Baba, S., *Sens. Materials*, **7** (1995), pp. 261–270.
[23] Iiyama, S., Suzuki, Y., Ezaki, S., Arikawa, Y., Toko, K., *Materials Sci. Eng.*, **C4** (1996), pp. 45–49.
[24] Oohira, K., Toko, K., Akiyama, H., Yoshihara, H., Yamafuji, K., *J. Phys. Soc. Jpn.*, **64** (1995), pp. 3554–3561.
[25] Fukunaga, T., Toko, K., Mori, S., Nakabayashi, Y., Kanda, M., *Sens. Materials*, **8** (1996), pp. 47–56.
[26] Oohira, K., Toko, K., *Biophys. Chem.*, **61** (1996), pp. 29–35.
[27] Ninomiya, T., Ikeda, S., Yamaguchi S., and Yoshikawa, T., *Proc. 7th Congress Sensory Tests* (1966), pp. 109–123 (in Japanese).
[28] Kurihara, K., Yoshii, K., Kashiwayanagi, M., *Comp. Biochem. Physiol.*, **85A** (1986), pp. 1–22.
[29] Roper, S. D., *Ann. Rev. Neurosci.*, **12** (1989), pp. 329–353.
[30] Kirin Brewery Co. Ltd. (ed.), *Biru no Umasa wo Saguru [Study of Deliciousness of Beer]*; Tokyo: Shokabo (1992), p. 81 (in Japanese).
[31] Ezaki, S., Yuki, T., Toko, K., Tsuda, Y., Isoe, A., Nakatani, K., Yamafuji, K., *Extended Abstracts (43rd Spring Meeting) Jpn. Soc. Appl. Phys.* (1996), p. 1114 (in Japanese).
[32] Wada, M., *Quark*, **13**(8) (1994), pp. 40–45 (in Japanese).
[33] International Coffee Organization (ed.), *Consumer-Oriented Vocabulary for Brewed Coffee*, (1991), *Report No. 3*, p. 4.
[34] American Dry Milk Institute (ed.), *Standards for Grades of Dry Milk Including Methods of Analysis*; Illinois: Am. Dry Milk Inst. (1971), p. 4.
[35] Tsugo, T., Yamauchi, K., *Gyunyu no Kagaku (Chemistry of Milk)*; Tokyo: Chikyu-sha, (1975), chap. 5 (in Japanese).
[36] Arikawa, Y., Toko, K., Ikezaki, H., Shinha, Y., Ito, T., Oguri, I., Baba, S., *J. Fermentation Bioeng.*, **82** (1996), pp. 371–376.
[37] Imamura, T., Toko, K., Yanagisawa, S., Kume, T., *Sens. Actuators*, **B37** (1996), pp. 179–185.
[38] Akiyama, H., Toko, K., Yamafuji, K., *J. J. Appl. Phys.* **35** (1996), pp. 5516–5521.
[39] Iiyama, S., Miyazaki, Y., Hayashi, K., Toko, K., Yamafuji, K., Ikezaki, H., Sato, K., *Sens. Materials* **4** (1992), pp. 21–27.
[40] Akiyama, H., Tsuzaki, T., Toko, K., Yamafuji, K., *T. IEE Japan*, **117-E** (1996), pp. 89–94.
[41] Ema, K., Yokoyama, M., Nakamoto, T., Moriizumi, T., *Sens. Actuators* **18** (1989), pp. 291–296.
[42] Lundström, I., Erlandsson, R., Frykman, U., Hedborg, E., Spetz, A., Sundgren, H., Welin, S., Winquist, F., *Nature*, **352** (1991), pp. 47–50.
[43] Hierlemann, A., Weimar, U., Kraus, G., Gauglitz, G., Göpel, W., *Sens. Materials*, **7** (1995), pp. 179–189.
[44] Weimar, U., Vaihinger, S., Schierbaum, K. D., Göpel, W., in: *Chemical Sensor Technology*, (Seiyama, T., ed.), vol. 3; Tokyo: Kodansha (1991), pp. 51–88.
[45] Cornila, C., Hierlemann, A., Lenggenhager, R., Malcovati, P., Baltes, H., Noetzel, G., Weimar, U., Gopel, W., *Sens. Actuators*, **B24–25** (1995), pp. 357–361.

[46] Gardner, J. W., Bartlett, P. N., in: *Olfaction and Taste XI* (Kurihara, K., Suzuki, N., Ogawa, H., eds.), Tokyo: Springer-Verlag (1994), pp. 690–693.
[47] Persaud, K. C., Qutob, A. A., Travers, P., Pisanelli, A. M., Szyszko, S., in: *Olfaction and Taste XI* (Kurihara, K., Suzuki, N., Ogawa, H., eds.), Tokyo: Springer-Verlag (1994), pp. 708–710.

PART 2

Sensor Applications

5 Ultrasonic Sensors for Process Control

PETER HAUPTMANN, Otto von Guericke Universität Magdeburg, and Institut für Automation und Kommunikation Magdeburg-Barleben, Germany, RALF LUCKLUM, Otto von Guericke Universitat Magdeburg, Germany, BERND HENNING, Institut für Automation und Kommunikation Magdeburg-Barleben, Germany

Contents

5.1 Introduction

5.1.1 Sensor Definition

Ultrasonic sensors actively transmit acoustic waves and receive them. They utilize ultrasonic transducers, which transform an electrical signal into an ultrasonic wave and vice versa. Piezoelectricity is the most commonly used physical effect. In practice it is often possible to use the same transducer for both transmitting and receiving. On its path from the transmitter to the receiver the acoustic wave is modified by the properties which are under investigation. The ultrasound signal carries the information about the parameters to be measured.

The task for ultrasonic sensors is not merely to detect the ultrasound. As intelligent sensors they have to extract and evaluate the information carried by the ultrasonic signal efficiently and with high accuracy and resolution by dedicated hardware. Software algorithms based on models for ultrasonic propagation and interaction between the ultrasonic wave and the physical or chemical variables of interest are employed for that aim. Techniques like fuzzy logic or neural networks are examples of highly sophisticated methods and are used more and more in ultrasonic sensor systems. Very often ultrasonic sensors are embedded into control systems that have access to additional sensors. Based on a microprocessor they combine information of the different sensors, handle the bus protocol and initiate actions.

Ultrasound covers a frequency range from 20 kHz to about 1 GHz. For technical applications the range 20 kHz to 10 MHz is the most important one. Dependent on application conditions and accuracy requirements different piezoelectric materials are used. Ultrasound enables instruments to be non-invasive because the acoustic wave can often penetrate the walls of vessels or pipes. The propagation of acoustic waves through multilayered structures depends on the acoustic impedance mismatch at each of the boundaries the ultrasound has to pass through, the attenuation of the ultrasound in the different materials and in some cases on the relationship between the wall thickness and the wavelength of the ultrasound waves. Longitudinal (compressional) ultrasonic waves have the highest relevance since shear waves are highly damped in liquids. Other kinds of waves are used e.g. for ultrasonic microsensors or for nondestructive testing (NDT) of solids. Their properties have to be known for the transmission of ultrasound at interfaces.

Ultrasonic sensors are applied in many fields. Their use has a long tradition in ultrasonic devices for NDT in material science, in medical applications like ultrasonic mammography or tomography or in weapons and military equipment. These sensors or ultrasonic methods are not the matter of this book. They are described elsewhere [1, 2]. A new field of ultrasonic sensors and ultrasonic systems is process monitoring and control, automotive techniques, chemical analysis, etc. These applications enjoyed a rapid increase of interest in recent years. The state of the art of this development will be discussed in the following.

5.1.2 Materials

Ultrasound can be generated in different ways [3]. Magnetostrictive, laser or capacitive techniques are of little importance at present. Most frequently the ultrasound is generated with piezoelectric materials. The conversion of electrical energy in acoustic energy and vice versa can be explained on the basis of the piezoelectric effect. Due to the reversibility of this effect the same material is applicable for the transmitter and receiver transducer. One transducer can fulfil both tasks. Most ultrasonic sensors act in this mode. For ultrasonic sensors the following piezoelectric materials are interesting:

- monocrystaline bulk material like quartz, lithium niobate ($LiNbO_3$) or lithium sulfate ($LiSO_4$);
- polycrystalline bulk materials or piezoelectric ceramics, e.g. lead zirconate titanate ($Pb(Zr,Ti)O_3$; trade name PZT), lead metaniobate ($PbNb_2O_6$) or barium titanate ($BaTiO_2$);
- polymer films like poly vinylidenefluoride (PVDF) or co- and terpolymers of difluoroethene (VF_2), vinylfluoride (VF), trifluoroethylene (VF_3) and tetrafluoroethylene (VF_4); as a new material poly hydroxybutyrate (PHB) may be interesting in the future;
- thin piezoelectric films, e.g. zincoxide (ZnO).

Criteria for using these materials as sensor materials are their electrical, mechanical and acoustic properties. ZnO plays a dominant role for microsensors. In Table 5.1 a few of these materials are compared.

Knowledge of these data is also necessary to use CAD systems for the design and optimization of ultrasonic sensors. Finite element (FEM) and boundary element (BEM) methods are often used for the optimization of mechanical properties. FEM software which provides piezoelectric finite elements also exists. CAD systems which combine successfully piezoelectric components with sonic field properties in gaseous, liquid media or in solids are rare. One exception is the FEM system CAPA (coupled acoustic and piezoelectric analysis) [5]. It is used for the design of piezoelectric sensors and actuators. For modeling of the dynamic behavior of ultrasonic transducers PSPICE, the well known software for circuit design, is applied [6].

5.1.3 Basic Physical Equations

The propagation of a plane elastic wave can be described by the sonic alternating pressure $p(x,t)$

$$p = p_o e^{j\omega(t-x/c)} e^{-\alpha x} \tag{5.1}$$

Table 5.1 Data of selected piezoelectric materials [4].

Material	Sound velocity c_L m/s	Relative permittivity ε_{rel}	Critical temperature T_k in °C	Piezoelectric coefficients $d_{11/33}$ 10^{-12} As/N	$e_{11/33}$ As/m^2	$g_{11/33}$ Vm/N	Coupling coefficient $k_{11/33}$
Quartz (X-cut)	5700	4,6	576	2.3	0.17	57	0.10
Lead zirconate Titanate	4350	1700	365	374	15.8	24.9	0.70
Lead metaniobate	3300	225	570	85	3.2	42.5	0.38
Zincoxid	6400	8,2		12.4	1.57	0.48	0.38
PVDF	2200	10	80	22	0.14	200	0.16

with c the velocity of sound and α the attenuation of sound. Both these parameters determine the wave propagation. They are strongly dependent on the properties of the media the wave is traveling through. This is used for many sensor applications. Molecular interactions, phase transitions, molecular rearrangements and others are responsible for the c- and α-behavior.

In an unbounded solid medium the longitudinal velocity of sound c_L and the transverse velocity of sound c_{Tr} are given by

$$c_L^2 = (K + \tfrac{4}{3}G)/\rho \tag{5.2}$$

and

$$c_{Tr}^2 = G/\rho \tag{5.3}$$

where K is the bulk modulus, G the shear modulus and ρ the density. In a rod the longitudinal wave travels with

$$c_{Lrod}^2 = E/\rho \tag{5.4}$$

where E is Young's modulus.

The attenuation in a solid is determined by several absorption losses. It cannot be described in a simple way. In a polycrystalline solid reflection and scattering effects occur at the grain boundaries. In polymers relaxation processes can cause the absorption of sound. Many other effects are also responsible for the attenuation of sound in solids. Despite these several reasons for attenuation the order of magnitude of the absorption coefficients in solids is comparable and similar in different materials. This explains the early development and broad application of methods for the nondestructive testing of solids on the basis of ultrasound. Only polymers present an exception.

Fluid systems show a more complicated behavior in the ultrasonic range in comparison to other materials. In liquids the velocity of sound is described by

$$c_L^2 = \frac{1}{\rho\beta} \tag{5.5}$$

with β the compressibility. β and ρ are substance specific, integral parameters. In electrolytes for instances β changes very strongly with slight variation of the ion concentration or the kind of ions.

The description of the attenuation of sound in liquid systems is relatively simple when homogeneous, so-called "classical" liquids are considered. For α_l equation (5.6) is valid:

$$\frac{\alpha_L}{f^2} = \frac{2\pi^2}{\rho c^3}\left(\tfrac{4}{3}\eta_s + \eta_v\right) \tag{5.6}$$

η_s is the shear viscosity and η_v is the bulk viscosity. The attenuation shows a quadratic dependence on frequency. This is important for most technological applications. At very high frequencies, e.g. higher than 10 MHz, the attenuation in many real liquid systems reaches such a value that the sound is absorbed at very small distances. The distance which can be passed by a sound wave is less than about 5 mm. On the other hand, at very low frequencies, lower than 100 kHz, the accuracy of most measurements is not sufficient. Remarkable attenuation effects only occur when the sound has passed a distance of about a meter.

Many fluid systems which are dominant in technical applications are multi-component systems such as suspensions, dispersions, emulsions, colloidal systems or aerosols. Their acoustic behavior cannot be described easily. It is characterized by the influence of different material parameters like density ρ, viscosity η, thermal conductivity χ, thermal capacity c_p, thermal expansion coefficient $\bar{\alpha}$ of the different phases and the particle radius R or diameter D in the dispersion or suspension. They can be temperature and pressure dependent. The attenuation in such a system is then caused by different losses:

$$\alpha = \alpha_0 + \alpha_{\text{vis}} + \alpha_{\text{therm}} + \alpha_{\text{sc}} + \alpha_{\text{relax}}. \tag{5.7}$$

α_0 is the basic absorption of the emulsion or dispersion medium, α_{vis} describes the viscous losses, α_{sc} the scattering losses, α_{therm} the thermal losses and α_{relax} losses due to relaxation processes.

The viscous losses are given by

$$\alpha_{\text{vis}} = \frac{\left[\dfrac{\omega}{2c}\right] k \left[\dfrac{\rho'}{\rho} - 1\right]^2 \dfrac{9}{4}\left[\dfrac{\delta}{R}\right]\left[1 + \dfrac{\delta}{R}\right]}{\left\{\dfrac{\rho'}{\rho} + \dfrac{1}{2} + \dfrac{9}{4}\left(\dfrac{\delta}{R}\right)^2 + \left[\dfrac{9}{4}\right]\left(\dfrac{\delta}{R}\right)\left(1 + \dfrac{\delta}{R}\right)^2\right\}} \tag{5.8}$$

where

$$\delta = \sqrt{\frac{2\nu}{\omega}}$$

and k is the concentration of the disperse phase and ν is the kinematic viscosity. The main effect for the losses is caused by the friction between the particles in the suspension or emulsion (primed quantities) and the carrier medium like water or oil.

As can be seen the parameters listed above also influence the acoustic behavior in a complex manner. In principle viscous losses could be very interesting for technological applications.

The scattering losses correspond to

$$\alpha_{sc} \approx kk_0^4 R^3 \left[\frac{1}{6} \left(\frac{\beta - \beta'}{\beta} \right)^2 + \frac{1}{2} \frac{\left(1 - \frac{\rho}{\rho'}\right)^2}{\left(2 + \frac{\rho}{\rho'}\right)^2} \right] \tag{5.9}$$

where

$$k_0 = \frac{2\pi}{\lambda},$$

and k_0 is the propagation constant. In dependence on the diameter and the concentration of the dispersion particles Rayleigh or Mie scattering is responsible for the measuring effect. Rayleigh scattering occurs at particle diameters which are small in comparison to the sound wavelength. It is assumed that there is no particle interaction. Mie scattering occurs at particle diameters comparable with the sound wavelength. Multiple scattering effects and particle interactions are observed in the case of Mie scattering. The complicated influence of particle radius and particle distribution on the attenuation can be observed. Also material parameters like the compressibility β play an important role. Scattering losses offer interesting applications.

The thermal losses are very often neglected in theoretical and practical considerations. This is wrong. They are described by

$$\alpha_{\text{therm}} = k \frac{3}{2R^2} cT\chi\rho \left[\frac{\bar{\alpha}}{\rho c_p} - \frac{\bar{\alpha}'}{\rho c_p'} \right] f(R, \omega, \chi, \chi', c_p, c_p') \tag{5.10}$$

where χ is the thermal conductivity, c_p is the specific heat, and f is an analytical function of the different parameters. The main loss effect is caused by the difference in the thermal expansion coefficients. The thermal losses can strongly influence the attenuation behavior of the system under investigation. It is difficult to know or to evaluate the values of the thermal parameters of the different components. Therefore thermal losses were very often not considered in the past.

In gaseous media the ultrasonic velocity and absorption are described by the classical thermodynamic relationships. For the velocity c for instance

$$c^2 = \left(\frac{\mathrm{d}p}{\mathrm{d}\rho} \right)_{\text{ad}} \tag{5.11}$$

is valid.

c and α react very sensitively in their dependence on changes of state of the gaseous medium.

The determination of velocity and absorption can be realized in different ways. In principle all methods are based on the following two equations

$$\alpha = \frac{1}{\Delta x} \ln \left(\frac{p_0}{p} \right) = \frac{1}{\Delta x} \ln \left(\frac{U_0}{U} \right) \tag{5.12}$$

and

$$c = \lambda f = s/t \tag{5.13}$$

where U and U_0 are the electrical voltages which are proportional to the sonic alternating pressures p and p_0. Δx or s are the path lengths traveled by the sound and t is the traveling time. Consequently the absorption measurement is reduced to an amplitude measurement. The attenuation of ultrasound is difficult to measure absolutely except in the laboratory. However, in principle this is a very interesting and important method to determine concentrations, particle diameters and particle size distributions, etc. The sound velocity is mostly measured by determination of the transit time of ultrasound along a known path length. Because time measurement can be done with high resolution the ultrasonic parameter sound velocity is often known or determined very precisely. This offers many sensor applications.

The acoustic impedance Z is a third acoustic parameter:

$$Z = \rho c \tag{5.14}$$

It determines the behavior at interfaces. For normal incidence of a compressional wave at an interface between two materials of different acoustic impedance the reflection coefficient of the acoustic wave, R, is given by

$$R = \frac{A_r}{A_i} = \frac{(Z_1 - Z_2)}{(Z_1 + Z_2)} \tag{5.15}$$

where A_r and A_i are the particle displacements or particle velocities of the reflected and incident waves and Z_1 and Z_2 are the acoustic impedances of the material the wave is traveling in and the material the wave is reflected by, respectively. The transmission coefficient T is deduced from

$$T = 1 - R. \tag{5.16}$$

R and T are determined by amplitude measurements and therefore underlie the same accuracy restrictions as the attenuation coefficient.

Beside c, α, and Z the wavelength, λ, and frequency, f, of the sound wave, its phase, φ, and the resonance frequency f_r of the transducer may be of interest for the sensor design or special applications. Summarizing it can be concluded that in most cases the ultrasonic parameters are experimentally determined by time measurement (c, f, f_r), amplitude measurement (α, Z, R, T) or more seldom by interferometric principles (λ, φ). The determination of the spectral amplitude distribution can play a role if scattering effects are used.

5.1.4 Ultrasonic Sensor System

An ultrasonic sensor alone makes no sense. For an efficient use of this sensor principle well-adapted transmitter and receiver electronics and intelligent data acquisition electronics are necessary. Whereas in the past many ultrasonic systems could be realized only as scientific instruments [7] due to modern microelectronics they are now feasible as compact devices at low cost. More recently the proliferation of control systems opened up high application potential for ultrasonic sensors [8].

The scheme of an ultrasonic sensor system and possible sensor configurations are shown in Figure 5.1. Two ways can be distinguished. First the ultrasonic sensors

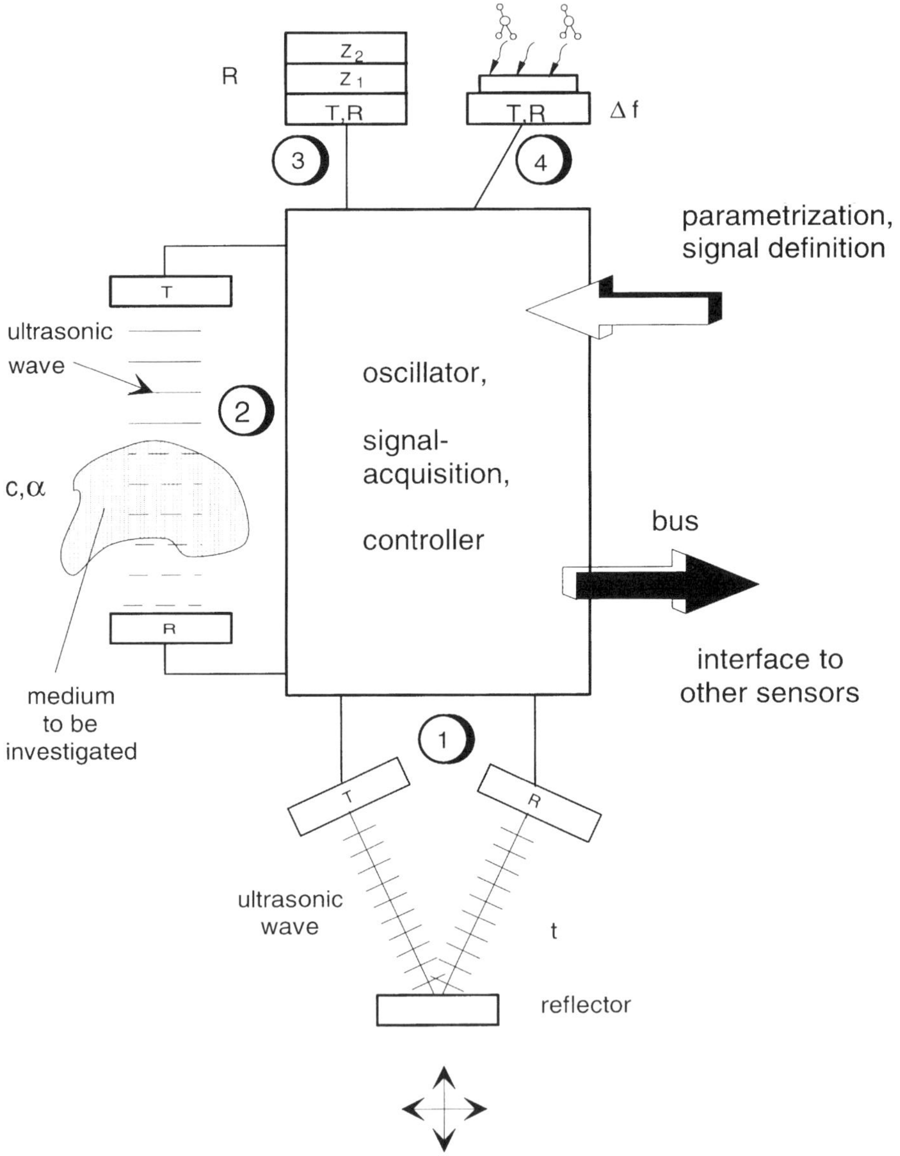

Figure 5.1 Ultrasonic sensor system.

actively transmit acoustic waves and receive them later. Usually piezoelectric transducers are exploited which transform an electrical signal into an ultrasonic wave and vice versa. In many applications it is possible to use the same transducer. This is the case in examples 1 and 2 (Figure 5.1). Depending on the situation along the acoustic path, characteristic parameters of the ultrasonic wave are modified. Thus the ultrasound signal carries information about the properties between transmitter and receiver. An intelligent sensor system now has to extract this information. To achieve this performance the signals are processed, demodulated and evaluated by dedicated hardware. A time resolution of less than 1 ns can be attained for the determination of propagation time. That is equivalent to an accuracy for the ultrasonic velocity of better than 1‰. The amplitude accuracy is normally measured to within 1%. Higher resolution of both parameters can be realized by special electronic circuits (e.g. 12 bit AD-converters and very fast circuits), averaging and specific signal evaluation. These measures only make sense when state parameters like temperature, pressure, etc., are measured simultaneously with the ultrasonic parameters with high accuracy. In the next step special algorithms are used to calculate the parameters of interest. They are based on the laws of wave propagation and include the interaction of the acoustic wave with the investigated matter. Knowledge of these relations is necessary for the development of new applications for ultrasonic sensors, e.g. in process control, for special distance measurements and others.

Two other ultrasonic sensor principles are shown in Figure 5.1, denoted by 3 and 4. In these cases one sensor is directly coupled with the medium under investigation and influenced by it. In example 3 the acoustic impedance or the reflection or transmission coefficient are measured by detection of the amplitude changes. The sensor acts both as a transmitter and receiver. Acoustic emission with only a receiver for detecting the emitted signals uses this principle. It is not discussed further in detail.

In example 4 the resonance frequency of an ultrasonic sensor is changed by deposition of a material on to the sensor surface or by interaction of a sensitive layer on the sensor with a surrounding species in a gas or a liquid. The last case is the basis for so-called mass-sensitive sensors. They play an increasing by common role on the sensor scene.

5.1.5 Sensor Classification

Depending on how an ultrasonic signal has been changed on its path from transmitting transducer to receiving transducer, or how the transducer properties were changed, ultrasonic sensors can be distinguished into four groups (Figure 5.2):

- distance sensors
- ultrasonic propagation path sensors
- impedance sensors
- mass-sensitive sensors.

Distance sensors use the travel time and amplitude of a received signal (e.g. the echo) to derive the presence, distance and type of the sound-reflecting object. Intelligent evaluation methods allow conclusions about objects to be recognized and classified. Ultrasonic propagation path sensors use the effect that the tested variables change the transmission of the ultrasound. The affected parameters are the ultrasonic velocity and attenuation, acoustic impedance, scattering coefficients, local changes of propagation (due to diffraction and refraction), etc.

Impedance sensors determine the change of amplitude to provide information about material properties like density or transmission coefficients. Mass-sensitive sensors have become increasingly important in the last few years due to progress in chemical layer synthesis and deposition methods. Their working principle is simple but they can achieve very high resolution. A quasi-digital signal is provided.

In Figure 5.2 the listed principles are summarized and selected applications are shown. Ultrasonic sensors can be applied in gaseous media, liquids, solids, and melts.

Ultrasonic distance sensors	• Distance, position, level • Path • Object structure • Multidimensional recognition • Presence of objects • Process monitoring • Motion, speed, path (Doppler effect)
Ultrasonic propogation path sensors	• Concentrations in multicomponent systems • Particle size distribution in suspensions, emulsions • Volume and mass flow velocity • Transmission properties • Temperature measurement • Dynamic pressure measurement
Impedance sensors	• Density • Acoustic impedance
Mass-sensitive sensors	• Mass • Viscosity • Specific chemical or biological species • Multicomponent analysis

Figure 5.2 Classification of ultrasonic sensors.

In homogeneous liquid systems the velocity of ultrasound can be used for their identification. In solutions the velocity of ultrasound usually depends on the concentration of the solution. In two-phase liquid systems, e.g. suspensions or emulsions, the attenuation changes with the concentration of the solid or emulsified phase. In mixtures of gases the velocity varies linearly between the velocity in one gas and the velocity in the other pure gas. In melts the attenuation is strongly influenced by the viscosity of the melt. In solids Young's and shear moduli determine the acoustic behavior.

The advantages of ultrasonic sensors are:

- ultrasonic techniques are noninvasive;
- in-line measurement;
- rapid response, usually a fraction of second;
- low power consumption;
- excellent long-term stability;
- high resolution and accuracy.

Some disadvantages are:

- increase in the amount of electronics for high accuracy and information (this does not mean high costs due to decreasing prices for electronic components),
- integration along the entire sound path,
- knowledge of the acoustic properties of the substances through which the sound travels is necessary.
- increase of the attenuation of sound with frequency.

The most interesting frequency range for technological applications of ultrasound lies between 0.1 and 10 MHz. At higher frequencies the measurement distance is very small. At low frequencies the resolution is reduced and higher sensitivity to acoustic interference noise occurs.

5.2. Ultrasonic Flow Sensors

5.2.1. Principles

Measurement of the volume flow velocity and of the mass flow velocity is very important for a variety of applications. In industrial processes flow sensors are one of the most frequently applied sensors after temperature and pressure sensors. Other examples where flow plays a fundamental role are human respiration measurements in medicine, intake air determination of internal combustion engines, the measurement of gas for heating homes and of drinking water consumption. For all these applications there is a trend to robust, accurate, bus-compatible measuring devices without moving parts and with a minimum disturbance of the flow. Besides ultrasonic flowmeters only thermal methods and vortex meters are in accordance with this trend.

Various methods have been used for ultrasonic flow meters. They are based on three acoustic effects [9]. These are:

- the drift effect;
- the Doppler effect;
- the attenuation or diffraction effect.

High accuracy results are achieved by application of the drift effect. It makes possible direct ultrasonic measurements of the flow velocity [10]. The ultrasonic waves are transported by the flowing medium. The measurement principle of the flow in a pipe by ultrasound is shown in Figure 5.3. The time required for sound to travel between two ultrasonic transducers set at a distance L from one another depends on the flow velocity v, the speed of sound c, and the angle α between the direction of flow and direction of sound propagation:

$$v = \frac{L}{2\cos\alpha}\left(\frac{1}{t_2} - \frac{1}{t_1}\right) = \frac{L}{2\cos\alpha}\left(\frac{t_1 - t_2}{t_1 t_2}\right) \tag{5.17}$$

The difference of the reciprocals of the propagation time in both directions is required for the calculation of the mean value of the flow velocity. The constants of the measuring tube have also to be known. These reciprocal times can be measured very accurately by different methods. Due to progress in microelectronics the traveling

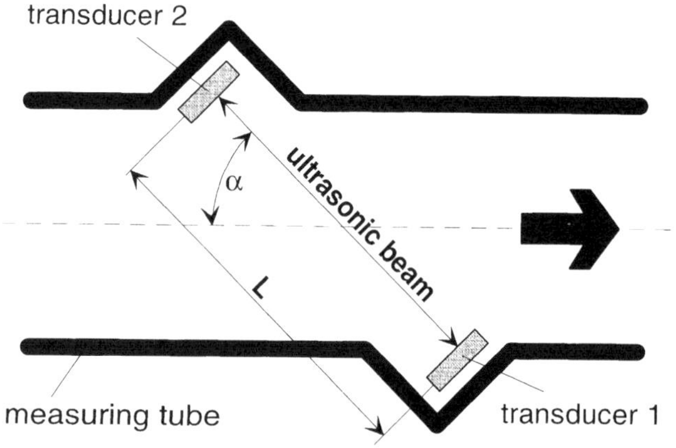

Figure 5.3 Principle of ultrasonic flow measurement.

time can be determined very rapidly with high accuracy and a resolution better than 1 ns. The following appropriate calculations are done easily by a microcomputer. An older classical principle is the so-called "Sing around method" where the reciprocal time is determined directly as a repetition frequency. At the moment of reception of an ultrasonic pulse the transmission of a further pulse is triggered and so on, thus giving the repetition frequency $1/t_1$, or $1/t_2$, depending on the direction of transmission. It works very well if the medium the ultrasound is traveling through does not change its ultrasonic attenuation behavior. Also other reasons for pulse failure have to be avoided.

The flow velocity measured by ultrasound is the volume flow velocity, i.e. the transported volume per unit time. The measured flow velocity is an integral value. It is therefore influenced by the flow profile. Tube geometry or inner surface structures of the tube have also to be considered. The acoustic flow measurement principle has many advantages. Examples are: no moving parts, no pressure losses, independence of viscosity and conductivity of the medium, and nondestructive measuring principles.

In spite of these important advantages ultrasonic flow measurements are often considered as unreliable and of low accuracy. One of the reasons for this evaluation is the insufficient knowledge of physical effects which are relevant for the ultrasonic flow measurement. Thus in pipes the flow velocity can differ locally as a result of turbulence. The measured value is an average of all local speeds along the path of the ultrasound signal. It is generally different from the actual mean speed in the pipe, particularly in the case of low flow velocities with laminar flow profiles. Other causes of improper function are refractive effects of the flow profile, like lenses, which affect the ultrasound. Combined with the wave-guiding properties of the measuring tube these refractions give improper results particularly when coaxial sound radiation is used. Simulation results of this effect are presented in [11]. They show the change of the sound pressure distribution in a coaxial measuring tube as a function of flow directions. A further effect which reduces the accuracy is the so-called "zero slow" error. Due to nonsymmetric properties of the ultrasonic transducers and electric circuits on the transmitting and receiving side and nonsymmetries between the two directions of the ultrasonic path even at nonflow there are slightly different values of the propagation time t_1 and t_2. This causes a rapidly increasing error at low flow velocity. Strict requirements for the uniformity of the transducers raise costs and cannot be maintained for long periods of time.

Characteristic data for an industrial liquid flow sensor system using the transit time difference principle are given in Table 5.2 (Schwing Verfahrenstechnik GmbH, Germany). They are representative of other producers.

To increase the accuracy of flow systems the influence of the velocity profile has to be eliminated. For measurement tubes of large nominal diameter several sound paths can be placed in such a way that the influences of flow profile differences and of flow disturbance are surpressed [10]. The problem is posed by the fact that sound paths near the middle of the tube measure too much flow in the case of a laminar profile, while paths near the walls measure too little. The sensitivity of other parts lies in between.

Table 5.2 Typical data for a liquid flow system

Specification of the system		
Pipe diameter: 6–300 mm Medium temperature: −200 °C to +150 °C (steel probe)		
Accuracies		
Reynolds number range	Maximum error at 30 cm/s or more	Largest range of flow change
> 100 000	± 1% of measured value (m. v.)	optional
10 000–100 000	± 1.7% of measured value	3 : 1
2000–10 000	larger than ± 10% of m. v.	
< 2000	± 0.2% of m. v.	optional

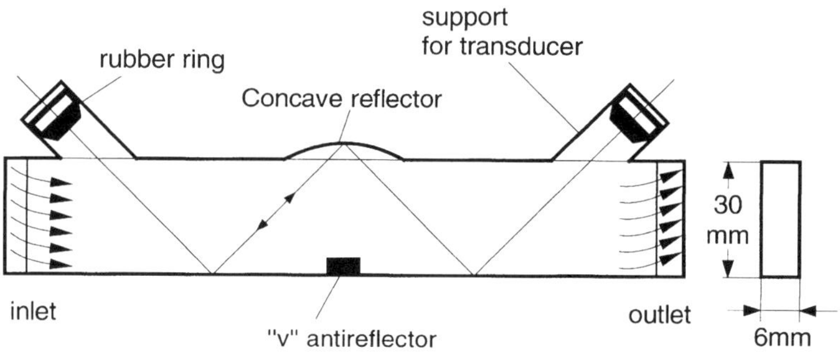

Figure 5.4 Measuring tube configuration for an ultrasonic gas meter (after [12].

For small diameters it is complicated to arrange several soundpaths side by side due to lack of space for the ultrasonic transducers and above all, because wall reflections and wave-guide effects prevent the formation of line-shaped paths. For small measuring tube diameters a broad sound beam can be generated instead, which fully fills the width of the rectangular tube. The aim is that all parts of the flow profile contribute to the result of the measurement. This integration over the complete flow profile maintains the ultrasonic flow measurement independent of the flow profile and flow disturbances.

As an application example a special configuration of a measuring tube for the measurement of gas consumption in households is presented in [12]. A w-shaped sound path in the measuring tube with a slim rectangular cross-section produces very good results (Figure 5.4). Selection of the intended sound path is achieved by using short pulses with high initial slope. With special design of the measuring tube parasitic sound paths were suppressed. Finally very low errors could be realized by application of ultrasonic transducers of high quality and uniformity. From 20 l/h to

20 000 l/h, this means that within a range of 1–1000 the error was less than 0.5 %. This is used in a SIEMENS ultrasonic gas meter [13].

Ultrasonic gas flow measurement methods are characterized by rapid response time and the ability of flow direction recognition. Therefore they are important for air mass meters which are necessary to measure the intake air flow of internal combustion engines. In pulsating and reversible flows ultrasonic sensors have proved to have a faster response time than hot-wire and hot-film sensors.

Equipments using the drift effect to measure the propagation time of sound in opposite directions is mostly employed in ultrasonic flow meters for gaseous and liquid media. The medium must have low sound attenuation and scattering. In the case of significant sound scattering, e.g. a medium such as a suspension, dispersion or aerosol, the Doppler effect can be utilized. In ultrasonic flow meters of the Doppler type the sound transmitter and receiver are immovable. Inhomogeneities in the flowing medium, e.g. bubbles or suspended particles, act as mobice sonic scatterers or reflectors. Under the assumptions that the diameter of the reflector is larger than 1/10 of the sonic wavelength and $v \ll c$ the following relationship between the Doppler frequency f_D and the flow velocity v is valid:

$$f_D = f_T - f_R = 2f_T \frac{v \cos \alpha}{c} \qquad (5.18)$$

where f_T is the transmitted frequency, f_R is the received frequency and α is the angle between the direction of flow and the direction of sound propagation. From equation (5.18) it can be concluded theoretically that the higher the transmitter frequency the better the resolution of the flow velocity v according to

$$v \sim \frac{f_D}{f_T}. \qquad (5.19)$$

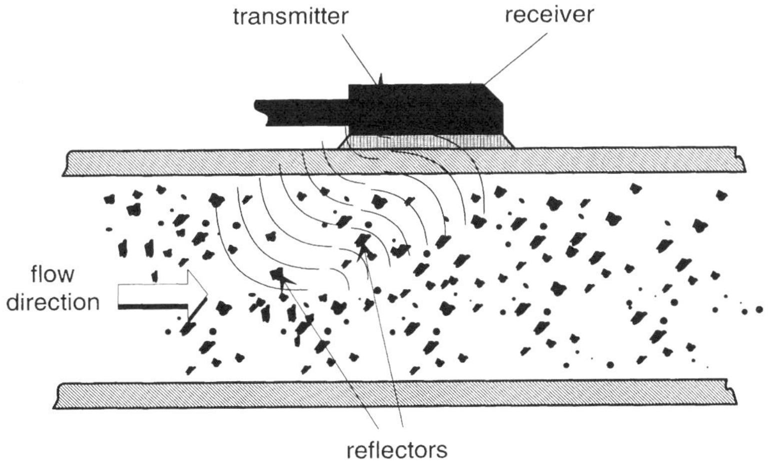

Figure 5.5 Sensor arrangement in a Doppler flow meter.

Consequently smaller reflectors come into force. In practice Doppler devices work at MHz frequencies and then the natural impurities in a fluid, e.g. in water, are sufficient to act as reflectors. One example of an ultrasonic sensor arrangement is given in Figure 5.5. Both ultrasonic sensors are located on the same side of the tube. Other configurations are possible and employed in industrial applications.

The third effect which is used in ultrasonic flow sensor configurations is the attenuation or diffraction effect. Normally the propagation of ultrasonic waves is influenced by wavelength- and medium-dependent attenuation and scattering. Its temporal attenuation pattern exhibits characteristic and thus identifiable features. This is the case when a fluid carries natural or artificial inhomogeneities. Their local distribution is constant over a certain time interval or path. It can be considered as the motion of a "frozen" pattern. This pattern is observed by two acoustic barriers which are placed at a distance. The technique is called correlation flow measurement. Figure 5.6 shows the scheme of this principle. The correlation function

$$R_{xy}(\tau) = \frac{1}{T}\int_0^T x(t-\tau)y(t)\,\mathrm{d}t \tag{5.20}$$

is calculated. x and y are the received signals at different points and different times. From the time τ_m at which the correlation function has a maximum and the known

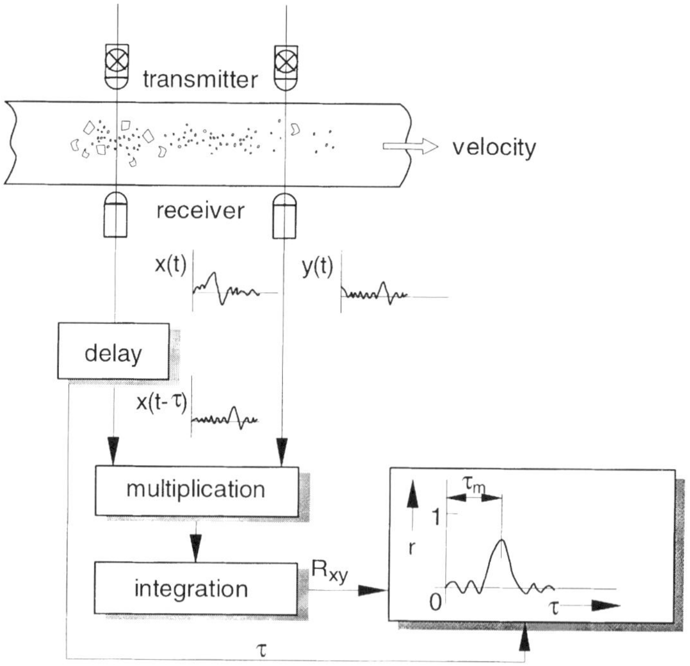

Figure 5.6 Principle of the correlation technique.

distance s of both ultrasonic sensors the flow velocity

$$v = \frac{s}{\tau_m} \tag{5.21}$$

can be calculated. The correlation technique is an old well-known but complicated technique. Due to developments in electronics this principle can be realized nowadays without problems. In the past it was mostly used in applications where multi-component or multiphase mixtures and very contaminated liquids were analyzed.

5.2.2 Industrial Applications

Ultrasonic flow meters have been well established on the market for about 20 years but in comparison to other flow meter principles they have played a secondary role. According to a market analysis in 1988 [14] they had only a few percent of the market share. The main reasons for this unsatisfactory situation were the high costs for the total equipment and insufficient accuracy and reproducibility of the principles used. Nowadays the situation has completely changed. Due to the development of electronics and software tools, the availability of new technologies and the synthesis and manufacture of new or modified piezoelectric materials, ultrasonic flowmeters with high accuracy and reliability are now offered. They combine many properties which are wanted by many users. These are:

- no pressure losses;
- high accuracy for moderate and high flow velocities, most of all for liquids;
- contactless measurement;
- application in opaque media;
- miniaturized devices for special cases;
- digital electric output signal.

The prices for ultrasonic flow meters are now comparable with other types.

In the following a few selected examples for a variety of application areas and typical parameters of industrial devices will be given. One of the most important applications is the measurement of flow in liquids. The company SCHWING (Germany) sells a system consisting of a measuring pipe with two ultrasonic sensors, pipe adapter and electronics (Mapco type flow meter). It can be applied for pipe diameters from 6 to 3000 mm, a temperature range from −200 °C to +150 °C and pressures up to 200 bar. The two sensors are arranged opposite to one another at an angle of 45° to the flow direction. A transit time difference system is used. Liquids like oil, petrol, organic solvents, drinking water, inorganic acids and alkaline solutions, waste water, corrosive solutions, emulsions, etc., with a broad viscosity range can be measured. Typical values for the accuracy of the velocity measurement

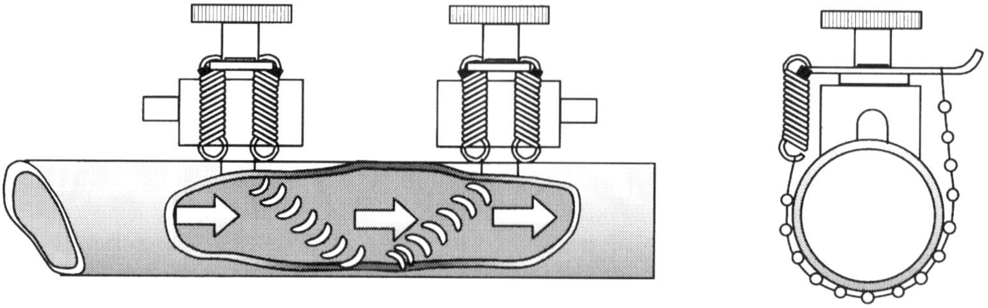

Figure 5.7 Clamp-on technique: mounting of sensors at the measuring point (1 – chain clasp; 2 – sensor; 3 – runner with sensor)

in their dependence on the Reynolds number are given in Table 5.2. Restrictions for the application of these devices occur in systems with high attenuation, e.g. fluids with higher concentration of bubbles or with scattering centers. PANAMETRICS (USA) offers very flexible equipment, e.g. Model 6068. A large number of measurement pipes with different building shapes (e.g. U-type for high resolution) and built-in sensor configurations (different ultrasonic paths, e.g. the w-path) allows the user to construct systems with very high accuracy (better than 1% in pipes with diameters from 2 mm to 2 m). SIEMENS (Germany) will bring on the market a new in-line ultrasonic flowmeter for fluids which is based on a helical sound path. High measurement accuracy is maintained independent of the fluid flow profile. The device consists of a tube in which two special reflectors are mounted. These influence the ultrasonic wanes in such a way that they subsequently touch the tube walls. This path is the reason for the uniform interrogation of the different flow velocities in the tube leading to a measurement independent of profile changes [15].

The disadvantage of measurement pipes with built-in sensors can be avoided by using devices based on the clamp-on principle. The sensors are attached to the pipe wall (Figure 5.7) and can make measurements at every point of a pipe. No special installation at a point in the pipe is necessary. The sound wave from the transmitter propagates through a wedge-shaped sonic feed pipe-body, the pipe wall and the medium to be measured. The ultrasonic pulses are transmitted at an angle into the pipe. Therefore refraction of the sound occurs at every interface with a different sonic velocity value. It is influenced by the material of the pipe wall and the properties of the measurement medium. Secondly the flow profile and the kind of flow determine the flow velocity. This has to be corrected. In the past all these factors meant that clamp-on devices played only a secondary role in flow measurement. Nowadays the situation is completely changed. Due to modern electronics a very fast on-line information processing and self-learning process can be constructed. The material parameters, flow characteristics and geometrical data are stored in a data bank. They can be used in solving the special measurement problem. Many companies offer very interesting products for liquid flow measurements in water and seawater management, environment protection, in the chemical industry, the food and beverage industry, power stations and energy supply, machine and plant construction, and semiconductor manufacturing plants. Thus FUJI ELECTRIC Co. (Japan) produces different clamp-on type ultrasonic flowmeters based on the transit-time measuring method (e.g. Portaflow X, Time-Delta Series). Thanks to microprocessor based electronics the flow meters can be easily configured from the keyboard to specific applications. Special types of flowmeters are suitable for liquid flow measurements for pipe sizes from 13 to 6000 mm, fluid temperatures from $-40\,°C$ to $200\,°C$, a flow range from 0.3 to 32 m/s and a turbidity of 10^4 mg/l or less. Accuracies of about 1% can be reached. The response time is 1 second. With the advanced Anti-Bubble Measurement (ABM) system the measurement procedure is less affected by bubbles. Similar devices are manufactured by other enterprises like NUSONICS (USA), PEEK Measurement (USA) or FLEXIM (Germany). One of the FLEXIM devices, the FLEXUS ADM, is not only used for flow measurement but also for heat quantity measurement. This special application is very important for heat flow distribution systems, e.g. in huge buildings, thermal power stations or processes (e.g. in the chemistry and food industries). The heat quantity can be determined by a difference method. The in- and out-going enthalpy flows are considered. This is done in practice by measurement of the volume flow V as well as the feed pipe (ϑ_f) and return pipe (ϑ_b) temperatures. The heat flow Q is given by

$$Q = k_i(\vartheta_f - \vartheta_b)V \tag{5.22}$$

with the heat coefficient k_i. k_i is specific for every substance and can be calculated from the specific enthalpy and density. Devices for measurement of the heat quantity consist of an ultrasonic flow system in combination with a temperature difference measurement system both supported by a computer or controller for the calculation procedures. Many other big and smaller companies sell heat quantity devices.

In 1996 PANAMETRICS (USA) brought a new clamp-on flow meter on the market (DF 868) which differs from other methods. It can measure in clean water as well as in mud. With this principle the whole range of liquid systems can be investigated. In liquids with bubbles or solid particles it uses the so-called transflection principle, a combination of transit time and reflection measurement. For each measurement the pure transit time mode or the transflection mode is in action. Another version of this principle (PT 868) unites three devices in one: flow measurement, heat quantity measurement and wall thickness inspection. This is an impressive example of progress in this field.

Doppler flow meters are used for fluid multicomponent systems such as emulsions, suspensions, dispersions or similar liquids. These systems are characterized by the existence of sound scattering centers like bubbles or solid particles which are necessary for the Doppler effect and they normally posses a high sound attenuation. Therefore the other known methods cannot be applied. Doppler flow meters very often use the clamp-on technique. The MAPCO Doppler flow meter (NUSONICS, USA) or the Polysonics DFF 3088 from PEEK Measurement (USA) are examples of commercial devices. The last one has the following parameters: accuracy about 2% for a flow range from 0.002 m/s to 15 m/s, temperature range from $-40\,°C$ to $175\,°C$. The pipe wall must be homogeneous and not thicker than 25 mm.

Transit time flow meters are also used for gas inspection. The SIEMENS ultrasonic household gas meter was already mentioned [13]. PANAMETRICS has the GP 68 on the market. In a range from 0.03 m/s to 45 m/s, gas flow can be detected with an accuracy of about 2%. There are many activities that use ultrasonic intake air flow meters for internal combustion engines, e.g. in cars. Ultrasonic flow measurement methods have two important advantages in comparison to other methods, e.g. the hot-wire and hot-film sensors. They have a faster response time and recognize the flow direction. These advantages open up a new technical potential for improved fuel injection systems that can manage individual cylinders and transient loads. This technique is an important tool on the way to cars with lower consumption and cleaner exhaust.

Which trends are typical for ultrasonic flow meters? Due to the development in electronics (faster processors, higher storage capacities, new digital processors, etc.) more data can be processed faster. The results are flow meters with higher accuracy, enhanced capabilities (three in one like the PANAMETRICS PT 868) or new application areas (one device for a broad spectrum of liquids or gases). New transducer materials, e.g. with a higher Curie temperature, will open up new high-temperature applications in gaseous media or melts.

5.3 Ultrasonic Distance Sensors

5.3.1 Principles

Ultrasonic distance sensors can be used for many tasks. Examples are the determination of distances or the presence of an object and object recognition. Often the propagation of ultrasonic waves in air is used. Level sensors can also be included in the category of distance sensors. In this case the propagation of ultrasound in liquids, solids or bulk goods can be exploited. Ultrasonic sensors evaluate the information from propagating ultrasonic signal parameters. Typical characteristics are the ultrasonic propagation time, signal amplitude or echo shape. Ultrasonic distance sensors mostly utilize the impulse-echo principle. A short ultrasound pulse is transmitted in the direction of the object that reflects the signal back to the ultrasonic transducer. The propagation time t of the sound signal is directly related to the distance d from the sensor to the object according to

$$d = \frac{ct}{2} \tag{5.23}$$

where c denotes the velocity of sound. Depending on the ultrasonic frequency (most devices work with ultrasonic frequencies from around 9 to a few hundred kHz) a measurement range from mm to a few meter can be realized. Distance sensors operating in air and using broadband ultrasonic transducers achieve a high local resolution in the direction of sound propagation. Accurate measurements down to

0.25% can be expected [16]. In conventional ultrasonic distance sensors simple comparator circuits are used to detect the arrival of the reflected signal in a specific time window. By considering whether the echo has been received within a time window, i.e. a time interval, which can be preset by the user, the distance range is given in which the sensor responds to the presence of an object. Time-dependent control of sensitivity is used to compensate the distance dependency of the echo amplitude. Different reflection properties are compensated by an automatic gain control which holds the average echo amplitude constant. Normally in commercial distance or presence sensors the echo amplitude has very little influence on the accuracy of the distance measurement. An essential advantage consists in the general insensitivity against dust, lighting effects and color, surface roughness or material properties of the measured object. Such distance sensors are employed in the automation of industrial processes, e.g. the positioning of workpieces, as well as in traffic control systems, e.g. for collision avoidance or monitoring of free parking places. Ultrasonic distance meters are used also for the measurement of the filling level in containers or the height of material on conveyor belts.

Problems occur when the distance is measured simultaneously to several objects. Good reflecting objects cause multiple echoes due to repeated reflections between sensor and object. They simulate the existence of further objects, which is very critical if the distance to more than one object must be determined. Figure 5.8 shows such a situation. Fluctuations in the speed of sound , turbulences and external noise can cause further disturbances that degrade the accuracy of the measurement and lead to significant changes of the received echo amplitudes. Conventional ultrasonic distance sensors which use a simple time window technique for the detection of the arrival of the reflected signal can solve these problems only insufficiently. For this reason there is a need for advanced ultrasonic sensors that have the capability to

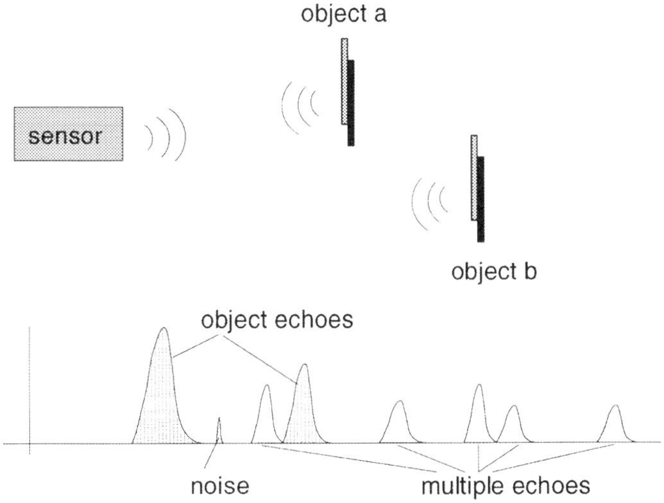

Figure 5.8 Echo profile with a number of interfering echoes (after [17]).

evaluate partial echoes in complex echo profiles. In [17, 18] the concept of signal processing for an intelligent distance sensor and its realization is described. The main point is to separate real target echoes from misleading echoes by evaluating a set of suitable rules. Because fuzzy logic allows partial membership and logic operations on distributions of membership values over an interval, it is much better suited for the required decision processes than comparator techniques (Boolean logic) which are based on fixed thresholds and allow only binary yes/no decisions. Suitable inputs for a fuzzy echo profile evaluation are for instance:

- features characterizing a partial echo;
- features describing the relations between several partial echoes;
- knowledge about the time history of measuring situation;
- a priori or expert knowledge.

The methods of data analysis and the classification algorithm are described in detail in [18]. The results are very important for level measurements in complex environment. Figure 5.9 illustrates typical echo profiles which were obtained in a bulk goods container after reflection of the transmitted ultrasound signal. In Figure 5.9c the results after fuzzy evaluation are shown. The set-up of the intelligent level meter is determined by a microcontroller and a SIEMENS fuzzy coprocessor (SAE

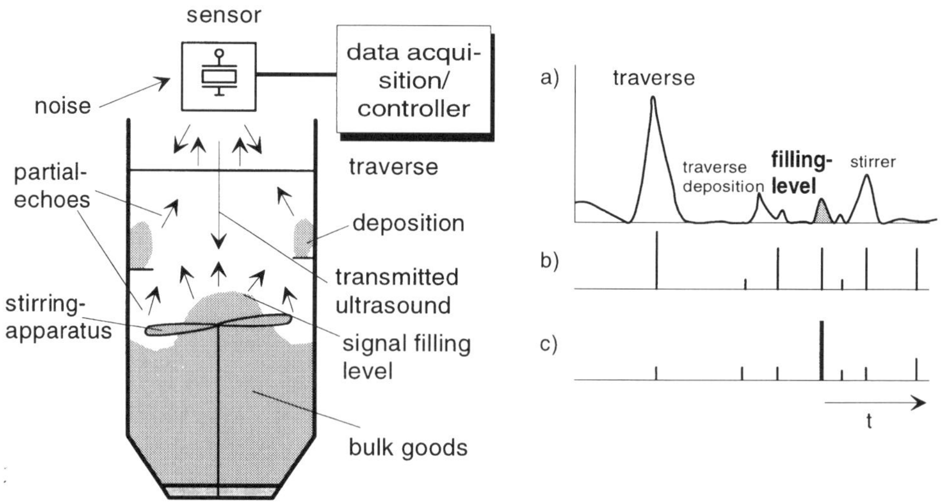

Figure 5.9 Typical measuring situation of a level meter and results of fuzzy-evaluation (after [17]): (a) echo profile; (b) surpression of multiple echoes; (c) evaluation of temporal echo development.

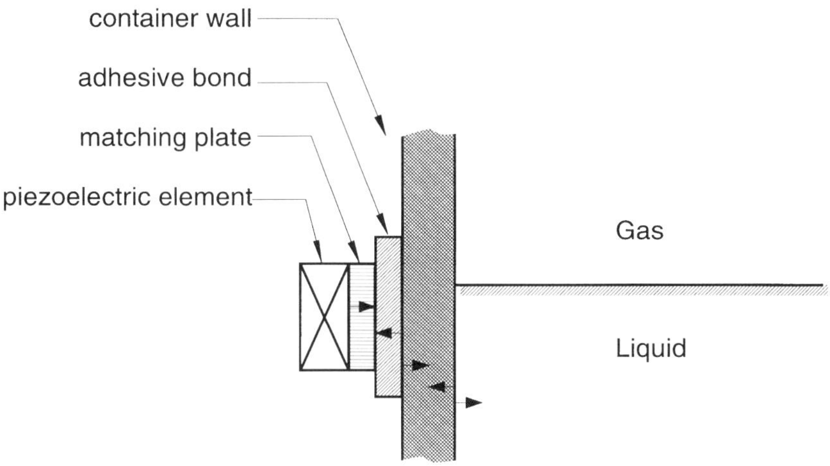

Figure 5.10 Basic structure of an ultrasonic low liquid level sensing system (after [24]).

81C99). Such intelligent distance sensors using fuzzy logic rather than Boolean logic will play a dominant role in different applications in the future.

The high local resolution of distance sensors in the direction of sound propagation can be used for object recognition. For this aim the whole high-frequency receiver signal is converted into a digital signal and stored as an echo profile in the memory of a microcomputer. By comparing the stored echo profiles of a reference situation with an actual situation it can be decided whether or not the actual situation corresponds to the reference. Techniques to get a so-called acoustic fingerprint are reported in [19]. In order to achieve rapid object recognition so-called inverse filtered excitation signals are transmitted, which are adapted to a specific object as a key is to a lock [20]. By using transducer arrays or a moving transducer, lateral structures can also be scanned [21]. Normally they would be indistinguishable for standard ultrasonic sensors. A resolution of about 1 mm in the axial and lateral directions can be achieved by using special composite sensors. Ultrasonic echo profiles pertaining to different parameters can also be processed by fuzzy logic or neural network techniques. The application of the latter technique for position-invariant recognition of small objects using a transducer array is shown in [22].

Ultrasonic level sensors can also operate on a principle which is based on the acoustic impedance variation caused by the presence (or absence) of a liquid above (or below) where the sensor is mounted on a tank or vessel. Figure 5.10 shows as an example the basic structure of a three-layered system (matching plate, adhesive bond, container wall). The functional relationship of time t_{ijk} and acoustic amplitude response A_{ijk} for such a three-layered structure can be written as

$$t_{ijk} = it_1 + jt_2 + kt_3 \tag{5.24}$$

where $t_m = 2d_m/v_m$. (d_m is the thickness, v_m is velocity of the medium m, and i, j, k are the numbers of reflections in media 1,2,3, respectively) and

$$A_{ijk} = A_0 \, T_{10} \, R_{10}^{i-1} \, R_{12}^{i+j} \, R_{23}^{j+k} \, R_{34}^{k} \, \mathrm{e}^{-2(i\alpha_1 d_1 + j\alpha_2 d_2 + k\alpha_3 d_3)}$$
$$\times \sum_{m=1}^{\min(i,j)} (-1)^{j-m} \binom{j}{m} \binom{j-1}{m-1} \left(\frac{1 - R_{12}^2}{R_{12}^2} \right)^m$$
$$\times \sum_{n=1}^{\min(j,k)} (-1)^{k-n} \binom{j}{n} \binom{k-1}{n-1} \left(\frac{1 - R_{23}^2}{R_{23}^2} \right) \tag{5.25}$$

where m and n are binomial parameters [24]. A_0 is the incident amplitude and $R_{ij} = (Z_i - Z_j)/(Z_i + Z_j)$ is the reflection coefficient of medium i to medium j; Z is the acoustic impedance. For applications the only variable is the amplitude of the acoustic signal from the container wall to the fuel or gas interface, expressed as the ratio of the kth echo amplitude for the container wall with a liquid backing (i.e. fuel) and without (i.e. gas). This can be expressed as

$$\frac{A_{ijk}}{A_{ijk}} = \left(\frac{R_{34}}{R_{34}} \right)^k \tag{5.26}$$

where the unmarked quantities are for a wall backed by a liquid, the marked quantites by a gas. The point is now to use the most sensitive signal of the echoes. This principle for level detection is used in industrial applications and for low liquid level sensors in space launch vehicles [24]. The sensor is able to withstand stringent environmental conditions, such as high pressure, extreme thermal cycles, high gravitational load, mechanical shock, and random vibrations.

The use of resonance and the Lamb wave principle for limit value measurements is reported in [23]. In the case of the resonance principle the container wall is excited in its fundamental thickness resonance mode by a piezoelectric transducer which is placed outside of the container. For steel containers with a wall thickness from 2 to 5 mm these frequencies lie between 0.5 and 1.5 MHz. The decay time of this mode depends on whether the inside of the wall is covered with liquid or not. The second principle uses Lamb waves. Special modes are excited in the container wall by electromagnetic transducers. A few of these Lamb wave modes are strongly attenuated. The frequencies lie in the same range as in the case of the resonance principle. In practice the resonance principle is used due to its highest sensitivity but special demands have to be met. The transducer must be damped optimally but posses high coupling factors. The coupling layers have to be long term and temperature stable, etc. This principle is realized in the product NIVOPLUS (ENDRESS & HAUSER, Switzerland) [23].

5.3.2 Industrial Sensors

Many different sensor types for measurement of distance, the presence of an object, level or object recognition are on the market. The big companies ENDRESS &

HAUSER (Switzerland) and HONEYWELL (USA) offer a wide variety of sensors for process monitoring and control. Distance or level sensors are produced also by many small and middle-sized companies because they satisfy very flexible special application demands or occupy niches. The spectrum of ultrasonic distance or level sensors reaches from relatively simple configurations to intelligent sensor systems. Mostly piezoceramic transducer materials are applied but also Polaroid based sensors (MESSRING, Germany) or PVDF arrays are used. As an example ENDRESS & HAUSER brought a big variety of liquid flow level and limit value sensor systems on the market. The sensor systems are called Prosonic T. They are widely used in the

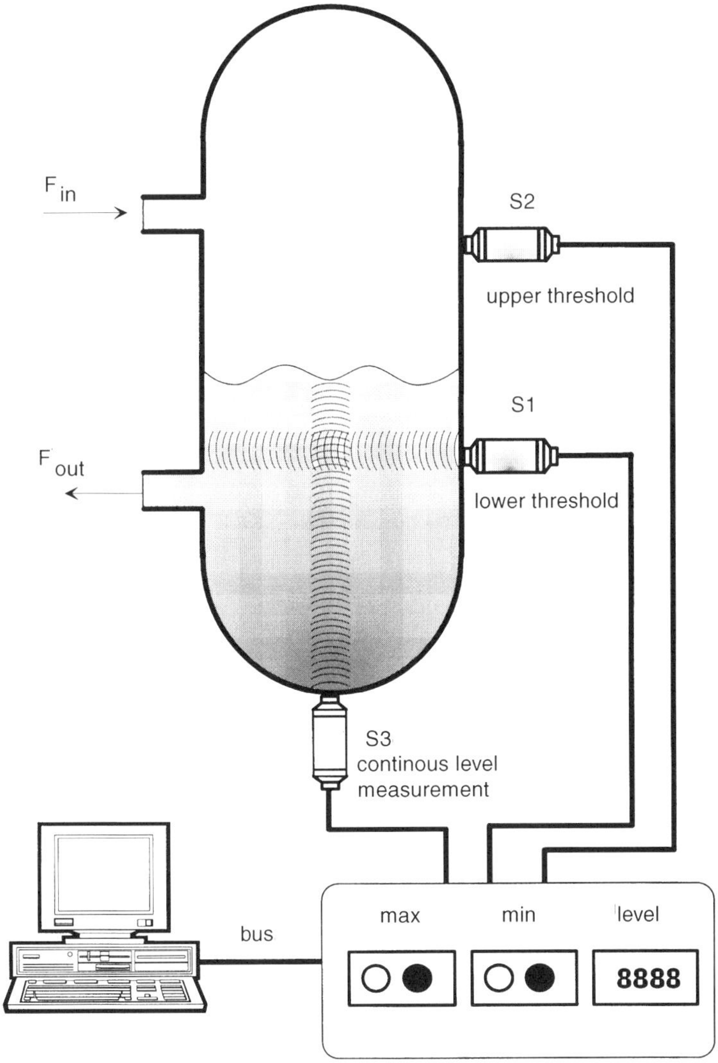

Figure 5.11 Principle of liquid level measurement with min/max limit switch.

chemical, food or oil industries. HONEYWELL offers distance and proximity sensors. LUNDAHL Instr. (USA) sells the DCU-10 programmable ultrasonic sensors or DCU-7 microprocessor controlled ultrasonic sensors. They can be used for most level and presence detection applications, e.g. for feeder bin level control, for monitoring the level of liquid in tanks during nickel and copper plating process, for a conveyor system beam counter or for portable level monitoring. IDEC GmbH (Germany) offers an analog distance sensor with a detection range from 10 to 1000 mm. Fuzzy logic is used for the compensation of temperature changes or variations of surfaces. MICROSONIC GmbH (Germany) manufactures ultrasonic sensors for safety control, passenger detection (wms types), proximity sensors (mic types) or low-cost sensors for obstacle detection of driverless transport vehicles (Trans-O-Prox). BAUMER ELECTRIC (Switzerland) offers a broad spectrum of ultrasonic distance sensors, two-point proximity switches and ultrasonic reflection barriers. UNISAT Sensortechnik GmbH (Germany) supplies level sensors for the food industry in a temperature range from $-10\,^{\circ}\mathrm{C}$ to $-600\,^{\circ}\mathrm{C}$, continuous level measurement up to 2000 mm and very short response time of about 0.1 s. In Figure 5.11 an example for a liquid level measurement system with a max/min switch is shown. This principle is used very often in many industrial areas.

Which developments can be predicted for distance sensors? It can be assumed that intelligent versions of sensors open up new application areas in industry and other fields such as safety techniques. This sensor type will play a more important role in object recognition due to more powerful sensor arrays and developments in data acquisition techniques. New materials will open up high temperature applications.

5.4 Ultrasonic State Sensors

5.4.1 Introduction

Ultrasonic techniques have become more and more attractive for process monitoring and control in many branches of industry such as chemistry, the oil industry, biotechnology, the food industry, agriculture, etc. Ultrasonic state sensors are defined as ultrasonic sensors for the detection of typical process parameters. This development has strongly accelerated in the last few years by progress in electronics and computer techniques. For a long time it was very well known what in principle ultrasound in process monitoring is able to do and what it cannot do. The book *Ultrasound and its application in Science and Technology* first published in the 1940s by Bergmann was a complete summary of the fundamental knowledge of his time [7]. A lot of examples for the application of ultrasound in different fields were given. Most of them are still valid but in the past they could not be implemented in practice due to the lack of a technical basis. Today microelectronics can handle data with high speed and resolution. Data can be stored and compared. By using special software new information can be extracted from experimental data. Thus new opportunities are given for ultrasonic techniques. An important advantage of ultrasonic techniques is that they can be used non invasively. The containment of the plant does not need to be breached. This results from the fact that containment materials, particularly metals, are usually "transparent" to ultrasound. Other advantages of ultrasonic techniques are that they normally do not involve "moving" parts, the response is rapid, the energy levels are low and nonhazardous equipments can be constructed, and they provide a range of mutually compatible techniques. Another important feature of ultrasonic techniques, which in one respect gives them the edge over optical techniques, is that the velocity of ultrasound is relatively low compared with that of light. Hence, the wavelength of an acoustic wave, e.g. in the order of MHz frequencies, is in the mm range. Disadvantages are that it is always the case that an integral information is received (ultrasonic techniques are not suitable for trace analysis), very often special knowledge of the substance properties are necessary, and sophisticated electronics is needed to develop and to handle the equipment.

The principal ultrasonic parameters which can be measured are: velocity, attenuation and impedance. From these parameters one can derive the chemical identity of pure liquids, the concentration of components in solutions, mixtures or dispersions, reaction conversions and under certain circumstances particle sizes or particle distributions of dispersed phases. Most ultrasonic state sensors use compressional ultrasound. The frequency is typically in the range 0.5–10 MHz except for attenuation particle size sensors. The ultrasound is normally generated by piezoelectric transducers of small and convenient size which are readily available commercially. Some form of "coupling" is used to transmit the ultrasound from the transducer into and back from the system under investigation. This impedance match is very often a serious problem which has to be solved. Electronic systems have been developed and are available which enable transducers to generate either continuous waves or pulses, or which detect and amplify the received signals, determine the time between a transmitted pulse and the received echoes and carry out data processing. Very high resolution and accuracy must be achieved, otherwise ultrasonic state sensors are not interesting for the applications listed above. That means the following conditions have to be satisfied:

$$\frac{\Delta c}{c} \leq 10^{-3}; \quad \frac{\Delta \alpha}{\alpha} \approx 0.5\% \quad \text{and} \quad \frac{\Delta Z}{Z} < 10^{-3}.$$

These conditions are equivalent to the demands of time measurements in the ns and ps range and of amplitude resolutions of 12 bits and higher. This expresses the high requirements for the electronics. It can be weakened only partly by using averaging procedures.

5.4.2 Ultrasonic Velocity Sensors

Ultrasonic measurement principles for sound velocity are well known. In Figure 5.12 one example with a fixed distance of transmitter and receiver is illustrated. It is the

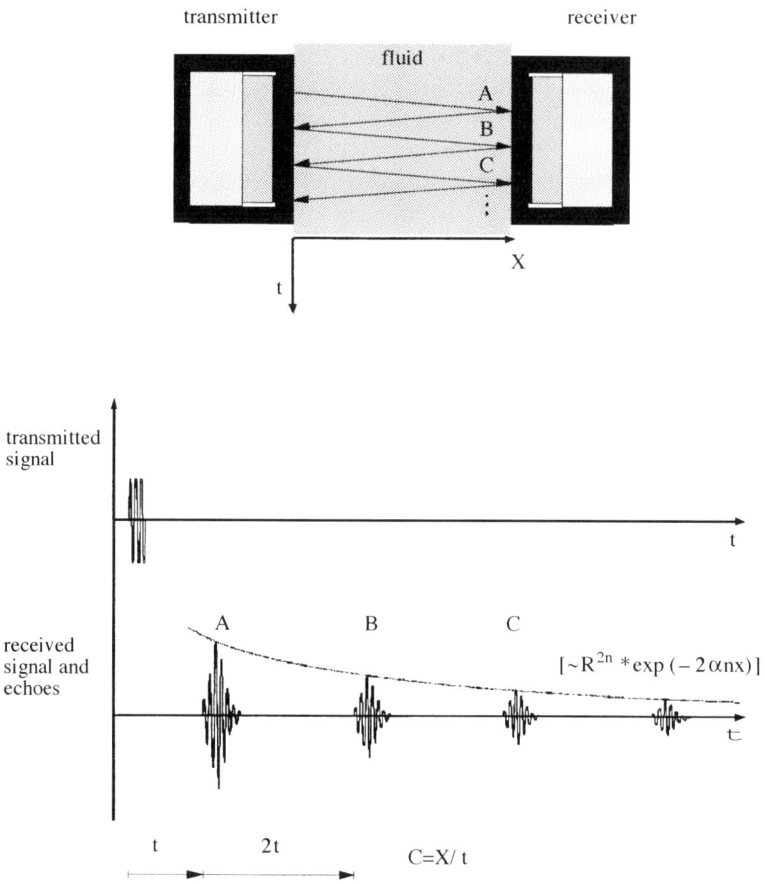

Figure 5.12 Measurement principle for ultrasonic velocity c and attenuation α.

most widely used arrangement in technical applications. The velocity, c, is calculated from the time difference between different echoes and the known distance, x. For very accurate measurements special tricks for the triggering of the echoes with different amplitudes are used and the thermal expansion coefficient of the container material has to be taken into account. The velocity, c, can also be determined by other techniques such as interferometric or resonance principles but they do not play a role in applications in industrial equipments.

At present the most fruitful ultrasonic parameter for use in process monitoring and control is the velocity of sound. Once the velocity of ultrasound in the medium is determined the data can be used in a number of ways. These are for identification of liquids, concentrations of solutions, behavior of mixtures of liquids and two phase liquid systems.

A most impressive example of this application is the continuous on-line-measurement of original gravity of beer [25]. Beer is considered to be a three-

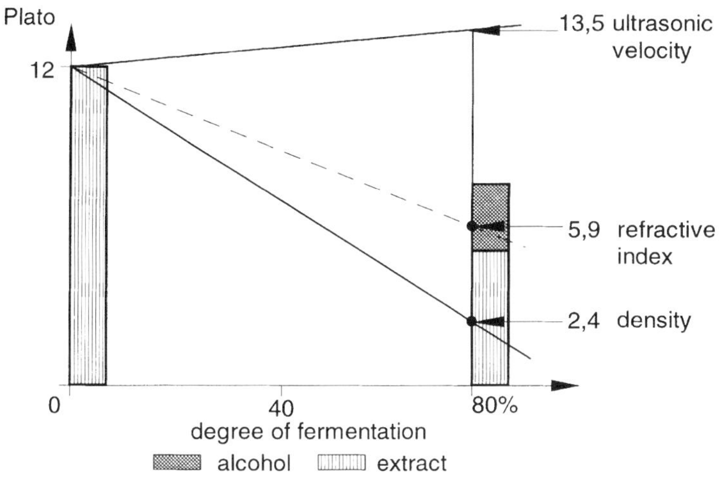

Figure 5.13 Dependence of sound velocity, density and refractive index on the degree of fermentation (12 Plato for a wort, 0 Plato for water).

component mixture consisting of alcohol, extract and water. Figure 5.13 compares the effect of the degree of fermentation upon the ultrasonic velocity, c, density, ρ, and refractive index, n. As the fermentation proceeds the values for density and refractive index fall rapidly while the value for sound velocity rises. The effect on the sound velocity can be seen to be much less significant than for the two other measurements. This important behavior can be explained as a result of the molecular conversion which occurs during fermentation. The effect of alcohol concentration on ultrasonic velocity is about twice that of sugar. Since fermentation converts two molecules of sugar to one molecule of alcohol, the resultant change in ultrasonic velocity is small. Ultrasonic velocity is suitable for continuous original gravity (OG) monitoring. In practice different error influences have to be taken into consideration. These are pressure dependence, carbon dioxide influence (in reality beer should be considered as at least a four-component mixture, the fourth component being carbon dioxide), extract error and dependence upon present gravity. After elimination or correction of these errors an accuracy of ± 0.2 OG can be reached. The first devices for on-line monitoring of OG during the brewery process were brought on the market by ANTON PAAR K.G. (Austria). The data processing system mPDS 4000 in connection with the ultrasonic sensor SPR is used for in-line monitoring of different processes. The ultrasonic velocity technique is also used for in-line applications in distilleries, to measure the performance stills and the alcoholic strength and the volume of bulk storage tanks and to monitor the blending of final products.

A schematic diagram of an ultrasonic concentration meter, the so-called FUD 1 model-41 (FUJI, Japan), is shown in Figure 5.14. It is possible to measure the concentrations of ternary ingredients in a solution by measuring the ultrasonic propagation velocity at different temperatures. The concentration k_n of the nth-component of the solution is a function of the temperature of the solution at the ith

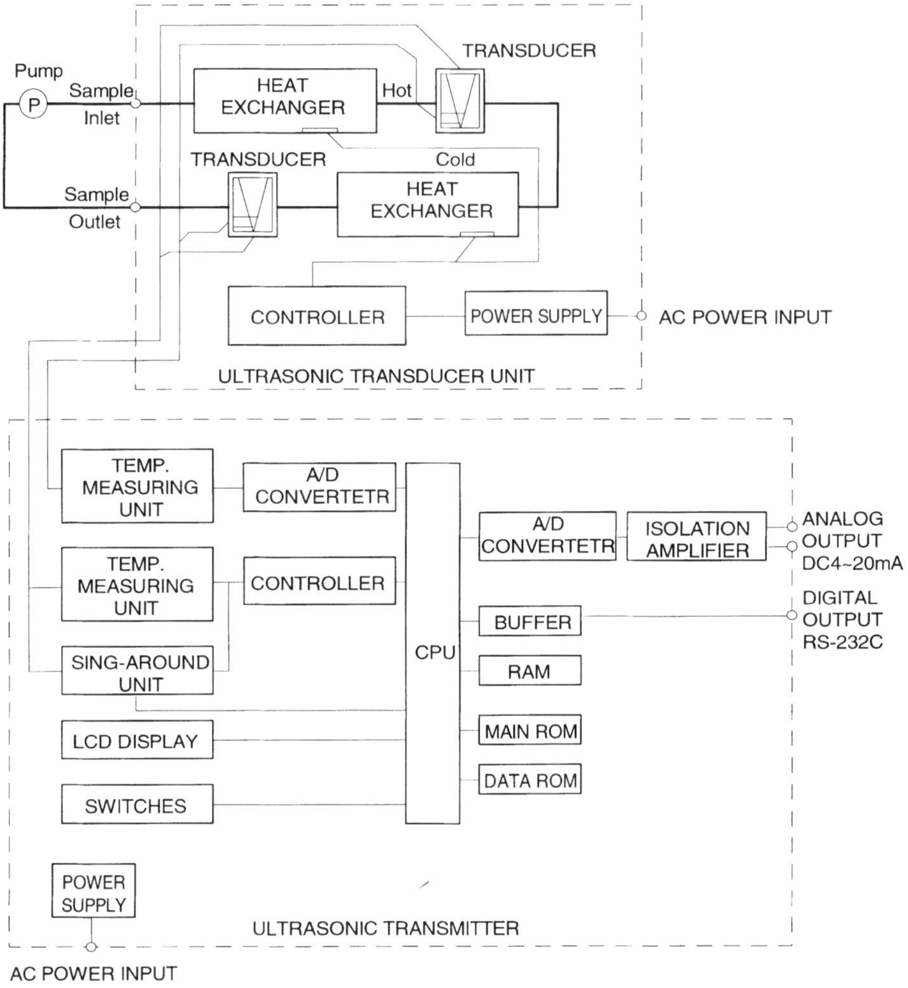

Figure 5.14 Schemetic diagram of a commercial ultrasonic concentration meter FUD 1 (FUJI)

temperature and the velocity of the solution at the ith temperature level

$$k_n = F(c_i, T_i). \tag{5.27}$$

It can be demonstrated that using a processor with internal ROM and EPROM the concentration of components of ternary systems were also obtained for such solutions as is shown in Figure 5.15 where no simple correlations between the components occur.

A more favorable case is illustrated in Figure 5.16. The change of ultrasonic velocity with sugar concentration is considerable. Therefore ultrasonic techniques are very interesting for applications in the food industry, sugar refineries etc. This was first pointed out in papers 20 years ago [26] but only recently have the corresponding

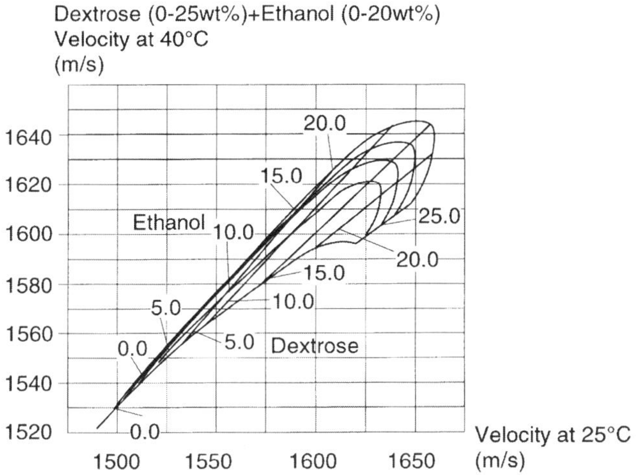

Figure 5.15 Ultrasonic velocity of the ternary system dextrose, ethanol, water at two temperatures.

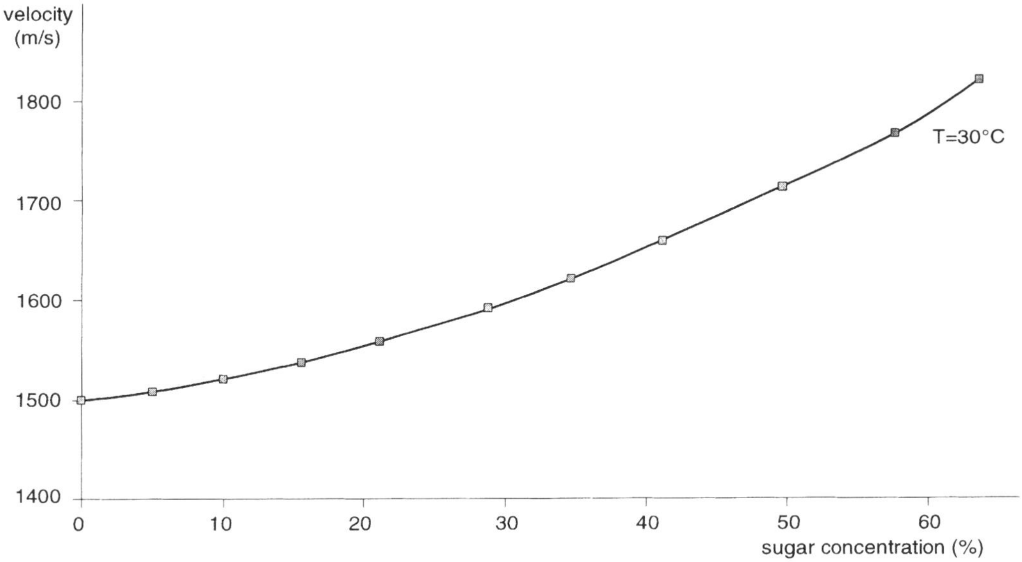

Figure 5.16 Dependence of ultrasonic velocity c on sugar concentration ($T = 30°C$).

techniques been available. This is expressed by many companies which are now on the market. Examples are BAILEY FISCHER & PORTER (Germany), PAAR (Austria), SCHWING (Germany), CANNONGATE (UK), RHOSONICS (Netherlands), FUJI (Japan) and SENSOTECH (Germany). The CIP (clean in process) qualified and under hazardous conditions applicable system SENSONIC 20 of the company SENSOTECH is shown in Figure 5.17. It is used in breweries as well as in

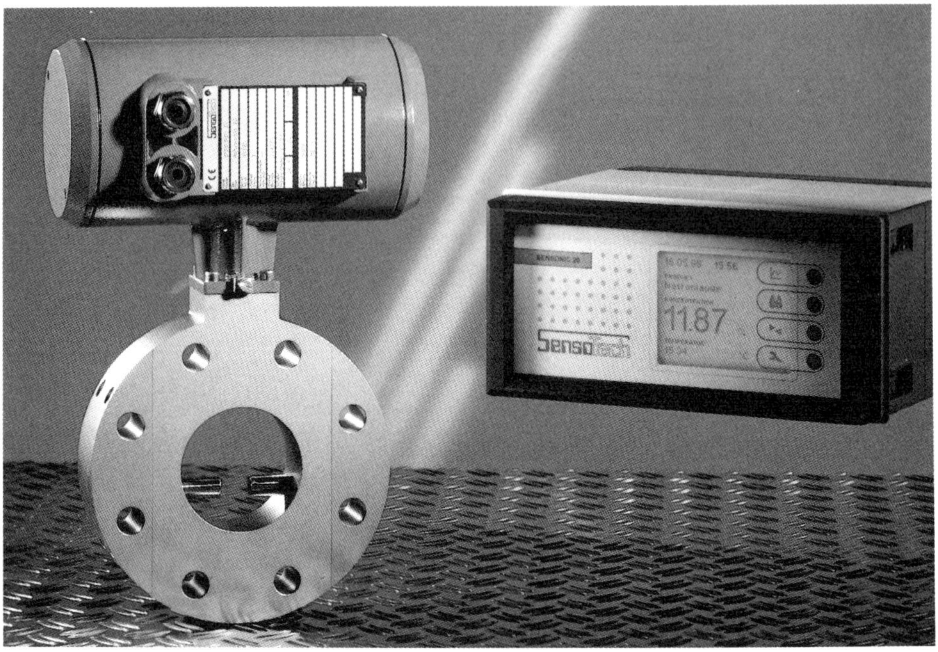

Figure 5.17 Commercial SENSONIC 20 device (courtesy of SENSOTECH, Germany).

chemical and pharmaceutical plants. An accuracy of about 0.1 ms^{-1} can be obtained. The temperature of the measured medium must be determined with an accuracy better than 0.1°C. These values are typical also for the products of the other suppliers.

In summary, one can say that systems which detect ultrasonic velocity have shown a rapid growth on the market. Liquids such as solutions and mixtures but also emulsions or suspensions are under investigation. This opens up the use of in-line applications in many processes for many substances. Besides the fields mentioned above these are product characterization in the food, chemical, pharmaceutical, and petrol industry, control of sewage treatment or polymerization processes, and monitoring of chemical etching, etc. [27].

Ultrasonic velocity sensor equipment is not applicable when the sonic attenuation of the medium the sound is propagating through is very high. This can occur in case of substances with a high content of bubbles or particles as scattering centers. No effects are measured also in multicomponent systems when the ultrasonic velocities of the components are not different.

5.4.3 Ultrasonic attenuation sensors

Attenuation of ultrasound is difficult to measure absolutely, especially under plant conditions. The basic reason is that ultrasonic signal strength decreases not solely as a result of attenuation in the medium under study but also because of other factors such as reflections at the interfaces, beam divergence, and changes in transmitter and receiver performances. These factors are not easy to keep constant. A problem of electronics is to achieve the same or comparable accuracy and resolution for attenuation determination in a wide range of received ultrasound signal amplitudes which can differ by a few orders of magnitude due to substance properties. Attenuation and velocity are mostly measured simultaneously. Thus following the electronic requirements are more strictly. In commercial systems for detecting ultrasonic velocity, e.g. the devices listed in Section 4.2., high accuracy is only achieved for a special class of liquids, e.g. two- or three-component solutions, special mixtures, etc. The question is now whether the measurement of attenuation supplies new or more information about the investigated liquid systems. The answer is the following:

1. Determination of attenuation is only useful in complicated, multi-component systems where the measurement of ultrasonic velocity is not very informative or very costly. Examples are the determination of the concentration of sewage slurry or the monitoring of fermentation processes. In Figure 5.18 the dependence of the ultrasonic parameters c and α and the biodrymass on the process time during a special fermentation is shown [28]. The calculated value of the absorption correlates very well with the biomass. This is an important result. New applications of the ultrasonic principle in biotechnology can be predicted because it can work under sterile conditions. In Figure 5.19 the block diagram of an ultrasonic device for simultaneous measurement of velocity and attenuation is depicted [29] which will come on the market as a commercial device in the near future. It works at a frequency of 1.5 MHz. The c accuracy is about 0.1%. The attenuation is measured with an error of about 1%.

2. Determination of attenuation is very useful when ultrasound is used for particle size determination. In Section 1.3 the fundamental equations for the description of attenuation in liquid multicomponent systems were introduced. One can see that the relationships between acoustic parameters and the influences of the several material data are very complicated. The equations were derived on the basis of different theoretical models [30–33]. These theories were developed under a few assumptions. Nevertheless one has to consider that the ultrasonic attenuation in emulsions, suspensions or similar systems is caused by different effects. These are thermal, viscous, scattering and structural effects. They are very sensitive to specific material parameters (thermal conductivity, viscosity, particle diameter, etc.) and the applied frequency. The complex relation of these variables and their influence on the attenuation are the reasons why only two commercial products were on the market. But this situation seems to be changing now. PEN KEM (USA) offers the new Acousto Phor 8000, a so-called particle analyzer on the basis of acoustic attenuation spectroscopy. A frequency range from 1 MHz to 100 MHz is used. Particle diameters and distributions from 0.01 to 10 µm can be detected. The most interesting property of this analyzer is that it is able to measure

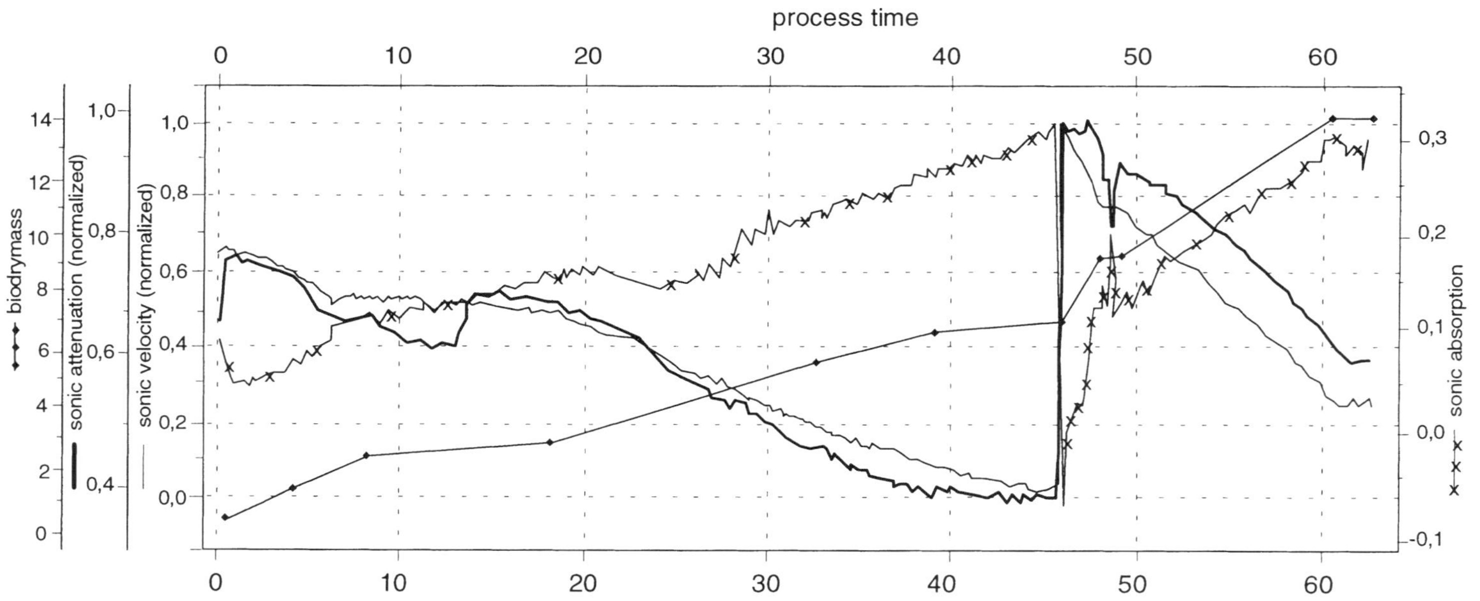

Figure 5.18 Measured and calculated ultrasonic parameters during a fermentation process.

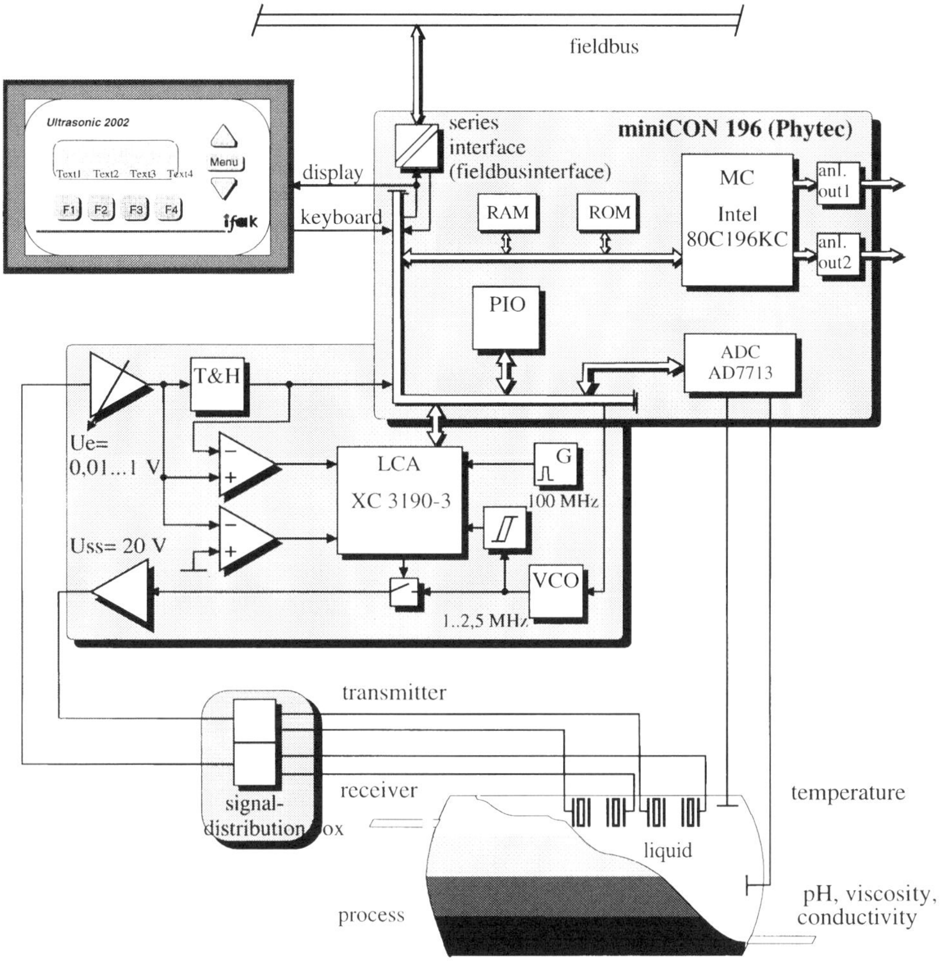

Figure 5.19 Block diagram showing an ultrasonic device for simultaneous measurement of ultrasonic velocity and attenuation.

volume fractions from 0 to 50%. The theoretical background for the explanation of this behavior is given in [33]. The SYMPATEC GmbH (Germany) offers the system OPUS. It is also an on-line particle diameter analyzer based on ultrasonic spectrometry. Diameters from 5 to 3000 μm can be investigated up to volume concentrations of 25%. MONITEK Technologies (USA) is on the market with an acoustic process particle measuring system (model AS 3/AT 3). Small particles and particle concentrations can be measured. In a specially configured cell the sound reflected by the particles is detected. MALVERN Instruments (USA) has just started manufacturing a new on-line particle size measurement system for process applications working on the basis of ultrasound.

5.4.4 Acoustic Impedance Sensors

The acoustic impedance is the product of the velocity of ultrasound in a medium and its density. The acoustic impedance mismatch at an interface between two media controls the relative amounts of ultrasound reflected and transmitted across the interface. On-line devices are available which determine parameters relating to this acoustic impedance mismatch between for instance the wall of a vessel containing a medium and the medium itself. They operate, in effect, by determining the amount of ultrasound reflected back at the interface between the wall and the liquid. In Figure 5.10 an example of using acoustic impedance measurement as an ultrasonic level sensor is illustrated. Acoustic impedance sensors have the very useful advantage of not demanding that the liquid is transparent to ultrasound. This greatly increases the range of materials which they can interrogate successfully. It opens up two possible fields of applications. First they can be used as a density sensor. In this case the ultrasonic velocity of the investigated liquid has to be known or to be determined simultaneously. Many attempts were made to develop sensors with high accuracy [34, 35]. They use an acoustic reference echo signal to make the measurement independent of the stability of the acoustic and electronic components. The effect of sonic diffraction usually limits the signal-to-noise ratio and thus the accuracy. This problem was solved by using PVDF as the sensor material [36] but this sensor is applicable only at low temperatures. Another way to reach high resolution is to construct a special sensor configuration as depicted in Figure 5.20. The piezoceramic and the buffer rods of the reference material have a cylindrical shape. The piezoceramic consists of lead-metaniobate, which is a compressional wave-only transducer; both buffer rods consist of quartz glass. It can now be shown that under certain geometrical conditions, this means that for a certain ratio of both the buffer rod lengths, the noise caused by edge waves is minimized [37]. An accuracy of the density measurement of about 10^{-3} is guaranteed. This sensor principle can be used for in-line process measurement. Density is an important quantity and every new technique is very welcome. The second field of applications is to offer the possibility

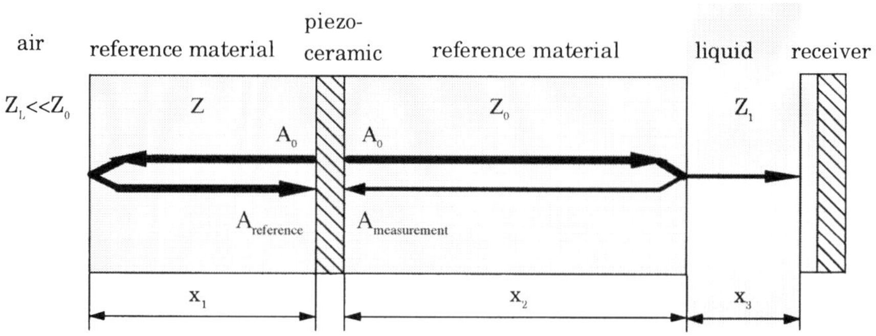

Figure 5.20 Schematic construction of the acoustic density sensor [37].

of determining the concentration of emulsions and, perhaps, solid dispersions, which cannot be analyzed ultrasonically by other techniques due to high sonic attenuation. No special devices are on the market at this time.

5.4.5 Conclusions

Determination of the velocity of ultrasound can be used to identify liquids, analyze mixtures of liquids, determine the concentration of components in multicomponent systems like ternary solutions, emulsions, dispersions and opens up new possibilities for in-line process monitoring. More and more producers of different sensor systems are on the market. It can be expected that in the next few years this sensor principle will be applied in many different processes. Ultrasonic attenuation cannot be determined easily in the absolute sense but, nevertheless, instruments for semi-quantitative determination of concentrations of solutions, emulsions or dispersions are appearing. They will have their chance where ultrasonic velocity determination is too insensitive or too time-consuming. Using ultrasonic attenuation losses like scattering effects, viscous or thermal losses in two- or three-component systems for the determination of particle sizes or particle size distributions is now very attractive. Not only systems for the laboratory but also process-compatible systems are available. Acoustic impedance mismatch determination is now used for level sensors. The application of this principle to determine the concentration of solutions and dispersions if they are not opaque to ultrasound has not yet been fully explored. The use of impedance sensors for on-line process density measurements is in progress.

5.5 Ultrasonic Microsensors

5.5.1 Principles

Microsensors that use acoustic waves comprise a very versatile class of sensors. Because they are highly sensitive to surface mass changes, they have many applications as chemical sensors. However, they can also determine a variety of other properties of solid or fluid media in contact with their surfaces, including liquid density, liquid viscosity, polymer modulus and electrical conductivity. Acoustic microsensors are interesting for the detection of mechanical quantities such as force, pressure or acceleration [38]. The use of acoustic waves for sensing applications requires a device to generate the waves, circuitry or instruments to operate the device, and methods to measure wave characteristics. Changes in wave velocity, frequency and/or amplitude indicate physical property changes occurring at the device surface. The typical acoustic microsensor consists of a piezoelectric material with one or more

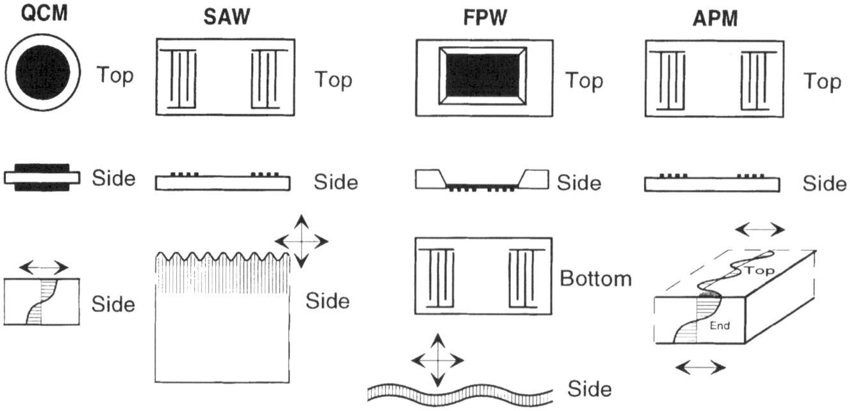

Figure 5.21 Views illustrating the structures of TSM, SAW, FPW, and APM (according to [39]); lower diagrams illustrate the wave motion.

metal transducers on its surfaces. These transducers launch acoustic waves into the material at ultrasonic frequencies, which may range from one to hundreds of MHz. Because of its temperature stability quartz is the most commonly used plate material but other materials may offer advantages in certain applications. Lithium niobate for example has a higher acoustoelectric coupling. Thin films of zinc oxide are used when very thin plates are desired or when the device must be created entirely by microfabrication or micromachining methods on a nonpiezoelectric substrate such as silicon. The crystal orientation and thickness of the piezoelectric material and the geometry of the metal transducers determine the type of acoustic wave generated and the device resonant frequency. The waves may travel through the bulk of the plate material, be guided by one surface as a surface wave or be guided by reflections from multiple surfaces, as in a plate wave. Transverse waves are most important in acoustic microsensor systems.

In Figure 5.21 the most commonly utilized acoustic microsensors are shown schematically, the thickness shear mode (TSM) resonator, the surface acoustic wave (SAW) device, the acoustic plate mode (APM) sensor and the flexural plate mode (FPM) device [39]. The acoustic waves generated in TSM devices are bulk transverse waves. They travel in a direction perpendicular to the plate surfaces. The resonant frequency of the fundamental mode is typically between 5 and 30 MHz. The frequency increases as the plate thickness decreases. SAW devices are composed of a relatively thick plate (e.g. ST-cut quartz) of appropriate crystal geometry and IDTs (inter digital transducers). These IDTs generate Rayleigh waves whose energies are confined largely to a zone at the surface approximately one acoustic wavelength thick. The wavelength is equal to the transducer periodicity p, defined as the center-to-center distance between two consecutive fingers of one comb of the IDT. With the wave velocity v, determined by the plate material and crystal cut, the "synchronous" frequency $f = v/p$ of a SAW device is determined by the finger spacing. Typical SAW sensor frequencies are 30–500 MHz. On thin plates, IDTs can generate two

Table 5.3 Comparison of acoustic waves and devices after [39]

Device type	Wave type	Particle displacement[a]	Transverse component displacement[b]	Media	Plate thickness	Factors determining frequency[c]	Typical frequency (MHz)	Example
TSM	Bulk	Transverse	Parallel	Gas, Liquid	$\lambda/2$	Plate thickness	5–30	Quartz $d = 277\ \mu m$ 6 MHz $\lambda = 554 \mu m$ $V = 3330$ m/s
SAW	Surface	Transverse, parallel[d]	Normal	Gas	$\gg \lambda$	IDT finger spacing	30–500	Quartz $d = 760\ \mu m$ 158 MHz $\lambda = 20\ \mu m$ $V = 3160$ m/s
FPW	Plate	Transverse, parallel[d]	Normal	Gas, Liquid	$\ll \lambda$	Plate thichness, IDT finger spacing	2–7	Zinc oxide $d = 3.5\ \mu m$ 5.5 MHz $\lambda = 100\ \mu m$ $V = 550$ m/s
SH-APM	Plate	Transverse	Parallel	Gas, Liquid	3–10λ	Plate thickness, IDT finger spacing	25–200	Quartz $d = 203\ \mu m$ 101 MHz $\lambda = 50\ \mu m$ $V = 5060$ m/s

[a] Particle displacement is relative to the direction of wave propagation.
[b] Transverse component displacement is relative to the sensing surface.
[c] Plate material also plays a role in determining frequency in all cases.
[d] Parallel component of particle displacement is in the direction of wave propagation and is also parallel to the sensing surface. The combination of transverse and parallel components produces a retrograde elliptical particle motion.

Table 5.4 Mass sensitivities of four types of acoustic microsensors according to [39].

Device type	Mass sensitivity increase with ...	Response to added mass[a]	S_m (unperturbed)[b]	S_m example (Hz/MHz)/(ng/cm^2)
TSM	Decreasing plate thickness (increasing frequency)	$\Delta f = -C_T F^2 \Delta m$	$S_m = -\frac{2}{\rho\lambda} = -\frac{1}{\rho d}$	0.014 (6 MHz, quartz plate)
SAW	Decreasing wavelength (increasing frequency)	$\Delta f = -C_S F^2 \Delta m$	$S_m = -\frac{K(\sigma)}{\rho\lambda} = -\frac{K(\sigma)}{\rho V_R}$	0.20 (158 MHz, quartz plate)
FPW[c]	Decreasing plate thickness (Decreasing frequency for constant wavelength)	$\Delta f = -\frac{1}{2M} F \Delta m$	$S_m = -\frac{1}{2M} = -\frac{1}{2\rho d}$	0.38 (5.5 MHz, composite plate)
SH-APM[d]	Decreasing plate thickness (Increasing frequency for constant wavelength)	$\Delta f = -\frac{J}{\rho d} F \Delta m$	$S_m = -\frac{J}{\rho d}$	0.019 (101 MHz, quartz plate)

Key: F, device frequency; Δf, frequency shift; Δm, change in mass per unit area; C_T, a constant dependent on the properties of the piezoelectric plate of the TSM device; C_S, a constant dependent on the properties of the piezoelectric plate of the SAW device; $K(\sigma)$, a factor dependent on the properties of the piezoelectric plate of the SAW device; M, the mass per unit area of the FPW membrane; ρ, the density of the plate material; d, the plate thickness; λ, the acoustic wavelength; V_R, the Rayleigh wave velocity; S_m, mass sensitivity factor.

[a] Assumed to be evenly distributed.
[b] Bare device in gas or vacuum.
[c] FPW composite membrane has zinc oxide piezoelectric layer. The expression for S_m in terms of ρ and d is for the special case of a single-layer isotropic FPW membrane.
[d] SH-APM expressions are approximations for an isotropic plate and are for individual modes, where $J = 1/2$ for $n = 0$ and $J = 1$ for $n > 0$. In the example, 101 MHz is the frequency for the $n = 0$ mode, and the mass sensitivity is for the $n = 1$ mode.

types of plate waves: Lamb waves and shear horizontal acoustic plate waves. In a plate of finite thickness Lamb waves generated by an IDT give rise to a series of symmetric and antisymmetric plate modes. The lowest-order antisymmetric mode has a wave velocity less than that of the SAW in the same medium. Its velocity decreases with decreasing plate thickness. This unique Lamb wave has a flexural character, hence the name "flexural plate wave" (FPW). The FPW device was microfabricated in a silicon substrate to create a very thin composite plate with a sputtered zinc oxide piezoelectric layer. The plate thickness is only a few percent of the acoustic wavelength, about 2–3 μm. The wavelength is determined by the IDT periodicity. By decreasing the plate thickness the frequency increases. Typical frequencies are 2–7 MHz. Shear horizontal acoustic plate mode, or SH-APM, devices are identical to SAW devices, except that their plates are thinner. Typical plates are a few acoustic wavelengths thick. In addition to generating Rayleigh waves, the IDTs on these thinner plates generate shear horizontal waves that propagate in the bulk at angles to the surface, reflecting between the plate surfaces as they progress down the plate between the IDT. Typical APM device frequencies are 25 to 200 MHz. In Table 5.3 typical properties of the sensor types are compared. It can be seen that SAW sensors of surface wave type can be used only in a gas whereas the other may be applied in gases as well as in liquids. Although mass loading is not the only mechanism of sensor response, it is one of the simplest response mechanisms. Therefore it is a good benchmark for characterizing and comparing devices. This is done in Table 5.4. The mass sensitivity S_m is defined in terms of oscillator frequencies as

$$S_m = \lim_{\Delta m \to 0} (\Delta f / F_0) / \Delta m \tag{5.28}$$

where F_0 is the unperturbed fundamental frequency of the oscillator in air and Δf is the frequency change that occurs on mass loading. S_m expresses mass sensitivity as the fractional frequency change per incremental change in mass loading. Table 5.4 shows that acoustic microsensors have very high mass sensitivities. Big differences occur between the different types. In this comparison the different working frequencies have to be taken into account. It can be concluded that acoustic or so-called resonant sensors are the most sensitive sensors in comparison to many other sensor types. Their frequency output signal can be processed very accurately and quickly. A lot of research activity has been undertaken in the last few years to built sensitive chemical, biological or physical sensors on the basis of these structures. A review on the state of the art is given in [39, 40].

5.5.2 Applications

Despite all the characteristics and advantages listed above, ultrasonic microsensors are not yet produced in huge amounts. There are many explanations for this situation. One is that the development of the first microsensors began only ten years ago. The second might be the design of these sensors. In most applications such as chemical sensors a sensitive layer on the sensor element is necessary. This layer or film must be uniform, thin, chemically and physically stable when in contact with its working medium (gas or liquid) and it must not electrically short-circuit the IDTs. The selected material must adhere to the acoustic wave device. Different techniques have been developed and used for depositing coatings on to the piezoelectric acoustic sensors. High selectivity and/or high sensitivity were achieved with polymer coatings, supramolecular materials or cavitands [40]. In many cases the long-term stability is unsatisfactory.

Acoustic microsensors are used in commercial "electronic nose systems" for analyzing the aroma or the odor of a sample. In many practical situations the human olfactory sense is used to assess subjectively for example foodstuffs, e.g. coffee, beer, tobacco, etc., or to design cosmetic products. The use of multiple odor sensors is helpful to produce "quantitative odor fingerprints". Sensors used in these systems are MOS (Metal Oxide Sensors), TSM- or QCM -(quartz crystal microbalance) sensors, SAW-devices and conducting polymer sensors. The signals from the sensors or sensor arrays can be passed to software package such as chemometrics or neural networking using both parametric and nonparametric pattern recognition techniques including principal components, discriminant or cluster analysis and artificial neural networks. The HKR-SENSORSYSTEME (Germany) enterprise offers a system with 6 QCM-sensors for gas analysis. ALPHA M.O.S. (France) is on the market with the FOX 2000 system. The basic detection system is made of semiconducting sensors and can be upgraded by SAW-sensors. The RESEARCH CENTER KARLSRUHE (Germany) has developed a gas detection system, SAGAS, on the basis of 8 SAW-sensors (fundamental frequency 433 MHz). It has now been introduced on to the market. The TOYO corporation (Japan) sells the TOYO SF-105 fragrance sensor as a single sensor with electronic unit. The newest product on the market with QCM-sensors is the modular system MOSES II made by the company LENNARTZ (Germany). These listed products complete other odor systems or are in competition with the NOSE 4000 system (NEOTRONICS Scientific, USA) or the products of AROMAS-CAN (UK). Both use conducting polymer sensor arrays.

To sum up it can be predicted that "electronic nose systems" in different versions on the basis of acoustic microsensors and other sensors will play an increasing role in the future because there is a need in many branches of industry, medicine, environmental techniques etc., for detecting odors. The main task nowadays is the improvement of the sensor properties.

5.6 Other Ultrasonic Sensor Principles and Applications

Since the quartz crystal is an anisotropic medium, the resonance frequency of a plate is highly dependent on the crystallographic orientation of the plate, the so-called angle of cut. This was already used in the early years of quartz crystal frequency control to produce crystal plates with vanishing temperature coefficient (AT- and BTcut). The frequency versus temperature behavior can always be approximated by a third-order polynomial

$$\frac{\Delta f}{f_0} = a_0 + a_1\Delta T + a_2\Delta T^2 + a_3\Delta T^3 \tag{5.29}$$

where $\Delta f = f - f_0$ and $\Delta T = T - T_0$, f and f_0 are the plate resonance frequencies at the temperatures T and T_0, respectively, and a_0, a_1, a_2, a_3 are the zero, first second, and third order temperature coefficients (TCs). In 1962 the first systematic approach to promote the temperature sensitivity of the frequency for a temperature sensor was published [41]. Commercial temperature sensors were first introduced by HEWLETT-PACKARD [42]; they used a Y-cut and the third order TC of about 35 ppm. Later a low-cost digital sensor system by [43] followed with a HT-cut quartz temperature sensor based on a slightly singly rotated Y-cut ($\Theta = -4°$). It worked at 16 MHz and had a frequency versus temperature sensitivity of 90 ppm/°C. This system was produced as the QuaT system by HEREUS (Germany) but it has not achieved great market dominance.

Quartz thickness shear resonator disks were also used for force, stress and pressure transducers [44] but in practice the tuning fork sensor elements have more significance.

The exploitation of nonlinear effects in quartz resonators is a new field. This is shown in a paper [45] in which the influence of the electric field on X-cut quartz resonator behavior in the MHz range was investigated. It opens up new applications in a high-voltage environment.

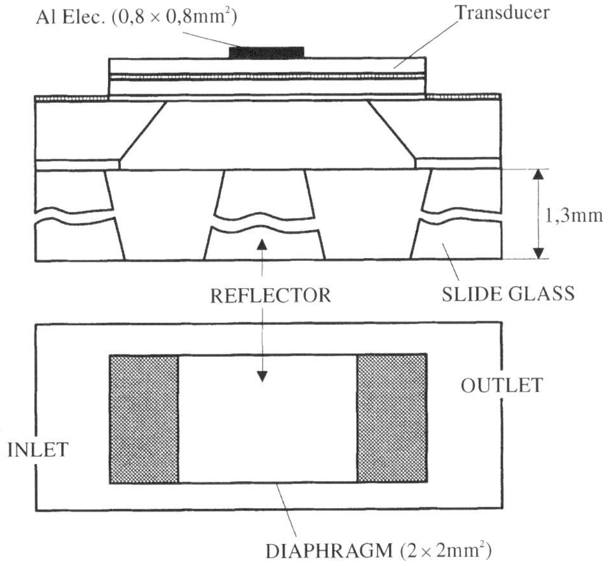

Figure 5.22 Miniaturized ultrasonic sensor system [46].

The main tendency in sensor development is miniaturization. First attempts have also been made to produce complete systems for ultrasonic velocity and attenuation measurements. In Figure 5.22 a miniaturized sensor structure is shown. It employs ZnO-film/Pyrex-glass composite diaphragm transducers. Note the low dimensions of the cell. The experimental results given in [46] are very imperfect but the attempt to design and build such a system is very remarkable. Such systems may be very interesting in the future for microreactors, medical applications, etc.

5.7 Summary

Ultrasonic sensors play an important role in many industrial applications as distance, proximity, level or flow sensors. By using new hardware and software techniques the accelerated development of new sensors and the improvement of known sensors on this field can be predicted. The use of ultrasonic state sensors is only beginning. New initiatives can be expected from new ideas for measurements in multicomponent systems. Acoustic microsensors are just starting to be used in industrial applications. Acoustic microsystems will play a very important role in the future.

References

[1] Krautkrämer, J., Krautkrämer, H., *Werkstoffprüfung mit Ultraschall*, Berlin: Springer Verlag, 1986.
[2] Wells, P. N. T., *Ultrasonics*, **30/1** (1992), 3–7.
[3] Kudruff, H., *Physik des Ultraschalls*, Stuttgart, Hirzel Verlag, 1988.
[4] Hauptmann, P., *Sensoren-Prinzipien und Anwendungen*, Munich, Hanser Verlag, 1991, p. 137.
[5] Lerch, R., Landes, H., Kaltenbacher, M., Lindiger, F., *Proc. SENSOR 95*, ACS Organisation, Wunstdorf, Germany (May 1995), 297–303.
[6] Püttmer, A., Krause, O., Lucklum, R., Henning, B., Hauptmann, P., *IEEE Transactions on Ultrasonics, Ferroelectrics, and Frequency Control*, **43**(1) (1997), pp 60–66.
[7] Bergmann, L., *Der Ultraschall und seine Anwendungen in Wissenschaft und Technik*, 6th edn. Stuttgart: Hirzel-Verlag, 1954.
[8] Lynnworth, L. C., *Ultrasonic Measurement for Process Control*, New York, Academic Press, 1989.
[9] Gätke, J., *Akustische Strömungs- und Durchflußmessung*, Berlin, Akademie Verlag, 1991.
[10] McShane, J. L., *Ultrasonic Flowmeter, Flow*, Vol. 1, Parts, Wendt, R. E., ed.; Pittsburg: 1974.
[11] Magori, V., *Proc. IEEE International Ultrasonics Symposium*, Cannes, France (November 1994), 471–481.
[12] Jena, A. V., Magori, V., *VDI Berichte*, **939** (1992), 165–170.

[13] Jena, A. V., Magori, V., Rußwurm, W., *Sensors and Actuators*, **37–38** (1993), 135–140.
[14] *Automatisierungstechnische Praxis*, **30** (1988), 459.
[15] Kroemer, N., Jena, A. V., Vontz, T., *Technisches Messen*, **6** (1997), pp. 180–189.
[16] McIntyre, J., *CS* (February 1992), 31–34.
[17] Kroemer, N., Vossiek, M., Eccardt, P.-C., Magori, V., *Proc. Sensor 93*, ACS Organisation. Wunstdorf, Germany (November 1993), 113–122.
[18] Eccardt, P. C., Vossiek, M., Magori, V., *Proc. EUFIT'95*, Aachen, Germany (August 1995), 1–5.
[19] Bernst, T., Löschberger, J., Magori, V., *NTG-Fachberichte*, **93** (1986), 253–263.
[20] Magori, V., *Proc. 3rd Annual European Computer Conference* (1989), 3-21–3-26.
[21] Schoenwald, J. S., Martin, J. A., Ahlberg, L. A., *Proc. IEEE'82 Ultrasonics Symp.* (1982), 945–949.
[22] Vossiek, M., Magori, V., Ermert, H., *Proc. IEEE Ultrasonics Symp.*, Cannes, France, (November 1994), 1–5.
[23] Müller, R., *VDI Berichte*, **1255** (1996), 247–252.
[24] Chern, E. J., Djordjevic, B. B., *Materials Evaluation* **48** (1990), 481–485.
[25] Stabinger, H., *Meßtechnik Nachrichten*, **6** (1985), Graz, Austria.
[26] Zacharias, E. M., Parnell, R. A., *Food Technology*, **26** (1972), 160–166.
[27] Benecke, F. W., *Chemie Technik*, **4** (1996), 42–45.
[28] Henning, B., Lachmann, H., Afschar, A., Bellgardt, K. H., *Proc. INCOM 96*, Düsseldorf, Germany (March 1996), 305.
[29] Püttmer, A., Hauptmann, P., Henning, B., Lucklum, R., *Sensors and their Applications VII*, Augousti, A. T., ed., Bristol/Philadelphia Institute of Physics, 1995, 141–145.
[30] Allegra, J. R., Hawley, S. A., *J. Acoust. Soc. Am.*, **51** (1972), 1545–1554.
[31] Harker, A. H., Temple, J. A. G., *J.Phys. D: Appl. Phys.*, **21** (1988), 1576–1588.
[32] McClements, D. J., Povey, M. J. W., *J. Phys. D: Appl. Phys.*, **22** (1989), 38–47.
[33] Pendse, H. P., Sharma, A., *Part. Part. Syst. Charact.*, **10** (1993), 229–233.
[34] Hale, J. M., *Ultrasonics*, **26** (1988), 356–357.
[35] McClements, D. J., Fairly, P., *Ultrasonics*, **29** (1991), 58–62.
[36] Adamowski, J. C., Buiochi, F., Simon, C., Silva, E. C. N., Sigelman, R. A., *J. Acoust. Soc. Am.*, **97**(1) (1995), 354–361.
[37] Püttmer, A., Hauptmann, P., Lucklum, R., Henning, B., *TRANSDUCER'97*, Chicago, USA, June 1997.
[38] Seifert, F., Bulst, W. E., Ruppel, C., *Sensors and Actuators*, **A44** (1994), 231–239.
[39] Grate, J. W., Martin, S. J., White, R. M., *Analytical Chemistry*, **65** (1993), 940A–948A, 987A–996A.
[40] Ballantine, D. S., White, R. M., Martin, S. J., Ricco, A. J., Zellers, E. T., Frye, G. C., Wohltjen, H., *Acoustic Wave Sensors*, San Diego, Academic Press, 1997.
[41] Wade, W. H., Slutsky, L. J., *Rev Sc. Instr.*, **33** (1962), 212–213.
[42] Hammond, D. L., Benjaminson, A., *Instr. Control Systems*, **38** (1962), 115.
[43] Ziegler, H., *Sensors and Actuators*, **5** (1984), 169–178.
[44] Dauwalter, C. R., *Proc. 26th Ann. Frequ. Contr. Symp.* (1972), 108–112.
[45] Bauerschmidt, P., *VDI Fortschrittsberichte 522*, Düsseldorf, VDI-Verlag, 1995.
[46] Hashimoto, K., Ienaka, T., Yamaguchi, M., *Proc. TRANSDUCER'93*, Yokohama, Japan (June 1993), 700–704.

6 Integrated Capillary Electrophoresis for Chemical Analysis

Holger Becker, Andreas Manz, Zeneca/Smithkline Beecham, Centre for Analytical Sciences, Department of Chemistry, Imperial College, South Kensington, London, SW7 2AY, UK

Contents

6.1 Introduction

The demand for analysis of samples with respect to their chemical composition and its quantification has risen enormously in the past few years. This demand does not have a single source, but comes from many different directions. Tighter regulations on pollution levels ask for online process control and environmental data acquisition, pharmaceutical and agrochemical research asks for rapid screening and sequencing methods and "smart machines" are to be equipped with artificial senses, for example drug or explosive sniffing robots need artificial noses. All this adds up to a big challenge for the field of analytical sciences, which tries to solve these problems with various interdisciplinary approaches, ranging from the physics of sensors together with the classical techniques of analytical chemistry to biochemical molecular recognition methods.

Amongst the separation methods of analytical chemistry, capillary electrophoresis (CE) has attracted much attention in recent years. Despite the fact that some principles of the behavior of molecules in a liquid under the influence of an electrical field have been known for more than a century [1] and the first CE apparatus was described in 1967 by Hjertén [2], it was not until the 1980s that CE was established as a standard technology in analytical sciences, and it took until 1989 before the first commercially available apparatus appeared. In recent years, it has evolved into a complementary method to high-performance liquid chromatography (HPLC) and gas chromatography (GC), the workhorses of most analytical laboratories. A particularly interesting feature of CE is its potential for a massive reduction in size, therefore becoming a central method for miniaturized total analysis systems (μ-TAS) [3, 4].

In this article, we will focus on this aspect of CE. We will give a short review of the principles of the method, with the following section devoted to an introduction to the main variables used in describing the performance of a CE system. We will then look into the properties and advantages of the miniaturization of analytical systems and describe the microsystem technology involved in the production of CE devices. Later sections are dedicated to the description of existing miniaturized CE systems and the experimental results obtained. In the concluding section we try to outline some of the future trends and developments in chip-based analytical systems.

The schematic setup of a CE experiment can be seen in Figure 6.1. Being a separation method, CE makes use of the fact that different species of the sample ions have different mobilities under the application of an electrical field. Thus, if a capillary connects two reservoirs of electrolyte buffer solution and has the sample injected, the application of an electrical field E will lead to a movement of the individual sample molecules, depending on their charge. The resulting linear velocity v_i of a given species i with the net charge $z_i e$ is given by

$$v_i = \mu_i E, \tag{6.1}$$

where

$$\mu_i = \frac{z_i e}{f_i} \tag{6.2}$$

is the electrophoretic mobility of the species (units: $cm^2\ V^{-1}s^{-1}$; typical values for singly charged ions are of the order 10^{-4} in these units) and f_i is its friction coefficient. The friction coefficient will generally depend on the size and shape of the molecule in its solvated state, which in a buffer solution can depend on variables such as ionic strength and the chemical nature of the buffer solution. In the limiting case

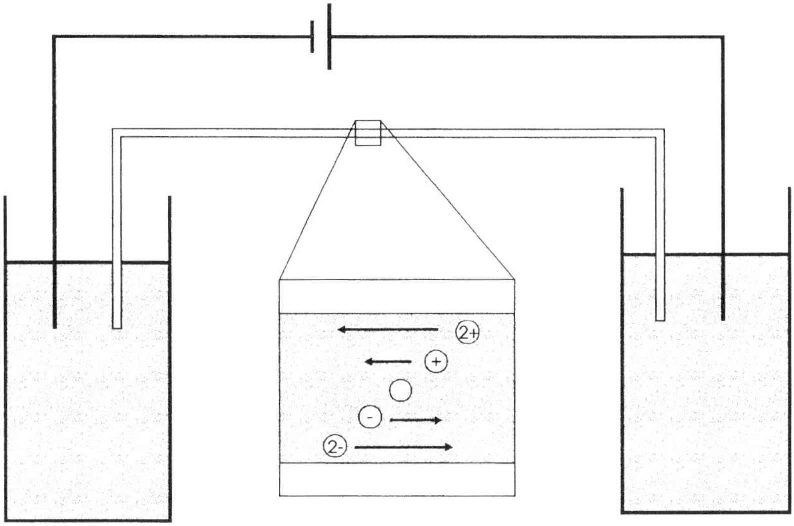

Figure 6.1. Diagram illustrating electrophoretic separation in a capillary. The circles represent molecules with their charges and their velocity vectors.

where Stokes law holds (a sphere with radius r moving in a homogeneous medium of viscosity η), it can be found to be

$$f_i = 6\pi r_i \eta. \tag{6.3}$$

As the forces acting upon the ions are the same in any place in the capillary, all ions of one species and charge move with the same velocity. They therefore move as a sample plug with a flat profile along the capillary. The dependence on the viscosity of the buffer medium can be eliminated by the use of gels, which act as a sieving medium, creating different migration velocities of different sized species due to their distribution in pore sizes.

A more detailed description of the principles of CE can be found in textbooks [5–8] and review articles [9–11].

A second phenomenon which can be observed in a capillary tube is electroosmosis. It is a macroscopic effect and results in a pumping action, when an electrical field is applied along the capillary. As shown in Figure 6.2, the ionized silanol groups (SiO^-) of the capillary walls form a charged surface layer and attract cations from the buffer. Thus, an electric double layer (the Helmholtz layer, or, more precisely, the Stern layer as the static layer and the outer Helmholtz plane as the mobile layer) is formed, consisting of the immobile negative charges at the wall and the positive ions in the solution. The thickness of this layer depends on the ionic concentration and is for low concentrations typically of the order of some hundred nanometers [6]. If an electrical field E is applied in the axial direction of the capillary, the positive ions in the Helmholtz layer will experience a force $F = \rho_e EV$ (V being the volume of the annular Helmholtz layer and ρ_e the charge density) and will start moving in a direction

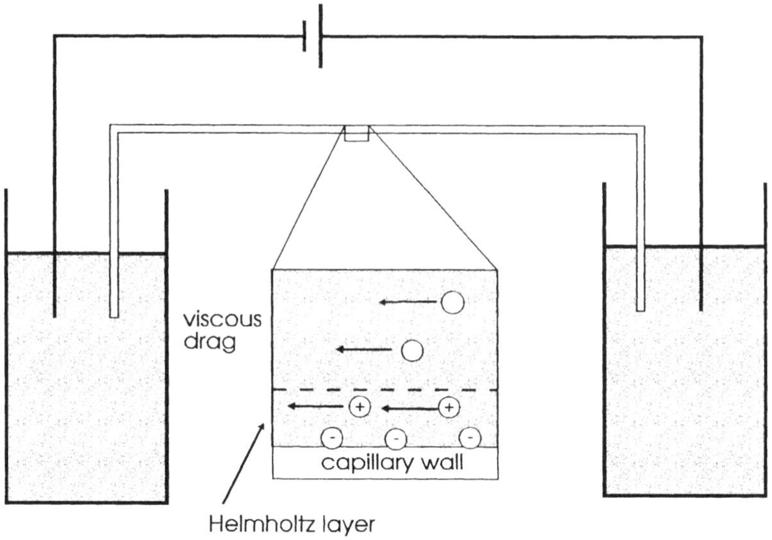

Figure 6.2. Diagram illustrating electroosmotic flow in a capillary. The circles represent molecules with their charges and their velocity vectors. The dotted line indicates the Helmholtz layer along the capillary wall.

according to this force. Due to viscous forces, the moving positive ions alongside the wall will start to drag the whole liquid column in the capillary along, leading within 100 µs to 1 ms [12] to the electroosmotic flow (EOF) through the capillary. Assuming the same viscosity η and dielectric constant ε in the buffer and the Helmholtz layer, the velocity v_{eo} of the fluid can be expressed as

$$v_{\mathrm{eo}} = \mu_{\mathrm{eo}} E, \tag{6.4}$$

with

$$\mu_{\mathrm{eo}} = \frac{\varepsilon\zeta}{4\pi\eta}, \tag{6.5}$$

where ζ is the zeta potential, the potential across the Helmholtz layer, which depends on the capillary material, the ionic concentration in the liquid and the pH value [13, 14]. In uncoated capillaries made out of fused silica or glass, v_{eo} is of the same order of magnitude as v_i, but can be manipulated over a wide range with capillary coatings or eliminated with gels. The total velocity v_i of a given species i is therefore given by

$$v_i = (\mu_i + \mu_{\mathrm{eo}}) E = \mu_{\mathrm{ce}} E = \mu_{\mathrm{ce}} \frac{U}{L}. \tag{6.6}$$

U is the applied voltage along a capillary of length L. Typical velocities lie in the range of 1–5 mm/s.

An important property of the EOF is its flat flow profile [15], as shown in Figure 6.3 in contrast to a parabolic flow profile in a pressure-driven flow system. Conventional

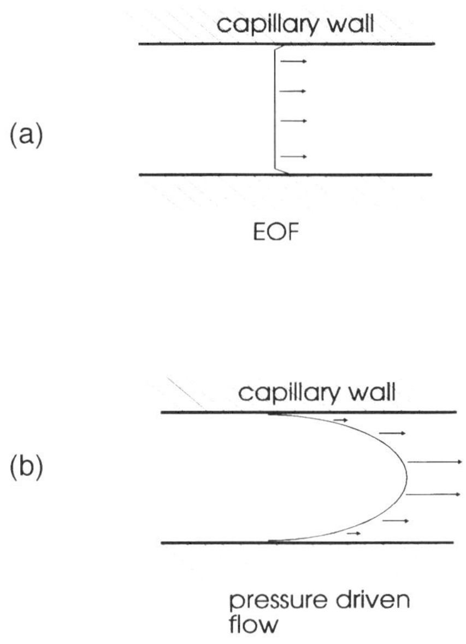

Figure 6.3. Flow profile (a) in a capillary in a system with electroosmotic flow (EOF) and (b) for comparison, a pressure driven system. The arrows indicate the velocity vectors.

pumps display a drastic increase in backpressure for channels with a diameter of some micrometers and some millimeters in length, while the EOF is almost independent of the capillary diameter. The existence of the EOF makes CE a particularly suitable candidate for miniaturization, as no pumps or other moving parts are necessary for liquid transport, the valve action can be substituted by switching voltages and the hydrodynamic dispersion is minimized [16]. An alternative method for valve-free liquid switching was proposed by Bevan and Mutton [17]. They use local rapid freezing of the capillary solution to control the liquid flow.

6.2 Theoretical Aspects of CE and its Miniaturization

If one assumes that the only contribution to band broadening is longitudinal diffusion, the sample concentration profile of a single species in a separation band is of Gaussian shape. In order to characterize the performance of the system, one can define variables for the separation efficiency (a measure for the total number of possible components) and for the resolution (i.e. the separation of the signals of different components). The third important parameter is the analysis time t, i.e. the time from sample injection until the information about its concentration is obtained.

The efficiency can be described in terms of the number of theoretical plates N, a description originally used in chromatography. To illustrate the concept, one can think of a beaker with water and an organic solvent. In a certain time after shaking the beaker, an equilibrium distribution between the two phases will be reached. This represents one theoretical plate. In chromatography, where a mobile phase is forced along a liquid stationary phase, many of these distribution equilibria are obtained along the column. In electrophoresis, this variable now has no such direct physical representation, but is a purely hypothetical value, defined by:

$$N = \frac{L^2}{\sigma_x^2}, \tag{6.7}$$

with σ_x^2 being the variance of the migrating zone width [6]. With the above assumptions, σ_x is given by:

$$\sigma_x = \sqrt{2D_i t} = \left(\frac{2D_i L}{\mu_{ce} E}\right)^{1/2}. \tag{6.8}$$

D_i is the diffusion coefficient of the component i. Inserting (6.8) into (6.7) and remembering that $E = U/L$ we obtain for N:

$$N = \frac{\mu_{ce}}{2D} U, \qquad \text{i.e.} \quad N \propto U. \tag{6.9}$$

In typical systems, N can reach values of several hundred thousands. But we have to keep in mind that the above estimate gives the upper limits of the plate number. Any other bandbroadening effect like injection, detection, wall interactions, interactions with micelles and particles, slow chemical reactions or complexation equilibria will result in an experimentally determined plate number which is lower than this theoretical limit. On the other hand there are possibilities to artificially increase the plate numbers for example by ways of sample stacking, isotachophoresis or isoelectric focussing. Other methods for expressing the separation efficiency are the plate numbers per unit time, N/t or the peak capacity n_c, the maximum number of separated peaks in an electropherogram. A variable to compare different CE systems is the height equivalent to a theoretical plane (HETP) H, which is the capillary length occupied by a single theoretical plate:

$$H = \frac{L}{N}. \tag{6.10}$$

The resolution R with respect to N, which can be expressed for a specified case, is given by [5]

$$R = \frac{\sqrt{N}}{4}\left(\frac{\Delta\mu_{app}}{\mu_{avg}}\right), \tag{6.11}$$

where $\Delta\mu_{app}$ is the difference in the apparent mobility of two separate components and μ_{avg} is the average mobility of the components.

We now have to look into the practical limitations of these quantities. By drawing a current I through the capillary, Joule heat is generated. This heat has to be carried away by active or passive cooling as otherwise the buffer will heat up, leading to bandbroadening due to a radial temperature gradient in the steady state, and in the end either to sample decomposition or column breakdown due to boiling. It is practical to describe the Joule heating P per length L of the capillary

$$\frac{P}{L} = \frac{UI}{L} = \frac{U^2}{LR} = C_{\text{max}}. \tag{6.12}$$

C_{max} is the maximum permissible power dissipated in the capillary and is typically of the order of 1 W/m. R decribes the resistance of the liquid column in the capillary and is given in the case of a circular capillary with diameter d and filled with a liquid of conductivity σ by

$$R = \frac{4L}{\pi d^2 \sigma}. \tag{6.13}$$

Substituting (6.13) in (6.12) and resolving the equation for U, we obtain an expression for the relation of the maximum applicable voltage along a capillary of given geometry (Length L and diameter d):

$$U = \sqrt{\frac{4\,C_{\text{max}}}{\pi\sigma}} \cdot \frac{L}{d}, \quad \text{i.e.} \quad U \propto \frac{L}{d}. \tag{6.14}$$

Recalling (6.9), we therefore get for the separation efficiency (number of theoretical plates):

$$N \propto \frac{L}{d}. \tag{6.15}$$

The analysis time t is given by:

$$t = \frac{L}{v_i} = \frac{L^2}{\mu_{\text{ce}} U}. \tag{6.16}$$

The second fraction has been obtained by inserting (6.6). Combining (6.14) and (6.16), we obtain for t:

$$t = \left(\sqrt{\frac{4\,C_{\text{max}}\,\mu_{\text{ce}}^2}{\pi\sigma}}\right)^{-1} \cdot L \cdot d, \quad \text{i.e.} \quad t \propto L \cdot d. \tag{6.17}$$

We now have the two important results motivating the miniaturization of CE systems: in order to increase the separation efficiency, one can either increase the capillary length or decrease its diameter. Looking at (6.17), it becomes clear that a decrease in capillary diameter also leads to a faster analysis. These two developments can be expressed by a single equation:

$$\frac{N}{t} \propto \frac{1}{d^2}. \tag{6.18}$$

The potential advantages of smaller size systems in terms of performance now becomes apparent: a faster analysis with an increase in separation efficiency is obtained by reducing the capillary diameter, something that can easily be done by modern microfabrication methods.

Obviously, several drawbacks arise from miniaturization as well. The problems of small volume injection and detection, where the sample volume is much smaller than the peak volume, have to be tackled. For most detection systems, the sensitivity decreases rapidly as the number of sample molecules is reduced, so the detector signal decreases proportional to d^3. With small size microchannels, the risk of clogging has to be counterreacted by a suitable sample pretreatment; usually some sort of filtering has to be involved.

6.3 The Concept of Miniaturized Total Analysis Systems

In addition to the above-mentioned theoretical improvements of a smaller analysis system there are further advantages of miniaturization. First of all, the amount of consumables needed, namely the buffer solution, is greatly reduced, leading to lower costs and minimized waste. Secondly, by having a system which can operate close to the location where the sample is originating, the overall cycle time for a complete analysis cycle (including sample extraction, possible pretreatment plus the aparatus analysis time itself) is greatly reduced. This can lead to a sensor-like application of the analysis system, but without the limitations connected with a chemical sensor. Chemical sensors are ususally designed to measure a single substance (as the chemical interface has to be specified), thereby lacking flexibility. Once a sensor is optimized, it is difficult to change to a new analyte or a new matrix. It is however well suited for a large number of analyses of a small number of analytes. Their lifetime is often limited as the active interface, membrane or other molecular recognition mechanism degrades and the selectivity achieved nowadays is often not satisfying. In terms of the efficiency as defined in Section 6.2, sensor extraction is a process with a theoretical plate number of $N = 1$. For industrial applications like online *in situ* measurements of chemical concentrations, extremly important for process control or monitoring purposes, "total analysis systems" (TAS), adapted from conventional instruments like gas chromatographs (GC), mass spectrometer (MS), total organic carbon measurements (TOC) or flow injection analysis (FIA) are used, where sensors are mostly integrated in the detection system. As stand-alone systems, sensors can be found in clinical applications as fiber-optic or pH-glass electrode sensors. With the idea of miniaturized total analysis systems (μ-TAS), introduced in 1990 [3], there exists a concept which combines the flexibility of the established benchtop systems of analytical chemistry with the advantages of the chemical sensor. Figure 6.4 graphically explains these concepts. A sensor, as depicted in Figure 6.4a, can ideally be immersed in the gas or liquid under investigation, and therefore needs no sample pretreatment and reacts directly to changes in the concentration of the detected substance. Its output is a simple optical (intensity) or electronic signal (e.g. voltage, current or frequency), therefore the data acquisition and interpretation is usually simple. With a modified standard stand-alone analysis system ("total analysis systems"), like a converted GC or MS apparatus (Figure 6.4b), a sample volume has to be taken from the location of interest. A suitable pretreatment is normally necessary and the output signal (e.g. from a fluorescence or an ionization detector) is

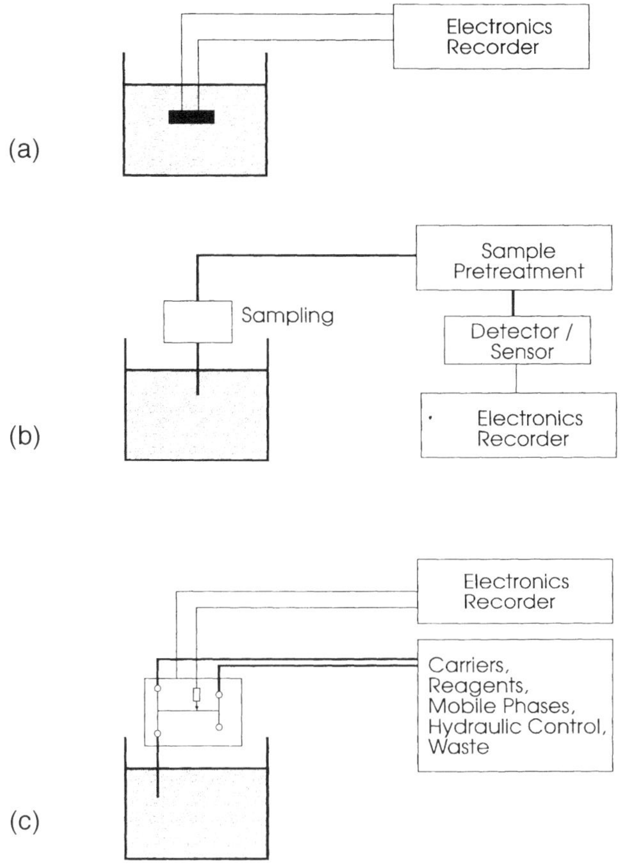

Figure 6.4. Conceptual comparison between (a) an ideal chemical sensor; (b) a total chemical analysis system (TAS);and (c) a miniaturized analysis system (μ-TAS).

more complex. A miniaturized analysis system (Figure 6.4c) would perform possible preparation and analysis steps like a sensor close to the sample, thus cutting the total analysis time due to a reduced sample transport length. Applications for such a system would include environmental monitoring and industrial process control, as well as fast screening sequences.

6.4 Microfabrication

The microfabrication process for μ-TAS chips makes use of planar chip technology, developed originally for the microelectronics industry [18]. Some of the original designs (see Section 6.5) have been made in silicon due to the familiarity with the

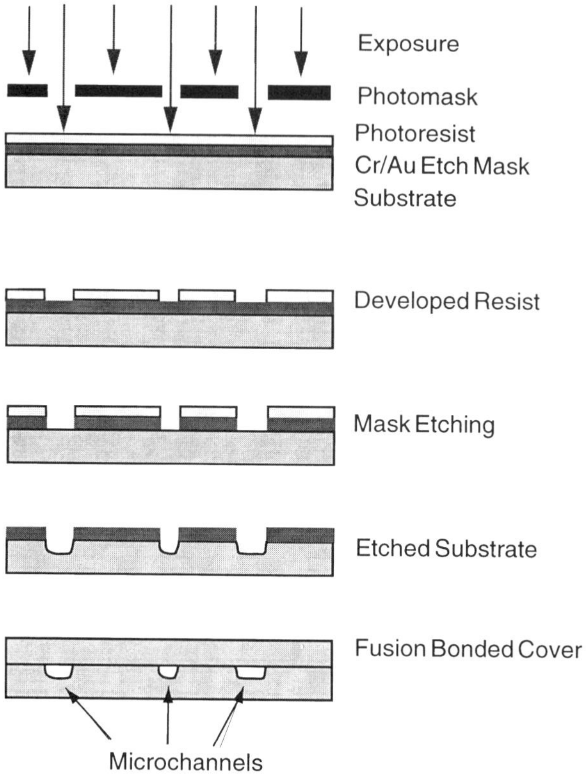

Figure 6.5. Process steps in a simple one-mask process for the micromachining of channels in a substrate.

machining processes as well as the easy availability of high-quality wafer material. CE chip fabrication relies on well-established processes including photolithography, wet etching or advanced micromachining technologies like reactive ion etching (RIE). In Figure 6.5, a typical fabrication process is depicted. Due to the applied high voltages in recent CE designs, usually glass (pyrex or borosilicate) or quartz is used as a substrate. It is covered with a metal film (usually chromium with a thin gold layer to promote adhesion) and a photoresist layer. With a photomask, produced with standard lithography processes, which contains the structural information of the device, the resist film is exposed and consecutively developed, removing the photoresist from the exposed areas. The substrate is then placed in an etching bath, which etches through the metal film in the areas not covered by photoresist. In a second etch step, the substrate itself is etched, typically in HF/HNO_3 or KOH. Depending on the etchant and the substrate, the microchannels form out. In glass or other amorphous material, one usually has isotropic etching conditions, yielding the same etch rate in any direction and therefore forming rounded channel edges. On a monocrystalline silicon or quartz substrate, anisotropic etching is possible with suitable echants, leading to a

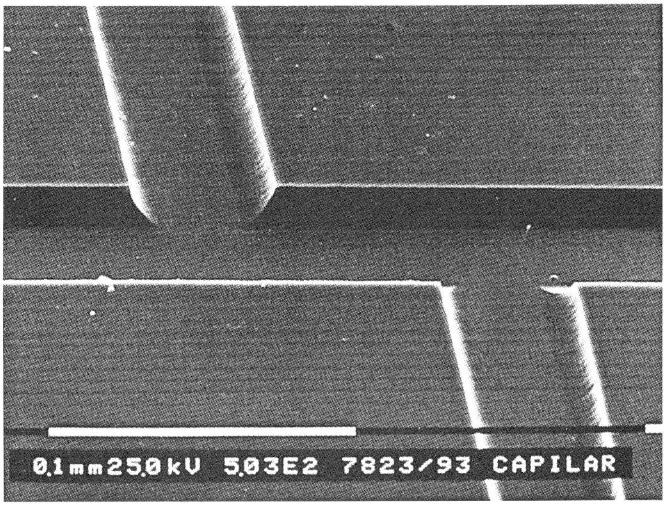

Figure 6.6. Electron micrograph of channels isotropically etched into glass.

channel profile which depends on the direction of the crystal cut. In the final step, a cover lid is fusion bonded on top of the substrate, closing the channels. This takes place in a furnace at temperatures above 500 °C for some hours, possibly assisted by an applied voltage (50–500 V, then called anodic bonding). Another possibility of closing the channels is to grow a top layer of SiO_2 on to the substrate. Typical channel dimensions are 20–70 μm width and 5–20 μm depth. An electron micrograph of a channel, isotropically etched in glass, is shown in Figure 6.6. Experiments show that the glass should be annealed (570 °C for about 6 h) prior to etching to ensure smooth channel walls, as otherwise internal stresses in the substrate due to its fabrication and polishing procedures lead to a higher etch rate along the stress lines, which then form crevasses at angles to the channel [19]. Further process steps, like the metallization of certain areas acting as electrical contacts or the formation of oxide or nitride layers, etc., can easily be included.

6.5 CE Device Designs

6.5.1 Early Developments

The concept of putting a CE system on a silicon wafer was first reported in a patent of Du Pont in 1990 [20]. In the patent, two separations of polynucleotides are shown but no subsequent publications from this source can be found. The first results published from a CE device on a planar chip were reported in June 1991 at the HPLC '91 conference in Basel and at Transducers '91 in San Francisco. The device was

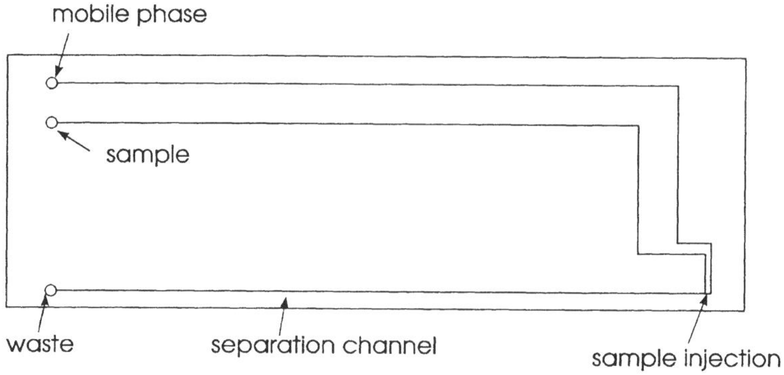

Figure 6.7. Layout of the first CE chip ("CETAS") (Ciba-Geigy). Adopted from [21].

designed by Manz et al. [21–23], produced at IMT Greifensee and experiments were carried out by Manz and Harrison at Ciba-Geigy, Basel, Switzerland [24, 25]. It was fabricated on glass and had an overall size of 15×4 cm (Figure 6.7). The flow system consisted of two channels with dimensions 30 μm (width) and 10 μm (depth) and one channel with a width of 1000 μm width and 10 μm depth, intersecting at one point. Detection was carried out by a laser fluorescence set-up at variable positions along the separation channel. The separation of two fluorescent dyes with the number of theoretical plates set at 18 000 within less than 8 minutes could be achieved.

An alternative CE chip was reported by Harrison et al. [26] at the University of Alberta. The chip layout is shown in Figure 6.8. The device was fabricated in ⟨100⟩ N-type silicon with various oxide and nitride layers for experiments to determine insulator breakdown measurements as a limit of applicable voltages in the separation column. The sample was introduced in one channel by applying voltages between points 1 and 2 and injected into the separation channel by switching on a high voltage

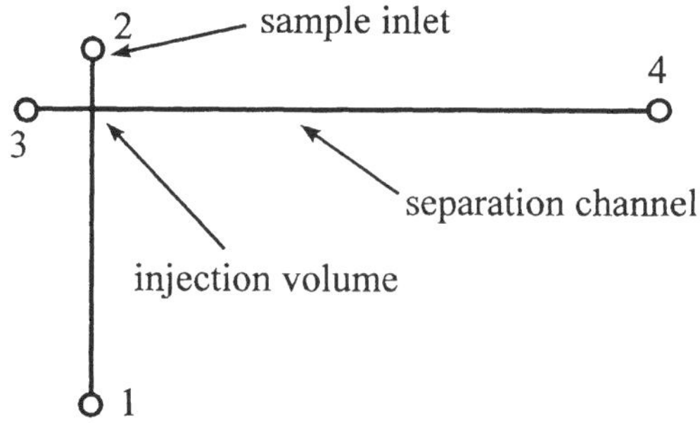

Figure 6.8. Layout of the JH-1 structure (University of Alberta). The numbers refer to liquid reservoirs. Adopted from [26].

between ports 3 and 4. By this means, a well-defined sample plug with the volume of the intersection region of the channels (approx. 300 pL) could be injected. Nominal channel width was 10 μm or 50 μm for all channels in one design type, or 10 μm for the separation channel and 50 μm for the injection channel in a second design. Channel depth was in the range of 5–15 μm. With this design, voltages of up to 1000 V could be applied to the channels, before a voltage breakdown in the silicon occured.

Pharmacia AB (Sweden) has produced a CE chip in a polymer material [27] by fabricating a silicon mold with the inverse of the desired structure, which can then be transferred into the polymer material. The channels with a length between 5 and 8 cm, width of 250 μm and depth of 50 μm form an S-shaped structure. The polymer slab containing the channel is clamped mechanically between two glass plates. With such a device, 6400–14 000 theoretical plates could be achieved for a separation of ϕX174 Hae III DNA restriction fragments in less than 30 minutes [28].

6.5.2 Advanced Designs

6.5.2.1 Ciba Geigy

Figure 6.9 shows a more recent design of a CE structure in glass as published in 1993 [29, 30]. The overall dimensions of the chip are 80 × 70 mm, the length of the separation channel is 50 mm, channel width is 50 μm, channel depth 12 μm. The design of the intersection region between the channels connection reservoirs 1–4 allows for three different sample volumes (sample plug length 150, 300 or 400 μm, corresponding to a sample volume of 90, 180 and 240 pl respectively) to be electroosmotically injected. To achieve well-defined sample plugs, the sample injection channel (connecting reservoirs 1 and 4) was homogeneously filled with the sample. Switching on the high potential along the separation channel between ports 2 and 5 and switching off the injection potential leaves the sample plug with the length

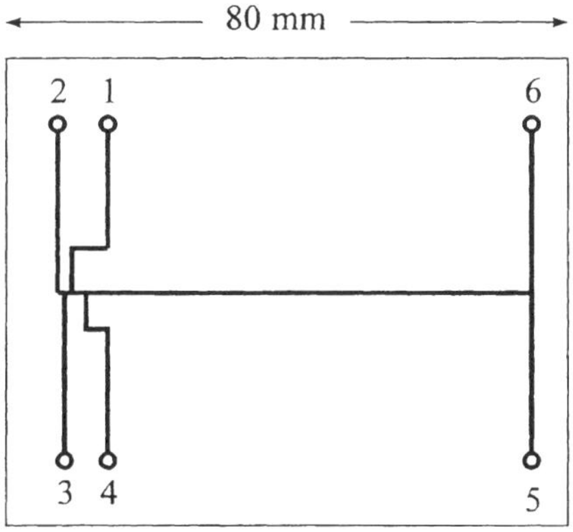

Figure 6.9. Layout of a CE chip with two different sample plug lengths (Ciba-Geigy). The numbers refer to liquid reservoirs. Adopted from [29].

of the lateral capillary distance in the separation channel ready for separation. A similiar structure was used for polyacrylamide gel electrophoresis [31].

6.5.2.2 Alberta

A design similiar to that of [24] (Figure 6.8), but fabricated in different types of glass (Pyrex, Corning, Tempax and Schott) has been presented in 1993 and was used to study the fabrication properties and the behavior of the liquid flow, particularly at the capillary intersections [32, 33]. Channel width was 30 μm for all channels in one design, or 30 μm for the separation channel and 70 μm for the sample inlet channel in a second design. Channel depth was always set to 10 μm. Another design developed from the one shown in Figure 6.8 is reported in [34], in which the channel width at the buffer inlet before the injection and for the separation waste (corresponding to ports 3 and 4 in Figure 6.8) after the detection zone has been increased to 300 μm to optimize the voltage drops along the capillaries, thus having a maximum field strength along the separation channel. In the same paper [34], a design with a meandring channel for mixing experiments is presented. Recently, a design in pyrex and fused silica with a postseparation reaction chamber was presented [35]. With this design it is possible to label the sample components after the main separation, directly before detection, instead of pretreating the sample in terms of fluorescence labeling.

6.5.2.3 Oak Ridge

In 1994 two main designs for separation chips, equipped partly with reaction chambers for pre- or postcolumn reactions, have been reported by the group at Oak Ridge National Laboratory. Figure 6.10 shows the central design elements for the CE glass chip [36, 37], which was modified to include an integrated precolumn reactor [38] or to allow for precolumn reactions [39]. The channels used have a width of approximately 75 μm at half depth and a depth of 7.5 μm with rounded channel edges due to the isotropic etching conditions. The overall chip dimensions are 50×25 mm. With these designs, high-speed separations and liquid flow charactersistics have been investigated. The second design, shown in Figure 6.11, makes use of a serpentine structure to achieve higher separation channel lengths, 165 mm [40–42] and 67 mm [43].

Designs nowadays have reached a stage where several of the larger instrument manufacturers are investigating this technology for possible commercialization.

6.5.3 Sychronized Cyclic Capillary Electrophoresis

A novel concept for repeated coloumn switching on a CE chip, named sychronized cyclic capillary electrophoresis (SCCE), was introduced by Burggraf et al. in 1992

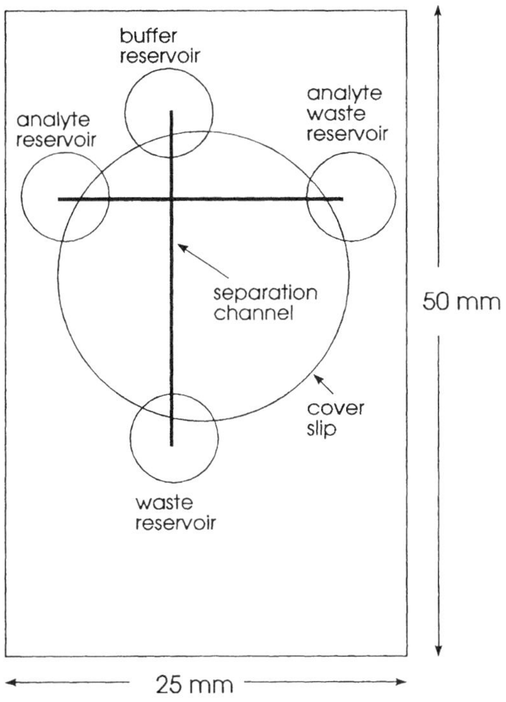

Figure 6.10. Basic layout of a CE chip from Oak Ridge National Laboratory. Adopted from [42].

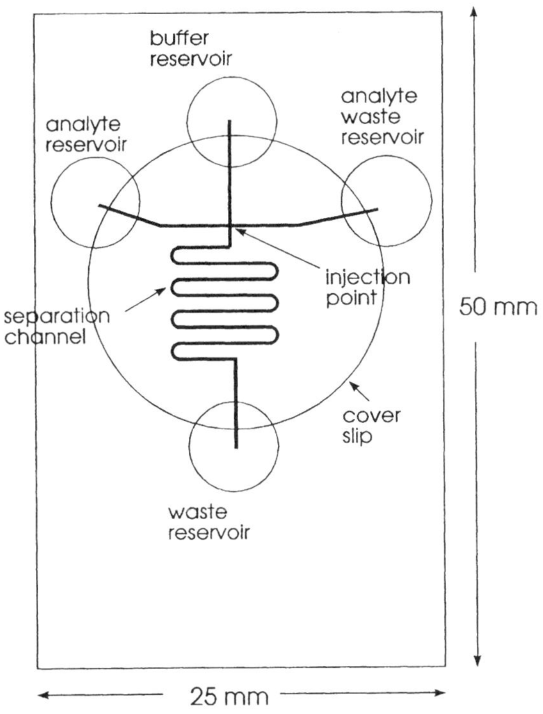

Figure 6.11. Design of a CE chip with a serpentine separation channel geometry for increased separation length (Oak Ridge National Laboratory). Adopted from [41].

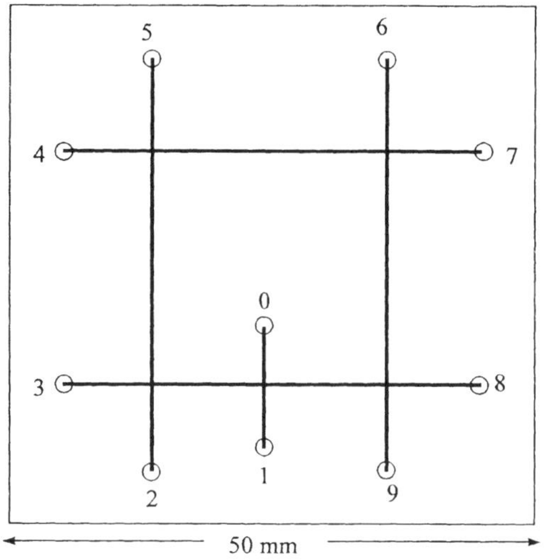

Figure 6.12. Layout of a SCCE chip. The injection channel is between reservoirs 0 and 1. Adopted from [44].

[44–46]. On a glass chip, microchannels (40 μm wide, 10 μm deep, 20 μm length of each channel) of a structure as shown in Figure 6.12 were fabricated. The basic idea is to eliminate unwanted sample components during the separation by synchronizing the HV switching of the different channels with the position of the component to be analyzed and at the same time, extract all other components into waste reservoirs [47]. In addition, the separation efficiency can be increased by the repeated passage of the applied voltage. The necessary steps are explained in Figure 6.13: A mixture of components (labelled 1, 2, 3 in the figure) is injected (Figure 6.3a) and starts to seperate after application of the electric field (Figure 6.3b). The HV switching is now synchronized with component 2, so as to remove the faster moving component 3 (Figure 6.3c). The separation of components 1 and 2 continues. Shortly before the slow component 1 reaches the lower right corner, the HV is swichted again, reversing the migration direction for component 1, but maintaining the direction of component 2. A third switching of the HV leads to a dumping of component 1 into one waste reservoir. If better separation is necessary, e.g. if the mobilities of the sample components are not very different, this cycle can be repeated until the necessary separation is achieved. This simulation of a capillary with theoretically infinite length leads to very high plate numbers (1200 000 after 20 cycles reported in [44]) even with a modest voltage (2 kV), as the ions are exposed twice to the applied voltage per cycle. The main problem of this design is a significant loss of material at the channel intersections after five to ten cycles, which leads to a signal decrease. With the advanced design by Burggraf, having T-shaped channel intersections, these losses could be reduced [48, 49].

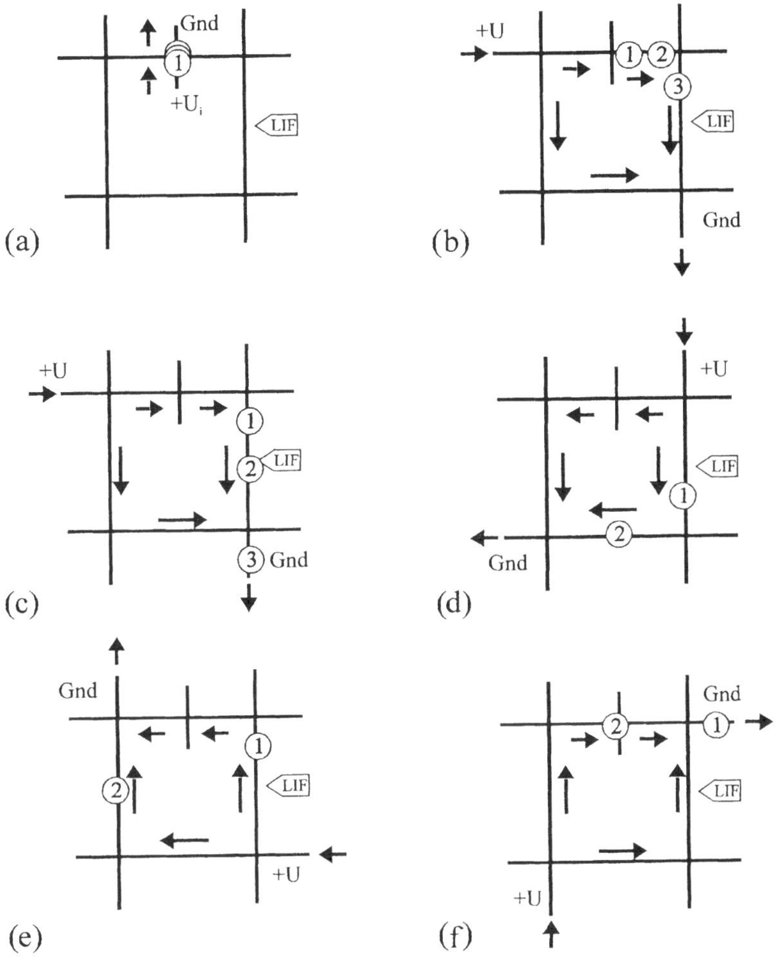

Figure 6.13. The principle of SCCE. Three sample components (indicated by the circles labeled 1, 2 and 3) are injected in (a), the voltage switching is synchronized to component 2. In the first phase of a cycle ((b) and (c)), component 3 is extracted, the separation continues ((d) and (e)), until component 1 can also be eliminated (f) and one cycle is finished.

6.6 Experimental Set-up

A typical μ-TAS CE experiment is depicted in Figure 6.14. Fluid reservoirs, usually plastic vials or pipet tips, are glued onto the chip. Through holes drilled into the cover plate (diameter typically about 1 mm) they are connected to the microchannels of the CE chip. Platinum electrode wires are placed into these reservoirs and connected to a HV power supply via a fast HV relay. For the sake of clarity, the

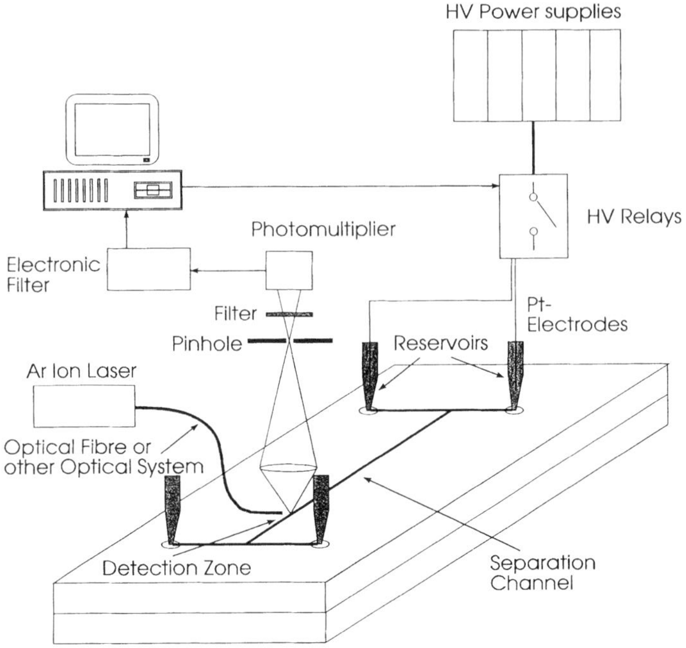

Figure 6.14. Setup principle of a miniaturized CE experiment. The main components are the fluid management with reservoirs and the chip itself, the detection unit with laser induced fluorescence detection and the computer-controlled HV units for injection and separation voltages.

electrodes connected to the lower reservoirs in the diagram are omitted. Detection is usually carried out with laser-induced fluorescence, where light from an Argon ion (514, 488 or 275 nm), He-Ne (594 nm) or He-Cd laser (325 nm) is coupled into the detection zone via an optical fiber. Recently, blue LEDs have also been used as a light source [50] for fluorescence excitation, indicating the possiblity of integrating the detection system on to the chip. The excited fluorescence is sampled, either directly with a microscope lens or indirectly by an optical fiber, filtered to suppress the light of the laser and registered in a photomultiplier tube or a CCD camera. Another detection method is realized by optical absorbance measurements, as used frequently in conventional HPLC systems. As the sample volumes in CE devices are of the order of one nanoliter or less, care has to be taken to optimize the coupling of the light in and out of the sample volume. Special flow cells have been designed to maximize the optical path length with respect to the detection volume, either by using multiple reflections of the channel walls [51, 52] or by probing the capillary in the longitudinal direction [53]. In early chip designs, platinum electrodes have been included in the chip [22] for electrochemical detection, but the requirements of floating the detector electronics on the HV level of the separation make this approach difficult.

6.7 Experimental Results

One of the prime motives for miniaturizing CE devices is the increase of analysis speed. In the examples given in this chapter, separations within seconds are shown, in contrast to typically 8–12 minutes analysis time for conventional CE.

In Figure 6.15 we show a separation of six fluorescein isothiocyanate (FITC) labeled amino acids, obtained by Manz and Effenhauser with the chip shown in Figure 6.9 [29]. The field strength was 1060 V/cm, corresponding to an applied voltage of 11 250 V between reservoirs 2 and 5. The sample consisted of a mixture of the aminoacids arginine (Arg), glutamine (Gln), phenylalanine (Phe), asparagine (Asn), serine (Ser) and glycine (Gly), each with a concentration of 10 µm in a buffer solution of 20 mM boric acid–100 mM Tris of pH 9.0. The sample volume was 100 pL. As can be seen from the figure, with the detector placed at a position corresponding to a separation length of 24 mm (top figure), all six of the amino acids

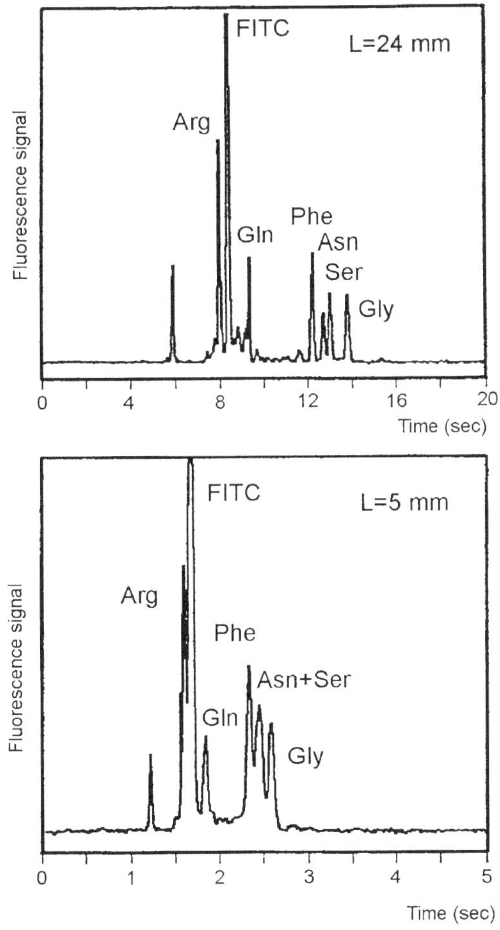

Figure 6.15. Electropherograms of a mixture of FITC-labeled amino acids obtained with a chip of the design shown in Figure 6.9. In the top figure, the detection took place after a separation length of 24 mm, in the bottom figure at a separation distance of 5 mm. Reprinted with permission from [29]. Copyright 1993 American Chemical Society.

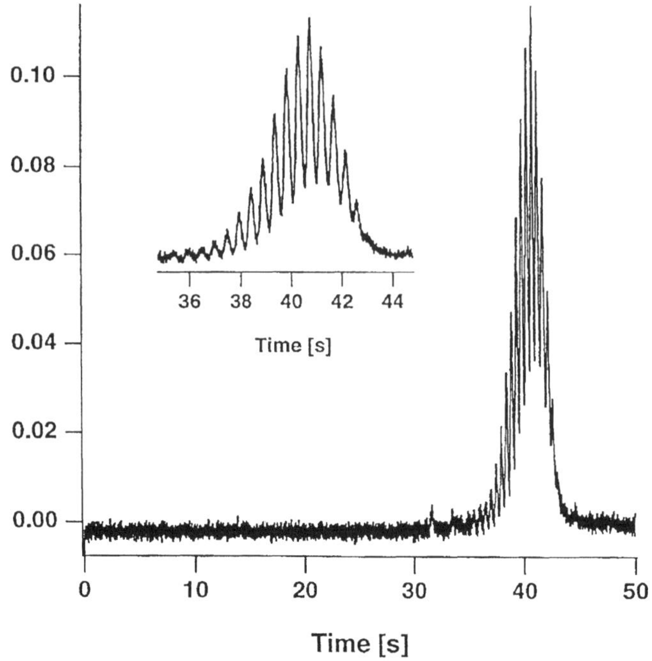

Figure 6.16. Electropherogram of a FITC-labeled oglionucleotide mixture. The inset shows the peaks with an expanded time axis. The different base lengths are clearly resolved. Reprinted with permission from [31]. Copyright 1994 American Chemical Society.

could be separated within 14 s. The plate numbers obtained for this experiment range from 45 000 for Gly-FITC to 75 000 for Gln-FITC corresponding to a plate height of 0.5 µm and 0.3 µm respectively. With the separation length set to 5 mm (bottom figure), baseline separation is not achieved for all peaks, nevertheless five of the six amino acids are resolved within an analysis time of 2.5 s. The maximum plate number in this case was 5800 for Gln-FITC. The highest plate number recorded with this device was 150 000 for Gly-FITC at a separation length of 50 mm. Another example is given in Figure 6.16, obtained with a similar device [31]. In this case, a synthetic mixture of fluorescent phosphorothioate oligonucleotides with lengths between 10 and 25 bases (PS $pd(T)_{10-25}$) in a buffer consisting of 100 mM Tris, 100 mM boric acid, 2 mM EDTA and 7 M urea with a pH of 8.5 was analyzed. The capillary was filled with a polyacrylamide gel, as used in conventional gel electrophoresis. With a field strength of 2000 V/cm, a separation efficiency of about 200 000 plate numbers in 45 s at a separation length of 3.8 cm could be achieved, corresponding to a plate height of less than 0.2 µm.

An example reported by Harrison et al. for separation in a pH 9.7 carbonate buffer, where the labeling of the amino acid mixture with *o*-phthaldialdehyde (OPA) took place after the separation, is shown in Figure 6.17 [35, 54]. From this figure, obtained from a sample volume of 60 pL, it is obvious that the efficiency of the separation is

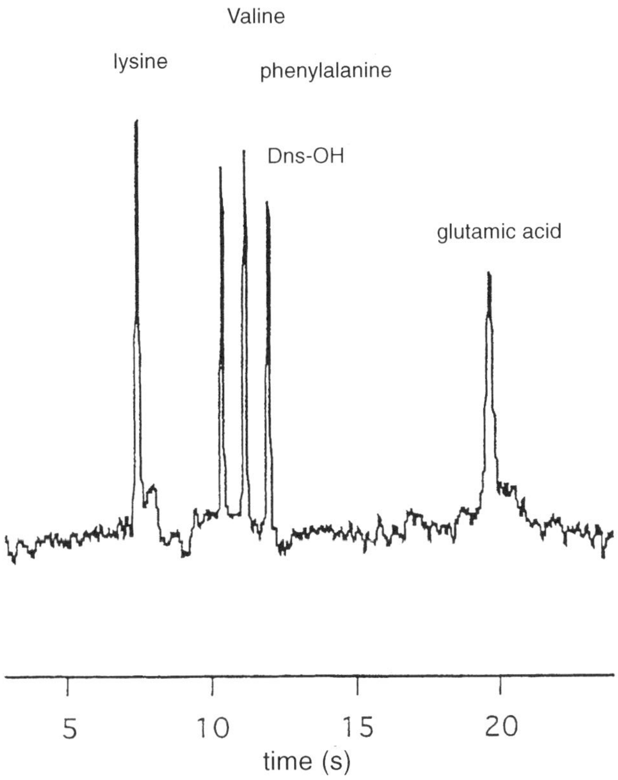

Figure 6.17. Electropherogram of amino acids which have been labeled after the separation process. From [54].

not significantly reduced by the following labeling reaction, considered an optimized design of the mixing/reaction chamber. The number of plates was estimated to be about 90 000, corresponding to a plate height of about 0.5 μm.

Very fast separations in the millisecond range have been reported by Ramsey and Jacobson [36] with a chip as shown in Figure 6.10. In Figure 6.18, a separation of the two dyes rhodamine B and disodium fluoescein with a concentration of 20 μM in a sodium tetraborate buffer (4 mM, pH 9.2) with a sample plug of volume of about 250 pL is shown. With an applied field strength of 1.5 kV/cm in less than 150 ms a separation could be achieved for a separation length of 0.9 mm. This corresponds to the number of plates equal to 130, increasing to 15 000 if a separation length of 11.1 mm is chosen, for which the separation takes about 1.6 s.

SCCE with the structure shown in Figure 6.12 has been performed in free electrolyte on fluorescein [46]; with a modified chip design micellar separations on a mixture of FITC labeled amino acids [48], and polyacylmide gel electrophoresis on oligonucleotides with base lengths of 2 to 10 bases [49] have been carried out. In Figure 6.19, the separation of FITC labeled amino acids over two cycles is shown.

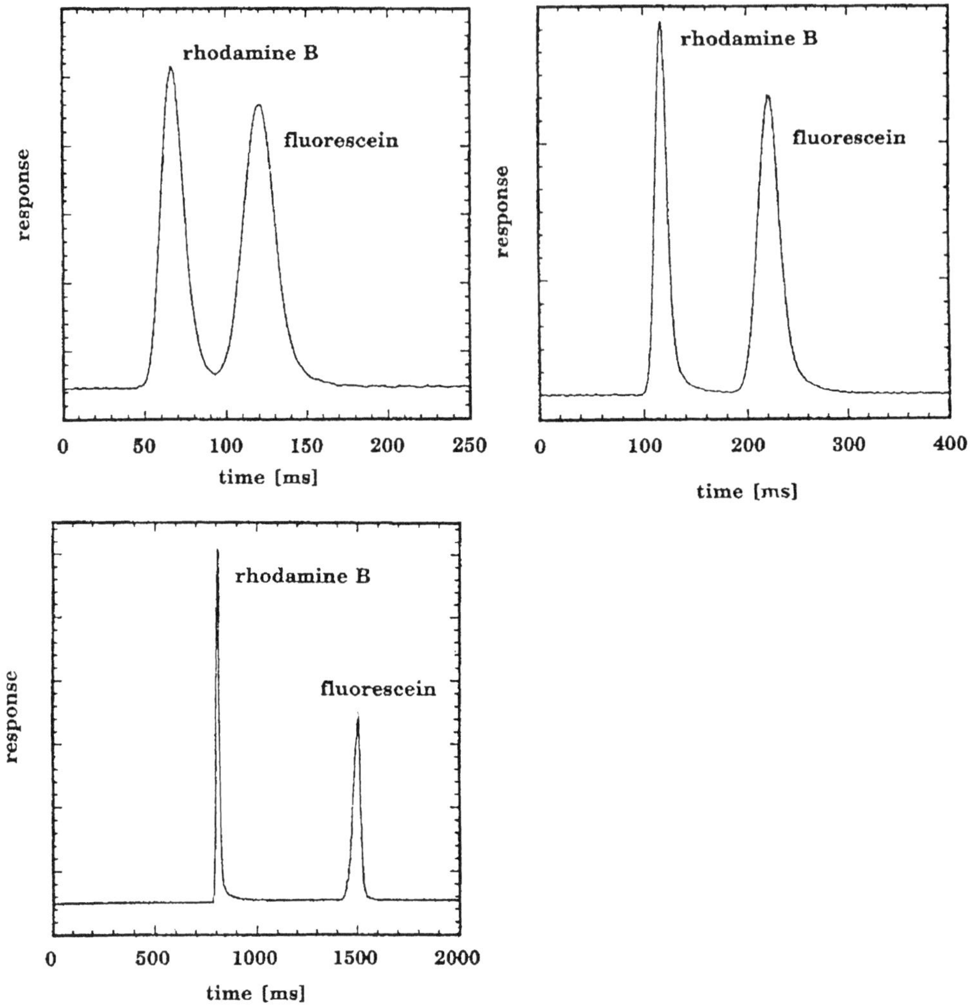

Figure 6.18. Fast separations of fluorescein and rhodamine B with a chip as in Figure 6.10. The separation lengths for the figures are 0.9 mm (top left), 1.6 mm (top right) and 11.1 mm (bottom). Reprinted with permission from [36]. Copyright 1994 American Chemical Society.

Two main features can clearly be seen. As expected, the separation efficiency increases with the cycle number. The peaks of the two amino acids arginine (Arg) and glutamine (Gln) are much more widely seperated in the second cycle than in the first. At the same time, however, the peak height is greatly reduced. This is due to leakage of the sample into the waste channels each time a corner of the structure is passed by the sample. With a different channel intersection layout and suitable countervoltages, this can be avoided [48, 49]. In these examples, SCCE is used in a sensor-like fashion

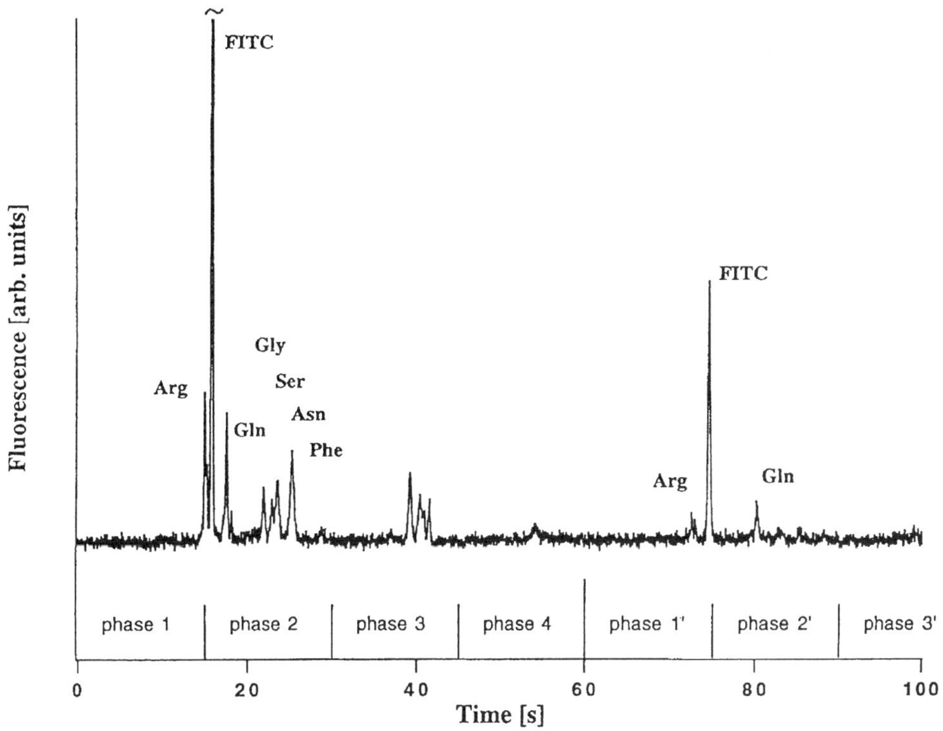

Figure 6.19. Separation of FITC-labeled amino acids with SCCE (for Chip design see Figure 6.12). The first cycle is finished after 60 s and two remaining amino acids enter a second separation cycle.

for analyzing a single component. While a chemical sensor features direct transduction and a fast response but lacks flexibility, SCCE is experimentally more complicated and slower but offers a high degree of flexiblity, as any component can be selected for measurement simply by switching the synchronization times.

In Table 6.1, a compilation of the separation properties of the various planar chip designs is shown.

6.7.1 Detection Limits

As we deal with detection volumes of the order of 1 nL to pL, the question arises of the feasibility of detection of sample molecules in such small volumes. For laser induced fluorescence detection (LIF) as the most widely used technology, detection volumes of 1 aL can be optically defined without too much experimental effort by focussing the excitation beam and using a confocal arrangement for collecting light for detection. The second limitation arises from the possibility of having small

Table 1. Separation properties of planar chips CE devices.

Reference year	Chip design	Separated mixtures	Field strength [kV/cm]	Separation length [mm]	Plate number	Plate height [μm]
Harrison, Manz, *et al.* [21], 1991	Fig. 6.7	Calcein fluorescein	0.2	65	18 000	3.6
Seiler, Harrison, Manz, *et al.* [55], 1993	Fig. 6.7	FITC-Arg FITC-Phe	0.655	96	70 000 (Arg)	1.5
		FITC-Gln			50 000 (Phe)	2
Effenhauser, Manz, *et al.* [29], 1993	Fig. 6.9	FITC-Arg, Gln, Phe, Asn, Ser, Gly	1.06	24	45–150 000	0.3–0.5
				5	5800	1
Effenhauser, Manz, *et al.* [31], 1994	Fig. 6.9	PS pd $(T)_{10-25}$	2	38	$\approx$200 000	0.2
Jacobson, Ramsey, *et al.* [36], 1994	Fig. 6.10	Rhodamine B	1.5	0.9	130	7.5
		fluorescein		1.6	450	3.6
				11.1	15 100	0.75
Jacobson, Ramsey, *et al.* [39], 1994	Fig. 6.10 +pre-column reactor	OPA-Arg	1.8	10	450	23
		OPA-Gly			280	36
Jacobson, Ramsey, *et al.* [38], 1994	Fig. 6.10 +post-column reactor	OPA-Arg	0.8	13.5	1320	10.2
		OPA-Gly			330	41
		OPA-Arg			1300	10.4
		OPA-Thr			2460	5.5
Moore, Ramsey, *et al.* [42], 1995	Fig. 6.11	Coumarin 440/450/460	0.047	21.3	$\approx$3750	$\approx$ 6
		(micellar CE)	0.5	13	3200–8800	1.5–4
Harrison, Fluri, *et al.* [54], 1996	Fig. 6.8 + pre-column reactor	Lys, Val, Phe, glutamic acid, dansylchloride	n.a.	n.a.	90 000	0.5
Burggraf, Manz, *et al.* [44], 1993	Fig. 6.12	FITC	0.5	80 per cycle	60 000 per cycle max. 1200 000	1.3

concentrations of species in the sample, as is the case of gel electrophoresis for DNA sequencing or many other biological applications [56]. In this case, despite a larger detection volume, the absolute number of molecules of the interesting species in the detection volume can be arbitrarily small, with an ultimate limit of a single molecule. As this problem is not limited to miniaturized CE systems, for conventional CE systems detection methods have already been designed to detect fluorescent light from a single molecule [57–61]. The main problem to overcome in these experiments is the differentiation between the molecular fluorescence and the background noise, mainly luminescence from the buffer molecules and Raman scattered light. This can be achieved by physically limiting the sample volume (which is automatically realized in CE devices by using microchannels), optically minimizing the detection volume by means of confocal detection and narrowband filtering of the emitted fluorescence. Other methods make use of time resolved LIF or fluorescence correlation spectroscopy.

For biological applications, the investigation of single cells is of particular interest. Due to its electroosmotic flow properties, CE is ideally suited for these experiments, as the injection of the cell can be realized easily. Successful separations of various components (e.g. hemoglobin [62] or catecholamine [63]) in single cells have been reported.

A second detection method makes use of potentiometric detectors. In this case, ion-selective microelectrodes are used as end-column detectors in CE, and ionic concentrations of down to 10 nM could be measured [64, 65].

6.8 Applications in the Life Sciences

In the biomedical sciences, the demand for rapid DNA sequencing technologies and new methods in screening has been enormous [66], fuelled particularly by the Human Genome Project. Conventional CE has already been proved useful for DNA sequencing, as it is possible to operate many capillaries in parallel to increase the analysis throughput [67–69]. Several approaches have been made to perform this task on miniaturized CE systems. At Perseptive Biosystems, Koutny et al. [70] used a CE chip made out of fused silica with a design similiar to the one described in Figure 6.8. It consists of three 2.5 cm long microchannels, two connected to the buffer and injection waste reservoirs and one acting as the separation channel, and one 0.5 cm long channel leading to the sample reservoir. The channels have a width of 68 μm at the top and a depth of 28 μm, with a rounded profile due to the isotropic etching conditions. The sample volume defined by this layout is about 700 pL. In Figure 6.20, repeated immunoassay separations of rabbit polyclonal anti-cortisol antiserum and fluoresein labeled cortisol are shown. Plate numbers of 6000 for the labeled antigen have been achieved within less than 30 seconds.

A complete microfabricated array of electrophoresis capillaries on a glass chip has been reported by Woolley and Mathies at UC Berkeley [71, 72]. On each chip, 15 CE systems were fabricated, with separation channels with widths of 30, 50 and 70 μm and a length of 45 mm. These channels were then filled with a hydroxyethyl cellulose sieving matrix. In Figure 6.21, a separation of ϕX174 Hae III DNA fragments is shown. The labels in the figure refer to the molecular weight of the fragments. For the 234-bp fragment, a plate number of 130 000 has been deduced.

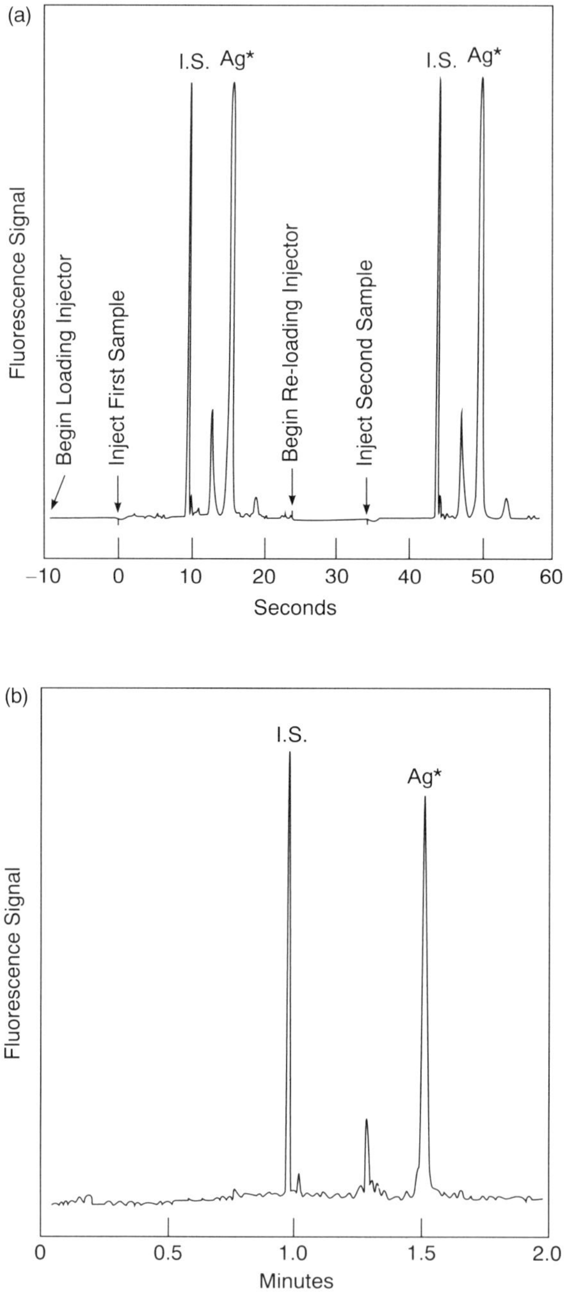

Figure 6.20. Immunoassay separations on a microchip. The labels stand for I.S.:internal standard; Ag*: labeled cortisol. The top figure shows two replicate sample injections, the bottom figure a single injection. Reprinted with permission from [70]. Copyright 1996 American Chemical Society.

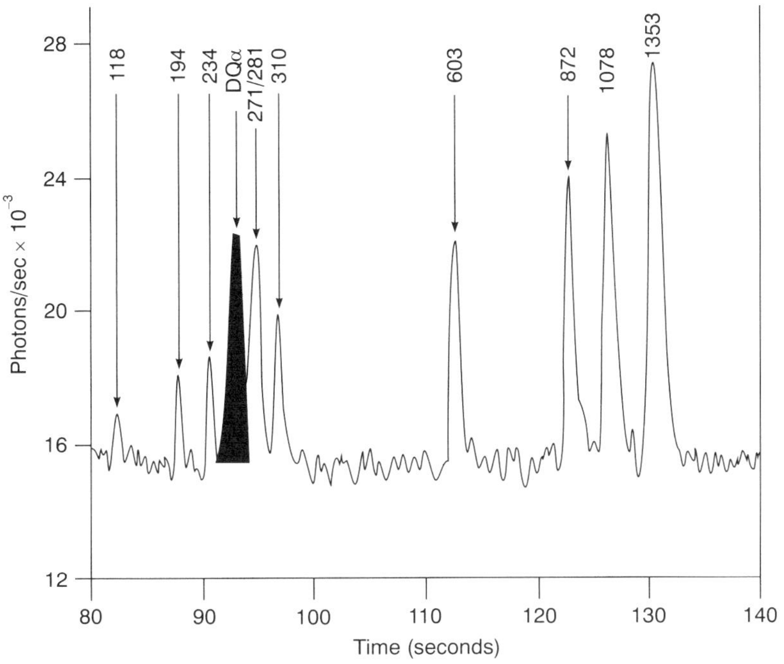

Figure 6.21. Electropherogram for high-speed DNA analysis on a CE microchip. The numbers indicate the lengths of the fragments, the shaded peak acts as a standard. From [72].

6.9 Conclusions

The availability of microfabrication technologies combined with the generality of the analysis system concept indicate a broad range of possible applications. In a commercial sense, microfabrication can promise low cost, high output figures for devices like those described above and therefore opening up new markets for analytical applications which so far have not been achievable for conventional desktop analytical instruments (due to their high prices and the necessary skills for operation) or have been looked upon as the domain of the chemical sensor (e.g. for on-line monitoring in process control or screening applications in clinical diagnostics). Obviously, the field of μ-TAS is still in its infancy, with the concepts and about a few dozen device designs published over the last six years. Many important areas await further development: for example, the interfacing of the microfluidics on the chip and the macroscopic world, sample handling with micromachined pumps, valves, mixers, etc., for optimized transport properties, advanced chip layouts with parallel or multidimensional channel arrays, on-chip integration of the detection system and advanced data acquisition techniques. But already the field of CE systems shows the competitiveness of the miniaturized systems with established technologies. Their performance in speed and selectivity, combined with the unique advantages of their down-scaled dimensions, such as reduced reagent consumption and therefore reduced waste generation, high flexibility and the potential to have mobile units as "on-chip laboratories" characterizes the miniaturized analysis systems as a verstatile and interesting field of research as well as a key product for any market linked with chemical analysis in the future. More than a dozen instrument manufacturers practice CE on a chip in their research laboratories. Performance should result in a commercial product being on the market by early 1998.

(Received June 30, 1996)

6.10 References

[1] Lodge, A., *B.A. Thesis*, 1886.
[2] Hjertén, S., *Chromatogr. Rev.*, **9** (1967), 122.
[3] Manz, A., Graber, N., Widmer, H. N., *Sensors and Actuators*, **B1** (1990), 244–248.
[4] van den Berg, A., Bergveld, P. (eds.), *Micro Total Analysis Systems*; Dordrecht: Kluwer Academic Publishers, 1995.
[5] Landers, J. P. (ed.), *Handbook of Capillary Electrophoresis*; Boca Raton: CRC Press, 1994.
[6] Li, S. F. Y., *Capillary Electrophoresis: Principles, Practice, and Application*; Amsterdam: Elsevier, 1992.
[7] Bruno, T. J., *Chromatographic and Electrophoretic Methods*; Englewood Cliffs: Prentice-Hall, 1991.
[8] Mosher, R. A., Saville D. A., Thormann, W. (eds.), *The Dynamics of Electrophoresis*; Weinheim: VCH Publishers, 1992.
[9] Gordon, M. J., Huang, X., Pentoney, S. L. Jr., Zare, R. N. *Science*, **242** (1988), 224–228.
[10] Karger, B. L. *Nature*, **339** (1989), 641–642.
[11] Grossmann, P. D., Colburn, J. C., Lauer, H. H., Nielsen, R. G., Riggin, R. M., Sittampalan, G. S., Rickard, E. C. *Anal. Chem.*, **61** (1989), 1186–1194.
[12] Dose, E. V., Guiochon, G. *J. Chromatogr.*, **652** (1993), 263–275.
[13] Hunter, R. J. and Wright, H. J. L. *J. Colloid Interface Science*, **37**(3) (1971), 564–580.
[14] Salomon, K., Burgi, D. S., Helmer, J. C. *J. Chromatogr.*, **559** (1991), 69–80.
[15] Stevens, T. S. and Cortes, H. J. *Anal. Chem.*, **55** (1983), 1365–1370.
[16] Manz, A., Effenhauser, C. S., Burggraf, N., Harrison, J. D., Seiler, K., Fluri, K. *J. Micromech. Microeng.*, **4** (1994), 257–265.
[17] Bevan, C. D., Mutton, I. M. *Anal. Chem.*, **67** (1995), 1470–1473.
[18] Moreau, W. M., *Semiconductor Lithography*; New York: Plenum Press, 1987.
[19] Fan, Z. H., Harrison, D. J. *Anal. Chem.*, **60** (1994), 375–377.
[20] Pace, S. J., U.S. Patent 4 908 112, 1990.
[21] Manz, A., Harrison, D. J., Fettinger, J. C., Verpoorte, E., Lüdi, H., Widmer, H. M., in: *Transducers '91, Digest of Technical Papers*; Piscataway, NJ: IEEE, 1991, pp. 939–941.
[22] Manz, A., Harrison, D. J., Fettinger, J. C., Verpoorte, E., Lüdi, H., Paulus, A., Widmer, H. M. *J. Chromatogr.*, **593** (1992), 253–258.

[23] Manz, A., Fettinger, J. C., Verpoorte, E., Lüdi, H., Widmer, H. M., Harrison, D. J. *Trends Anal. Chem.*, **10** (1991), 144–149.

[24] Harrison, D. J., Manz, A., Glavina, P. G., in: *Transducers '91, Digest of Technical Papers*; Piscataway, NJ: IEEE 1991, pp. 792–795.

[25] Harrison, D. J., Manz, A., Fan, Z., Lüdi, H., Widmer, H. M., *Anal. Chem.* **64** (1992), 1926–1932.

[26] Harrison, D. J, Glavina, P. G., Manz, A., *Sensors Actuators B*, **10** (1993), 107–116.

[27] Ekström, B., Jacobson, G, Öhman, O., Sjödin, H., International Patent WO 91/16966, 1990.

[28] Manz, A., Harrison, D. J., Verpoorte, E., Widmer, H. M., in: *Advances in Chromatography*, **vol. 33**, Brown, P. R., Grushka, E. (eds.); New York: Marcel Dekker, 1993, pp. 1–66.

[29] Effenhauser, C. S., Manz, A., Widmer, H. M. *Anal. Chem.*, **65** (1993), 2637–2642.

[30] Effenhauser, C. S., Manz, A., Widmer, H. M. *Anal. Chem.*, **34** (1995), 2284–2287.

[31] Effenhauser, C. S., Paulus, A., Manz, A., Widmer, H. M. *Anal. Chem.*, **66** (1994), 2949–2953.

[32] Fan, Z. H., Harrison, D. J. *Anal. Chem.*, **66** (1994), 177–184.

[33] Harrison, D. J., Fluri, K., Seiler, K., Fan, Z., Effenhauser, C. S., Manz, A., *Science*, **261** (1993), 895–897.

[34] Seiler, K., Fan, Z. H., Fluri, K., Harrison, D. J. *Anal. Chem.*, **66** (1994), 3485–3491.

[35] Harrison, D. J., Fluri, K., Chiem, N., Tang, T., Fan, Z., in: *Transducers '95, Digest of Technical Papers*; Stockholm: IEEE 1995, pp. 752–755.

[36] Jacobson, S. C., Hegenröder, R., Koutny, L. B., Ramsey, J. M. *Anal. Chem.*, **66** (1994), 1114–1118.

[37] Jacobson, S. C., Moore, A. W. Jr., Ramsey, J. M., *Anal. Chem.*, **67** (1995), 2059–2063.

[38] Jacobson, S. C., Koutny, L. B., Hegenröder, R., Moore, A. W. Jr., and Ramsey, J. M., *Anal. Chem.*, **66** (1994), 3472–3476.

[39] Jacobson, S. C., Hegenröder, R., Moore, A. W. Jr., Ramsey, J. M., *Anal. Chem.*, **66** (1994), 4127–4132.

[40] Jacobson, S. C., Hegenröder, R., Koutny, L. B., Warmack, R. J., Ramsey, J. M., *Anal. Chem.*, **66** (1994), 1107–1113.

[41] Jacobson, S. C., Hegenröder, R., Koutny, L. B., Ramsey, J. M., *Anal. Chem.*, **66** (1994), 2369–2373.

[42] Moore, A. W. Jr., Jacobson, S. C., Ramsey, J. M., *Anal. Chem.*, **67** (1995), 4184–4189.

[43] Jacobson, S. C., Ramsey, J. M., *Anal. Chem.*, **68** (1996), 720–723.

[44] Burggraf, N., Manz, A., de Rooij, N. F., Widmer, H. M., *Anal. Methods Instrumentation*, **1** (1993), 55–59.

[45] Burggraf, N., Manz, A., Effenhauser, C. S., Verpoorte, E., de Rooij, N. F., Widmer, H. M., *J. High Res. Chromatogr.*, **16** (1993), 594–596.

[46] Burggraf, N., Manz, A., Verpoorte, E., Effenhauser, C. S., Widmer, H. M., de Rooij, N. F., *Sensors Actuators B*, **20** (1994), 103–110.

[47] Gordon, G. B., U.S. Patent 5 328 578, 1994.

[48] Von Heeren, F., Verpoorte, E., Manz, A., Thormann, W. *Anal. Chem.*, **68**(13) (1996), 2044–2053.

[49] Von Heeren, F., Verpoorte, E., Manz, A., Thormann, W. *Microcolumn Sep.*, **8**(6) (1996), 373–381.

[50] Bruno, A. E., Maystre, F., Krattiger, B., Nussbaum, P., Gassmann, E. *Trend Anal. Chem.*, **13** (5) (1994), 190–198.

[51] Verpoorte, E., Manz, A., Lüdi, H., Widmer, H. M., van der Schoot, B. H., de Rooij, N. F., in: *Transducers '91, Digest of Technical Papers*; Piscataway NJ: IEEE, 1991, pp. 796–798.
[52] Verpoorte, E., Manz, A., Lüdi, H., Bruno, A. E., Maystre, F., Krattiger, B., Widmer, H. M., van der Schoot, B. H., de Rooij, *Sensors Actuators, B* **6** (1992), 66–70.
[53] Liang, Z., Chiem, N., Ocvirk, G., Tang, T., Fluri, K., Harrison, D. J., *Anal. Chem.*, **68** (1996), 1040–1046.
[54] Flui, K., Fitzpatrick, G., Chiem, N. Harrison, D. J., *Anal. Chem.*, **68** (23) (1996), 4285–4290.
[55] Seiler, K., Harrison, D. J., Manz. A., *Anal. Chem.*, **65** (1993), 1481–1488.
[56] Chen, D. J., Harke, H. R., Dovichi, N. J., *Nucl. Acids Res.*, **20** (1992), 4873-4880.
[57] Shera, E. B., Seitzinger, N. K., Davis, L. M., Keller, R. A., Soper, S. A., *Chem. Phys. Lett.*, **174** (1990), 553–557.
[58] Whitten, W. B., Ramsey, J. M., Arnold, S., Bronk, B. V., *Anal. Chem.*, **63** (1991), 1027–1031.
[59] Soper, S. A., Mattingly, Q. L., Vegunta, P., *Anal. Chem.*, **65** (1993), 740–747.
[60] Lee, Y. H., Maus, R. G., Smith, B. W., Winefordner, J. D., *Anal. Chem.*, **66** (1994), 4142–4149.
[61] Castro, A., Shera, E. B., *Anal. Chem.*, **67** (1995), 3181–3186.
[62] Lillard, S. J., Yeung, E. S., Lautamo, R. M. A., Mao, D. T., *J. Chromatogr. A*, **718** (1995), 397–404.
[63] Chang, H. T., Yeung, E. S., *Anal. Chem.*, **67** (1995), 1079–1083.
[64] Haber, C., Silvestri, I., Roosli, S., Simon, W., *Chimia*, **45** (1991), 117–121.
[65] Nann, A., Silvestri, I., Simon, W., *Anal. Chem.*, **65** (1993), 1662–1667.
[66] *National Research Council Report on the Committee on Mapping and Sequencing the Human Genome*; Washington, DC: Commission of Life Sciences, National Academy Press, 1988.
[67] Huang, X. C., Quesada, M. A., Mathies, R. A., *Anal. Chem.*, **64** (1992), 2149–2154.
[68] Mathies, R. A., Huang, X. C., *Nature* (1992), 167–169.
[69] Ueno, K., Yeung, E. S., *Anal. Chem.*, **66** (1994), 1421–1431.
[70] Koutny, L. B., Schmalzing, D., Taylor, T. A., Fuchs, M., *Anal. Chem.*, **68** (1996), 18–22.
[71] Wooley, A. T., and Mathies, R. A., *Proc. Natl. Acad. Sci. USA*, **91** (1994), 11348–11352.
[72] Wooley, A. T., and Mathies, R. A., *Anal. Chem.*, **67** (1995), 3676–3680.

7 Sensors in Biomolecular Interaction Analysis and Pharmaceutical Drug Screening

A. Brecht, G. Gauglitz, Göpel, W., Institut für Physikalische und Theoretische Chemie, Universitäte Tubingen, Germany

Contents

7.1 Introduction

Molecular recognition and interaction is of key importance in the metabolism and control of living systems. The interaction of macromolecules like proteins and DNA with other macromolecules or with low molecular weight compounds governs and directs metabolic activity, growth and differentiation, information storage and processing, and even movement and perception at almost any step and any level. Figure 7.1 illustrates typical sizes of the molecular species involved. The dramatic achievements in the field of molecular genetics now give access to the complete genetic information of entire organisms [1], i.e. to the "blueprints" of the molecular machinery of living organisms. The ambitious *Human Genome Project* in the USA and matching activities in other countries target sequencing of the entire human genome (with a total of about 4×10^9 basepairs, which encode for about 150 000 actively transcribed genes). This is a finite amount of information, which will be available within less than a decade.

Access to the genetic information allows the expression and manipulation of the encoded molecular structures. Growth, differentiation and diseases are hence increasingly understood at the molecular level. This knowledge results to a great extent from fundamental research. It starts to alter paradigms and approaches in

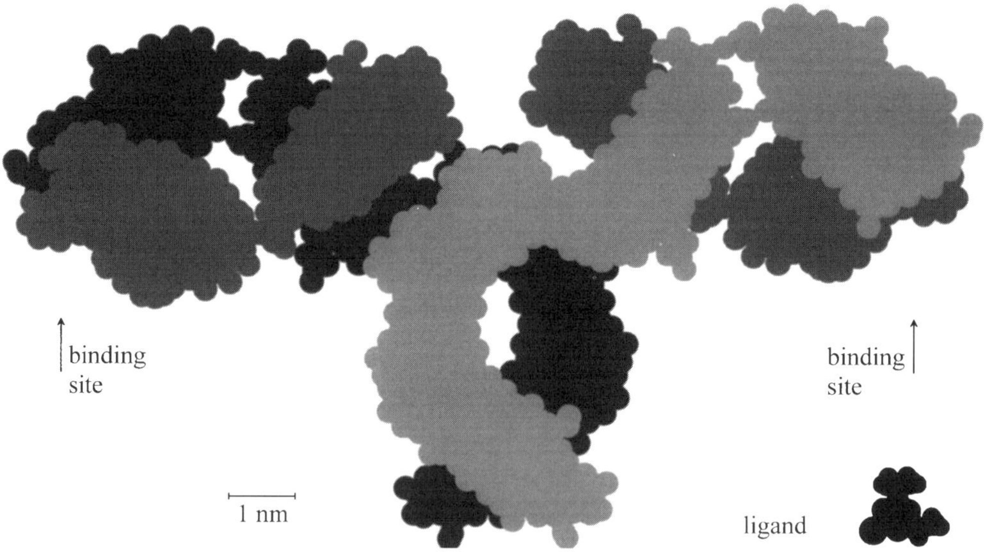

Figure 7.1. An antibody molecule (150 000 g/mole) from the immunoglobulin G class and a low molecular weight ligand (s-triazine derivate, app. 200 g/mole) which is bound by the antibody. The binding sites of the antibody molecule are indicated by arrows. The four polypeptide chains of the antibody are shaded differently. The scale bar is approximately 1 nm (modified from [2]).

biological research, in agriculture and in medicine. Generic approaches to the study of aspects related to biological function or activity are required to complement and to exploit the sequencing megaprojects which in the first instance deliver DNA sequences only. Sensors and sensor-related techniques will therefore support further investigation and exploitation of molecular function and molecular interaction.

Agriculture and medicine in particular constitute large markets and technologies involved with these developments will hence receive increasing attention. As an example, today no causal therapy is available for most known diseases. Research in pharmaceutical drugs will therefore continue to grow and involve all sources of new information. A significant proportion of the investments in drug search and research may be redirected to follow novel, promising strategies to find new drugs.

This contribution discusses possible impacts of these trends. Two emerging fields of application are discussed in which sensor technologies already earn revenues or may do so in the near future, i.e. basic research on molecular interactions and therapeutic drug screening.

Biochemical compounds may be used as part of a complex sensor system as already described in volumes 2 and 3 of this series [3, 4]. These cover the fundamentals of chemical and biochemical sensors, a range of transducer principles, and important fields of application. Improved understanding of molecular recognition and interaction in biological systems will now supply novel and improved recognition elements for the general development of chemical and biochemical sensors. Further to the use of existing biomolecular recognition compounds [5, 6], the development of biomolecules with new structures and specifities [7, 8] is pursued. Biomimetic host-guest-chemistry tries to mimic the selective biomolecular recognition by entirely synthetic receptor structures. Biosensor research is reviewed extensively in [9] and also covered in previous volumes of this series [3, 4] and is not the subject of this paper.

Most of the fundamentals of biomolecular recognition processes is textbook knowledge (e.g. [10, 2]). Therefore in Section 7.2, only a brief review of the basic principles is given, with the main emphasis on the application of sensor technologies. A survey of measurement approaches and transducer technologies currently adapted or adaptable to the interrogation of biomolecular interaction closes this chapter. In Section 7.3 actual and emerging applications from two fields are discussed with particular emphasis on the investigation of biomolecular interaction processes with sensor systems and the use of sensors in pharmaceutical drug screening.

7.1.1 Typical Bioaffinity Interactions

The metabolic machinery of organisms as represented by enzymes recognizes low molecular weight building blocks and fuels by spatial complementarity. The higher the specificity of a given molecular machine or enzyme is, the more perfect is the match between the binding pocket at the active site of the molecule and the target substrate. This process is often known as *key–lock* interaction or *induced-fit*.

Table 7.1. Typical examples of biomolecular recognition processes and the molecular species involved.

Receptor (mol. weight)	Ligand (mol. weight)	Effect of binding	Typical ligand concentration
Lactate dehydrogenase (≈ 35 000 g/mole)	Lactate (≈ 90 g/mole)	Conversion of lactate to pyruvate.	up to mM
Galactose-repressor (≈ 40 000 g/mole)	Lactose/allo-Lactose (≈ 342 g/mole)	Repressor dissociates from the operator sequence of the Gal-operon.	up to mM
Neuropeptide-Y-1 receptor, monomeric, 7-helix-protein class (≈ 70 000 g/mole)	Neuropeptide Y (≈ 40,000 g/mole)	G-protein coupled receptor. Involved in control of feeding behavior.	nM
GABA-A-receptor neuroreceptor consisting of multiple subunits (≈ 250 000 g/mole)	γ-amino-butyric-acid (≈130 g/mole)	Chloride channel of the GABA-A-receptor opens. Inhibitory effect on postsynaptic neuron.	nM to μM
Desoxyribonuclease I (≈ 30 000 g/mole)	DNA-double strands	Degradation of DNA.	—

Molecular recognition is also of high importance in the discrimination of self from non-self (immune system), in the control of growth and differentiation, and even in perception. Here the interaction of messenger molecules (like hormones, neurotransmitters or cytokines) with the matching receptor structures is the key event. A few examples are given in Table 7.1. The molecular weights of the species involved range from less than 100 g/mole to more than 100 000 g/mole. Actually multimeric proteins, multiprotein complexes, and membrane-integral structures may reach molecular weights of above 10^6 g/mole. The concentrations encountered start from millimolar concentrations at the high end (nutritives) to micro- and nanomolar concentrations (many messenger compounds), but even values below picomolar concentrations are encountered with some hormones (e.g. triiodothyronine: 10^{-13} M).

The most obvious process involving specific biomolecular recognition is the biocatalytic activity of enzymes in metabolism. Substrates are selectively bound and converted. The initial step of enzyme action is recognition and selective binding of substrates and cosubstrates. The modulation of enzyme activity by competitive inhibitors (i.e. substances that bind to the active site of the enzyme and thereby inhibit access of the substrate) or allosteric binding (i.e. binding of modulators to regulatory binding sites, apart from the active site) also involve specific recognition processes. Typical clinically relevant examples include the inhibition of the blood coagulation cascade, therapeutic blood pressure decrease by inhibitors of the "angiotensinogen converting enzyme", and inhibitors of the reverse transcriptase or the retrovirus protease from HIV.

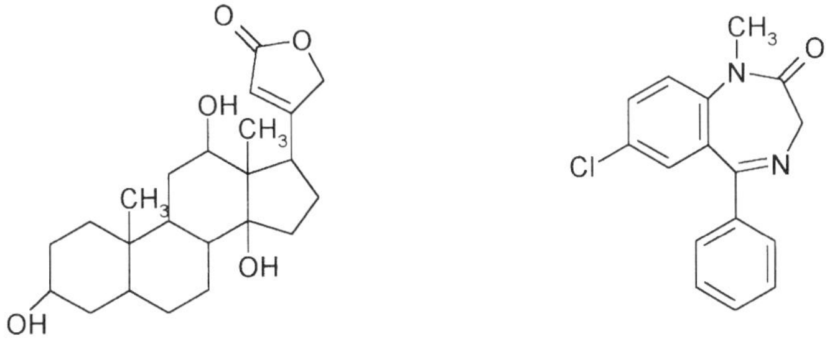

Figure 7.2. Two typical therapeutic drugs. Left: Digoxigenin – the active moiety of various drugs used in the treatment of cardiac dysfunction. The strophantin alkaloids to which this substance belongs are derived from natural compounds. Right: a benzodiazepin (Valium™), an entirely synthetic structure used in the treatment of depression as an anxiolytic agent.

Biological systems have to transmit and process information over distances ranging from the thickness of a biological membrane (a few nm) to entire organisms (a few m) to control metabolic and cellular activity. In many cases information is transferred by specific molecular information carriers (neurotransmitters, hormones, cytokines). Often the information carrier is a comparably small molecule that binds selectively to target sites which are typically receptor proteins, and in many cases located in a cell membrane. Binding of the information carrier causes a conformational change, that is further transmitted by protein-based cascades (e.g. protein G [11]) or by change in the ion conductivity of membrane channel forming proteins. [12]. Two clinically relevant examples are anxiolytics from the benzodiazepin-class (which bind to GABA-receptors in the brain) and strophantin alkaloids (which are used in the therapy of heart insufficiency and bind to a potassium channel in the heart muscle, see Table 7.1 and Figure 7.2).

Higher organisms have developed a sophisticated system for the discrimination between self and non-self in their defense against micro-organisms and other parasites. This system consists of cellular and molecular compounds. Again specific binding between recognition molecules and foreign structures is the first step in the activation of defense systems [13]. Antibodies represent the best-known recognition structures. They have become a versatile tool in bioanalytical chemistry [5, 14].

Finally, the storage and translation of genetic information also involves mechanisms of molecular recognition. The double strand formation of complementary DNA strands is of prime importance, as is also the interaction of DNA and regulatory factors that control the flow from genetic information to actively transcribed proteins [2].

Therefore, the investigation of biomolecular recognition processes is of very general and hence high interest, not only for fundamental research, but also for all practical fields in which control of metabolism is required. In particular this is the case in pharmacotherapy. Concepts from sensor research that allow us to optimize understanding (fundamental research) and control of metabolism (pharmacotherapy) will become of considerable commercial interest. An overview of biological fundamentals and the importance of molecular recognition processes with respect to biosensors is given in volume 3 of the series [15].

7.1.2 Biomolecular Interaction Analysis

Biomolecular interaction analysis (BIA) is a term that was introduced a few years ago by Pharmacia Biosensors of Sweden (now BIACORE AB) to characterize the study of molecular interactions between compounds of biological relevance. The study of biomolecular interactions evidently dates back much further, but Pharmacia introduced for the first time a commercially available sensor-based concept that did not rely on any specific molecular property (like intrinsic fluorescence or biocatalytic activity). This made the approach highly generic. This new technique met a situation in which the isolation and purification of biomolecules had already reached a high standard. In addition advances in molecular biology, especially the *polymerase chain reaction* and recombinant expression of proteins made possible the access to virtually any protein compound encoded in the genome of an organism.

Biomolecular interaction studies (BIA) answer questions related to the interaction of biomolecules. Typical questions include:

- Do two given compounds interact or bind and what conditions affect the interaction?
- What is the (thermodynamically controlled) strength of a biomolecular interaction?
- What are the kinetic on-rate(s) and off-rate(s) of a biomolecular interaction process?
- How do structural modifications affect biomolecular interaction?
- How do environmental factors affect biomolecular interaction?
- What sequence of interaction is required for the assembly of complex supramolecular entities?

An overview of commercial BIA devices is given in Section 7.3.

7.1.3 Pharmaceutical Screening

Traditionally novel drugs are developed by a systematic variation of a known structural motif (a lead) or by a random search of new substance classes. Strophantin alkaloids which are used in the treatment of cardiac dysfunction (derived from a natural compound) or benzodiazepins used as anxiolytics (a synthetic compound) may serve as two typical examples (Figure 7.2).

New therapeutic drugs have been and are still synthesized individually by classical organic chemistry. Tests are based on the functional evaluation of the compounds, e.g.

by testing the pharmaceutical potency in an animal model. A multistep selection and screening process is applied to select effective compounds. Typically more than 10 000 compounds are synthesized before a new pharmaceutical compound reaches the market. The development of a new drug from its first synthesis to the market involves a typical investment of a few 100 Mio US$. Synthesis, isolation and initial testing of candidate substances accounts for 10% to 25% of the total cost. Despite the high investment by pharmaceutical companies, the number of entirely new structures introduced per year is surprisingly low.

The considerable costs of drug search and research limits the number of companies world wide that are capable of participating in this activity. This favors concentration processes and mergers, even between already large companies. The hope of designing new drugs rationally, starting from known structure and sequence information, has not yet been demonstrated to be successful. Therefore, in recent years new approaches have been made to "widen" two of the bottlenecks in traditional drug search. These are:

(1) the availability of new compounds, and

(2) methods for their screening in a fast and cost-effective way to identify new potential drugs.

The number of available compounds is significantly increased by several routes. Traditional homogeneous phase approaches in drug synthesis are replaced by solid phase chemistry. The latter were originally introduced by Merrifield for the synthesis of polypeptides [16]. Other than homogeneous phase chemistry, solid phase approaches are well suited to automation strategies. Second, the concept of combinatorial chemistry [17], where a multitude of oligomeric compounds is synthesized from an alphabet of monomeric compounds gives access to an increasing number of substances. Understanding of the factors governing the pharmacological potency of a given set of substances is used to improve the selection of compounds and the strategies for their synthesis [18]. Third, pharmaceutical companies activate their internal substance bases for rescreening them for new applications. This involves reformatting of compounds into a condition that can be handled and distributed automatically. Sources are internal substance bases of companies, but also vendors, that offer substance libraries for screening on a commercial basis.

The capacities for testing new compounds are limited if complex test systems are required. Simple and rapid test systems are therefore a prime requirement for High Throughput Screening (HTS). Throughput is optimized by multistep screening procedures, where a primary screening with high throughput is used to identify suitable candidates for a more complex secondary screening procedure. Secondary screening often is carried out with well-established techniques, that have been commonly used in drug research for a considerable time (e.g. animal models). Primary screening techniques, which must serve to pick potential candidates from very large substance pools, are still a matter of research and development. Typical requirements as reflected by recent developments of screening approaches and screening technology [19] are given in Table 7.2.

Table 7.2. Requirements for techniques in High Throughput Screening.

Requirement	Typical Goal
High throughput	> 1 Mio samples/year
Simplicity	Wash-free assays, single step protocols
Broad applicability	No restriction to certain classes of target molecules
Cost effectiveness	Preferably < 1 US $ per screened substance
Low substance consumption	< 1 μg per assay, preferably < 1ng

In the screening for novel drugs two principal approaches are possible. Either the desired biological activity is tested or (as the more basic approach) the molecular interaction is probed between the drug and the molecular target as a precondition for biological activity. The related assay concepts are *functional screening* and *binding assays*.

Functional screening can be carried out at the level of individual molecules or at the level of more complex systems (organelles, cells, tissues). A wide range of cell-based assays is already available or under development. The major effort here is to implement a cellular system that responds unambiguously to the therapeutic agents under investigation. Sensor-based concepts have been introduced recently to that field [20, 21]. A considerable throughput can be achieved with cell-based assays, but the review of these techniques is beyond the scope of this contribution. At the molecular level functional screening is restricted to components with a distinct metabolic activity – typically enzymes. Functional assays detect directly the metabolic activity desired–a typical example is the proteolytic activity of the angiotensinogen converting enzyme (ACE), which creates angiotensin from its precursor angiotensinogen and plays a role in hypertension. An assay which detects the enzymatic activity of ACE on a model substrate, allows us to identify ACE-inhibitors, which constitute an important class of anti-hypertensives today. Assays based on a chromogenic or fluorogenic enzyme substrate are popular [22]. However, the effort to establish a new enzyme assay is considerable and in consequence the related costs are high.

Binding assays focus at the primary event of molecular recognition. A binding assay, which could detect the binding of a substance to ACE (see preceding paragraph) would certainly detect active site inhibitors of ACE, but may also respond to components which just bind to ACE but do not alter its catalytic activity. However, binding assays allow us to identify potential agonists and antagonists of receptors without any catalytic activity, like neuroreceptors or hormone receptors. Drugs are typically administered at a few mg/kg body weight daily intake. Depending on molecular weight, distribution to various compartments, and metabolism, the active levels are typically micromolar with few drugs reaching millimolar levels. At these concentrations a drug must interact significantly with a biological target molecule (receptor) to exert an effect. Therefore, tentative drugs for a given receptor can be recognized by specific binding at the concentration level targeted.

Table 7.3. Typical methods used in High Throughput Screening (primary screening).

Type of assay	Principal effect, remarks	Lit
Enzyme inhibition assay The activity of the enzyme is modulated by potential pharmaceuticals	Enzyme activity is detected, often photometrically or fluorimetrically. Typically endpoint assay with single readout.	[22]
Membrane receptor binding assay A labeled compound binds to membrane receptors in competition with test substances.	Equilibrium binding assay with a labeled competitor. Physical separation (filtration) of bound and free label is required before quantification of the bound label.	[47]
Scintillation proximity assay™ (Amersham) A radioactively labeled compound binds to receptors in competition with test substances.	Wash-free proximity type binding assay. Use of low-energy radioisotopes restricts excitation of luminescence to labels bound to the surface of scintillating beads.	[23]
Fluorescence polarisation assay A (preferably) low molecular weight fluorescence labeled compound binds to receptors in competition with test substances.	Fluorescence labeling of low molecular weight compound required. Changes in fluorescence anisotropy indicate binding. Wash-free assay.	[32, 49]
Fluorescence correlation spectroscopy A low molecular weight fluorescence labeled compounds binds to receptors in competition with test substances.	Fluorescence labeling of low molecular weight compound required. Translational diffusion coefficient changes due to binding. Wash-free assay. Low detection volume, short assay time and low working concentrations. Complex equipment.	[50]
Reporter gene assay Cellular assay with recombinant cells that respond to potential drugs by activation of a reporter gene.	Cell responds to stimulus by change in intrinsic luminescence, altered activity level of reporter enzyme, etc. Typically applied as endpoint assays. Development of system may be complex.	[95]
Extracellular acidification–Cytosensor™ (by Molecular Devices, USA) Test substances change the metabolic state of cells.	Change in extracellular acidification (excretion of "protons") is detected by solid state pH-transducer. Established assay works in microvolumes. Dynamic assay–dose response curves can be obtained.	[20]
Cytotoxicity and growth inhibition assays Test substances affect growth and viability of natural or recombinant cells.	Effect of test substance on growth behavior are investigated. Typically imaging techniques are employed. Cell culture over extended periods of time is common (days to weeks).	[96]

Binding assays may be simple in design and – due to their very general approach – broadly applicable. As *primary screening runs* will be followed by a more specific *secondary screening*, binding assays meet the requirements given in Table 7.2. Today binding assays are often carried out with radioactively labeled compounds. Specific assay formats like the Scintillation Proximity Assay™ from Amersham [23] allow detection in a wash-free single step assay. Table 7.3 lists typical assay formats and approaches currently used in HTS projects. The most common techniques will be explained in more detail below.

Irrespective of the nature of the assay, the required sample throughput ranges roughly from 10 000 assays per month to a few 100 000 assays per month. The further development of throughput requirements is not easy to predict, as synthesis capacities will grow, while improved selection strategies may at the same time reduce the input to a screening run. Costs encountered in a screening project include costs for establishing the assay for a given target and the costs for the screening itself. The costs for establishing an assay for a new target are highly dependent on the complexity of the approach. Costs encountered in screening assays range from below one US$ per substance tested up to 10 US$ per substance. The magnitude of libraries screened ranges are from a few 1000 (e.g. natural compound libraries) up to several 10000 (company-owned substance bases in the pharmaceutical industry). Special assay formats are developed to screen very large ($>$ 100 000) bead oriented libraries. However, it seems likely that strategies to increase and optimize molecular diversity in libraries will favor medium sized libraries. A careful estimate of the number of targets investigated in screens today (1996) amounts to a few hundred. The number of targets for screening will increase with the number of proteins available in recombinant expression systems. Finally the number of targets may be a considerable fraction of the number of actively transcribed genes from the human genome (estimated number 150 000). It is evident that the primary costs of a screening technology (hardware) are of limited importance in comparison with the actual costs per assay. Reductions in cost may be achieved by simplified establishment of new targets, by dispensing with costly reagents, and last but not least by reductions in assay volumes or test concentrations. It is already recognized that approaches based on affinity or immunosensors may be of value in HTS [24]. The subsequent review of fundamental and practical aspects of affinity sensor technology will point out routes, that will become important in BIA or HTS.

7.2 Fundamentals of Bioaffinity Interactions

7.2.1 General Aspects

Any approach towards probing biomolecular interactions with a sensor device must observe the fundamentals governing these interaction processes. Only very few biomolecular recognition processes are understood at a molecular level.

7.2.1.1 Interaction Forces

In many cases at least one partner involved in biomolecular interactions is a protein. Only part of the protein takes part in the binding event. This binding site or even binding pocket can readily be identified from crystallographic structures of the bound state. The bound compound can be either a low molecular weight compound or another macromolecular structure. Complex formation occurs due to structural complementarity of the reacting structures. In the complex both molecules assume a configuration, that enables a maximum of attractive interactions without generating repulsive forces. The unbound conformation of the reactants may be less defined than and significantly different from the bound state. The binding process is therefore best described as an induced fit of both compounds. The major forces involved are given in Table 7.4 [25].

The overall change in free enthalpy ΔG during binding contains the cumulative effect from various weak interactions distributed over the entire interacting surface. Interactions with a strong distance dependence will lead to a higher selectivity of interaction than interactions that decay over an extended distance. The repulsive interactions have by far the strongest distance dependence. Any misfit between the binding site and the bound structure will lead to a more significant change in binding constants than the lack of a particular attractive interaction. The hydrophobic interaction refers to an entropy effect. The exclusion of water from hydrophobic parts of a binding site will allow the water molecules to take up a less restricted structure. This leads to an increase of the entropy ΔS in the system. Due to the relation

$$\Delta G = \Delta H - T\Delta S \tag{7.1}$$

this increases the strength of the binding and makes the interaction of two hydrophobic spots in an aqueous environment energetically favorable. Hydrophobic interactions are related to the hydrophobic surface area of a molecule and may account for up to 50% of the free energy change in biomolecular interactions.

Table 7.4. Potentials and distance dependencies of interactions relevant in biomolecular recognition (for survey see [25]).

Interaction potential	Distance dependence
Coulomb interaction	attractive: $1/r$
Hydrogen bound	attractive $1/r^{10}$
Van der Waals (induced dipole)	attractive $1/r^{6}$
Hydrophobic interaction (exclusion of water from hydrophobic surface areas)	attractive: entropy effect
Repulsive forces (forbidden overlap of orbitals)	repulsive: $1/r^{12}$ or e^{-kr}

7.2.1.2 Equilibrium and Kinetics

The simplest quantitative description for a ligand L binding to a biological receptor R is given by the law of mass action for a simple bimolecular reaction. The binding equilibrium is determined by a dynamical balance of reaction rates. The binding constant K [L/mole] is given by the ratio of the on-rate k_1 [L mole^{-1} s^{-1}] and the off-rate k_{-1} [s^{-1}] respectively.

$$R + L \underset{k_{-1}}{\overset{k_1}{\rightleftharpoons}} RL \quad \text{and} \quad K = \frac{k_1}{k_{-1}} = \frac{[RL]}{[R] \cdot [L]}. \tag{7.2}$$

The simplest description of binding between a substrate and an enzyme for catalytically active molecules is based on a similar treatment, where the formation of the enzyme-substrate complex ES from enzyme E and substrate S under steady state conditions is described as an equilibrium reaction, too. In this context the Michaelis–Menten constant K_M corresponds to the equilibrium binding constant K [26].

Many biomolecular interactions, especially in the field of membrane receptors, are more complex and often involve multiple states of the receptor with widely differing binding constants.

Binding constants K found for simple biomolecular interactions range from 10^3 L/mole up to 10^{15} L/mole, with many important interactions ranging between 10^3 L/mole and 10^6 L/mole. The change in free enthalpy ΔG (in kJ/mole) during a binding event at ambient temperature can be estimated from the binding constants K (with $K_0 = 1$ L/mole) by the equation

$$\Delta G = -RT \ln(K/K_0). \tag{7.3}$$

Values for ΔG range from less than 20 kJ/mole to more than 80 kJ/mole. The equilibrium coverage of binding sites of a receptor with increasing concentration of a specific ligand shows saturation. One half of the binding sites of a given receptor are covered under equilibrium conditions, if the equilibrium concentration of the unbound ligand amounts to the inverse of the binding constant K. A decrease or an increase of equilibrium ligand concentration by one order of magnitude changes the occupancy of binding sites to 10% or to 90% respectively. To ensure occupation of a significant fraction of binding sites of a given receptor, e.g. in screening, the ligand must be present at a concentration greater than or equal to the inverse of the binding constant ($c_{\text{Ligand}} \geq K^{-1}$) [2]. Typical values of K targeted in screening are 10^6 L/mole and higher and in consequence concentrations of test substances in primary screening are about micromolar.

The kinetics of the binding process involve at least one binding step and at least one dissociation step. In simple cases the interaction may be described satisfactorily by two related rate constants (k_1, k_{-1}). Values found for the association rate constant are in many cases close to the diffusion limited values of about 10^7 (L mole^{-1} s^{-1}) for small molecules. Depending on the binding constant, values for the dissociation rate constant range from values above 1/s^{-1} up to values exceeding 10^{-6} s^{-1}, which is under practical conditions an almost irreversible process.

7.2.2 Physicochemical Aspects

A brief overview of the physicochemical effects of a molecular binding event are given in Table 7.5. Sensor concepts suitable for the detection of these effects are discussed subsequently.

7.2.2.1 Change in Enthalpy ΔH

As already discussed above changes in enthalpy (ΔH) and in entropy (ΔS) may drive molecular binding processes. In many cases the change in ΔH accounts for at least half of the change in free enthalpy (ΔG) which is up to some 10 kJ/mole. The change in ΔH can be measured directly by calorimetric techniques. A prominent example is the thermometric determination of enzymatic activities. Thermometric measurement techniques based on microbimorphs have achieved a high level of performance [27, 28, 29].

The thermal energy released and the resulting temperature increase during a binding reaction is readily estimated. Assuming a homogeneous phase interaction at a concentration of receptor c_{rec} of 1 μM and a molar reaction enthalpy Δh of 10 kJ/mole,

$$\Delta H = c_{rec}\,\Delta h = 1\,\mu\text{mole/L} \times 10\,\text{kJ/mole} = 0.01\,\text{J/L} \tag{7.4}$$

for complete turnover. In an aqueous medium (thermal capacity $\sim 4\,\text{J}\,\text{K}^{-1}\,\text{g}^{-1}$) the corresponding temperature increase is 2.5×10^{-6} K.

The reaction heat per unit surface released in a heterogeneous phase assay can be estimated likewise. Assuming a rather high surface coverage c_{surf} of 1 pmole receptor/mm^2 and again a molar reaction enthalpy Δh of 10 kJ/mole and complete turnover, the change is

$$\Delta H = c_{surf}\,\Delta h = 1\,\text{pmole/mm}^2 \times 10\,\text{kJ/mole} = 10\,\text{nJ/mm}^2. \tag{7.5}$$

The detection of such small thermal effects is not out of reach. However, typically reactions will proceed over a time span ranging from seconds to hundreds of seconds. This adds the requirement of very low drift and noise to the required high sensitivity. Therefore currently the feasibility of thermometric approaches in the direct monitoring of bioaffinity interactions is limited.

7.2.2.2 Conformational Changes by Induced Fit

The formation of a bimolecular complex may lead to a considerable change in the conformation of the reactants (Figure 7.3). A few examples are hexose kinase, DNA binding proteins [2], antibodies [13], and chemotactical proteins from bacteria [30], where structural changes may lead to displacements in the nm-range.

Structural (conformational) changes of a macromolecule will in most cases not significantly alter its macroscopic physicochemical properties. However, conforma-

Table 7.5. Physicochemical effects of a biomolecular recognition event (QCM = quartz crystal microbalance, SAW = surface acoustic wave, RIFS = reflectometric interference spectroscopy, LOD = limit of detection).

Phenomenon	Observed Effect	Transducer, Detection principle	LOD	Lit.
General effects (important for homogeneous phase detection)				
Enthalpy change (typically 20–50 kJ/mole)	Temperature change, heat flow	Microcalorimeter, microbimorph	50 μ J/K	[27, 28, 29]
Change in conformation (Movement of groups over distances up to 1 nm)	E.g. change in fluorescence characteristics of reporter fluorophore.	E.g. fluorescence spectroscopy	Depends on system investigated and on experimental conditions	[31]
Change in molecular weight due to complex formation (affects related molecular properties, e.g. diffusional behavior)	Change in diffusion coefficient (D_{tran} or D_{rot})	Fluorescence polarization spectroscopy, Fluorescence correlation spectroscopy, Electrochemical detection	Depends on system investigated and on experimental conditions (LOD 50 nM for change in conc. of redox label)	[49, 50, 57]
Surface effects (important for heterogeneous phase detection)				
Change in surface mass loading (app. 5 ng/mm^2 for a protein monolayer)	Change in frequency	Mass sensitive transducer (QCM, SAW, Love wave device)	10 pg protein/mm^2 (QCM) 3 pg protein/mm^2 (SAW), 1 pg protein/mm^2 (Love wave device)	[97, 98, 62]
Change in interfacial impedance	Interfacial dielectric properties, capacity effects or conductance	(Miniaturized) electrode structures	Depends on experimental conditions	[38, 39]
Change in surface refractive index (Δn app. 0.18 within a protein monolayer of 5 nm thickness)	Change in propagation characteristics of guided waves or surface plasmons	Microrefractometers (e.g. surface plasmon resonance, waveguide based refractometry).	10 pg protein/mm^2 (grating coupler, prism coupler), <0.3 pg protein/mm^2 (surface plasmon resonance)	[69, 74, 73]
Change in adlayer thickness (app. 5 nm for a protein monolayer)	Thin film interference	Interference spectroscopy (RIFS)	2 pg protein/mm^2	[81]

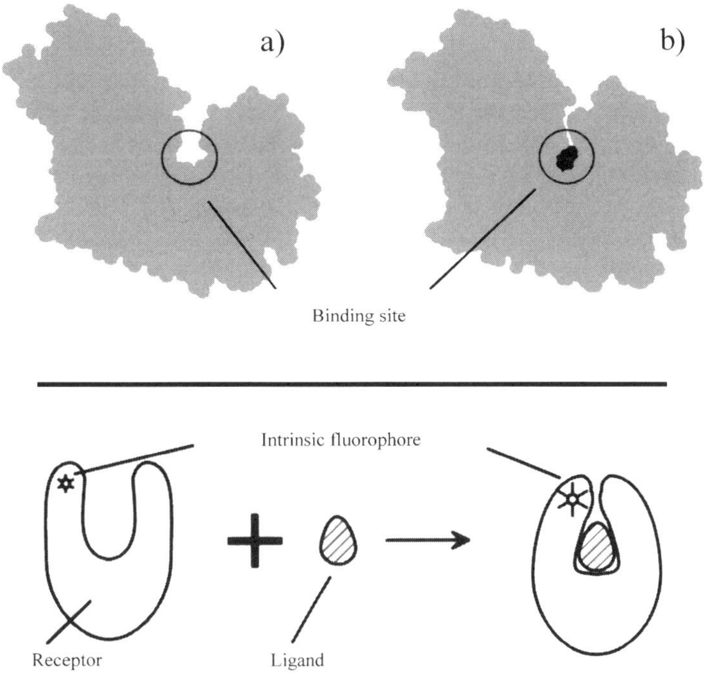

Figure 7.3. Change of molecular conformation during ligand–receptor interaction. Top: contourplot of the enzyme *hexokinase* which undergoes significant conformational change during binding of the substrate. Left (a): without ligand, right (b): with ligand. Binding pocket indicated by a circle. The bound ligand is indicated in black. Bottom: (schematic) altered fluorescence properties of an intrinsic fluorophore (*) in a receptor protein. Conformational changes of the receptor protein induced by a binding process influence the local environment of the fluorophore (modified from [2]).

tional changes may alter the microenvironment of groups with distinct spectroscopic properties. It is a common method in receptor research to monitor the fluorescence characteristics (quantum yield and/or lifetime) of intrinsic tryptophane chromophors of protein receptors. If more than one chromophor is involved, or if the material used is of limited purity the overall changes may be quite small. Recombinant protein techniques can be used to couple an external reporter fluorophore to a receptor protein [31] which leads to improved detection of binding events. However, the procedure is tedious, involves modification of the target and is not highly generic.

7.2.2.3 Molecular Weight and Diffusion Coefficient

In many bioaffinity interactions the molecular weights of the reacting species show a significant difference (see Table 7.1, Figure 7.1). The changes in molecular mass during the binding event may be massive for a low molecular weight ligand, but modest or even insignificant for the receptor compound. Techniques that take advantage of this fact may employ simple size exclusion (dialysis, gel filtration chromatography). Any change in the mass of a molecule or a complex will affect its diffusion coefficient. For a spherical particle, which is a reasonable approximation for many molecular targets, the translational diffusion coefficient D can be estimated by the Stokes–Einstein equation (solvent viscosity η, particle diameter a):

$$D = \frac{kT}{6\pi\eta a}. \tag{7.6}$$

Under identical experimental conditions, D is a function of the reciprocal of the particle diameter, which is roughly a function of the cubic root of the mass of the complex formed. Under practical conditions a change in D of up to one order of magnitude may result from a binding event. The rotational motion of a particle with the volume V is described by its rotational relaxation time ρ, which is the average time the molecule requires to rotate through and angle whose cosine is 1/e [32]:

$$\rho = \frac{3\eta V}{RT}. \tag{7.7}$$

Under the same assumptions as given above, the rotational relaxation time is a direct function of the particle volume or the particle molecular weight. An increase of ρ by up to three orders of magnitude may result from a binding event. Typical translational diffusion coefficients of some biologically relevant substances are given in Table 7.6.

While the translational diffusion coefficient of a small molecule may change by more than ten fold during binding to a protein, the diffusional properties of the protein will change only very slightly. If a ligand is labeled by a suitable marker, e.g. a fluorophore or an electrochemical label, changes in the diffusion characteristics allow one to probe a binding process. The translational diffusion coefficient D can be quantified by the electrochemical determination of limiting currents or by fluorescence correlation spectroscopy. Changes in the rotational relaxation time ρ may be quantified by static or dynamic measurement of fluorescence anisotropy. Technical approaches to the readout of binding assays by diffusional properties of a labeled ligand are discussed in Section 7.3.3.

Table 7.6. Molecular weight and diffusion coefficient of a few biologically relevant molecules (from [26]). Binding between a low and a high molecular weight compound will lead to a complex showing almost the diffusion coefficient of the high molecular weight compound alone.

Compound	Molecular weight (g/mol)	Diffusion coefficient D (cm^2/s)
Saccharose	342	4.586×10^{-6}
Lysozym	14 100	1.04×10^{-6}
Hemoglobin	68 000	6.9×10^{-7}
Urease	480 000	3.46×10^{-7}

7.2.3 Bioaffinity Interaction with Immobilized Compounds

Immobilization of one reactant from a bioaffinity system opens more routes to interaction monitoring (Table 7.5). Compounds can be immobilized directly at a transducer surface or in an immobilizing matrix [33]. Only if an extended three-dimensional matrix is used for immobilization can the density of biochemical compounds (mostly proteins) at a device surface exceed monolayer coverage [34]. The use of an extended three-dimensional coupling matrix at the device surface may restrict access to immobilized compounds and is therefore not very popular in binding studies. The physicochemical properties of a "prototype" protein monolayer are given in Figure 7.4. Immobilization techniques are available from affinity chromatography, biocompatibility studies and biosensor research [35, 33].

Binding processes at a surface involve diffusional transport of the reactants to the surface and the reaction kinetics itself. Therefore a bimolecular reaction with one reactand immobilized at a surface will in many cases proceed slower than the same reaction in the homogeneous phase. If kinetical investigation of binding processes is required, it must be ensured that the diffusional transport of reactants to the transducer surface does not become the rate limiting step. If binding sites are deliberately immobilized at a surface in high density, the maximum binding rates will be limited by diffusional transport. The binding rate is then linearly related to the concentration of the binding molecule in solution. Problems can be caused by the fact that at a high degree of surface coverage binding processes may be affected or hindered.

7.2.3.1 Mass Loading

A typical spherical protein molecule (50 000–100 000 g/mole) has a diameter of about 5 nm. A monolayer of protein molecules at a surface will therefore have a thickness of a few nm and, assuming dense packing, a surface coverage of about 10^{-7} moles/m^2 ($\Delta 10^{17}$ molecules/m^2). This corresponds to a mass loading of about 10 mg/m^2 or 10 ng/mm^2. These values correspond well to surface coverages found in experimental studies [36, 37]. Transducers that respond directly to mass loading or a related physical quantity can be used to measure changes in surface coverage during a

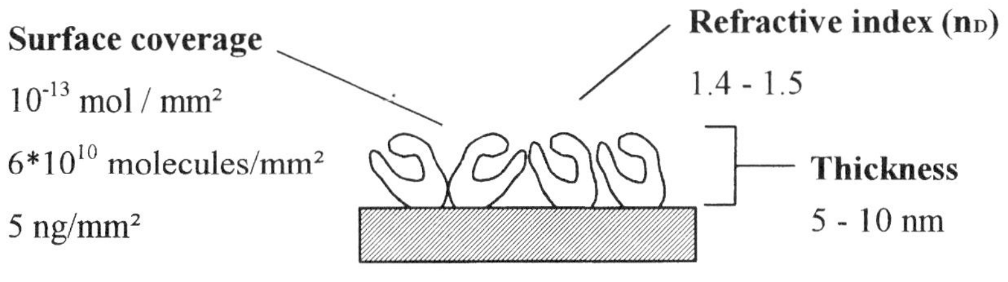

Figure 7.4. Prototype protein monolayer and important physicochemical characteristics observed in an aqueous environment (protein molecular weight app. 60 000 g/mole).

heterogeneous phase binding assay. The changes found for the binding of a low molecular weight compound (ligand) to a surface covered with receptor molecules will depend on the density of specific binding sites available on the surface. Under practical conditions only a few 10 $\mu g/m^2$ or 10 pg/mm^2 of a small ligand will be bound by a dense monolayer of receptor molecules. This is at the lower limit accessible by transducer principles currently available.

7.2.3.2 Conductivity

The impedance of an electrode in contact with an electrolyte is significantly affected if one or more adlayers with distinct dielectric properties are formed. Impedance spectroscopy serves to characterize interfacial properties and has been used successfully to monitor surface bound affinity interactions. The addition of a redox couple (redox marker) to the solution allows us to monitor "shielding" of the electrode surface by protein binding [38, 39, 40]. Limitations are the highly non-linear response of an electrode to increasing surface coverage. The impedance behavior of an electrode surface is primarily determined by highly conducting defects altering the real part of the impedance, so that changes in fractional sub-monolayer coverage do not exert major changes. Also the formation of secondary or tertiary layers on a precoated electrode leads only to minor changes as compared to the formation of the first layer.

7.2.3.3 Refractive Index

Organic matter shows a refractive index which in the UV-VIS range is higher than that of water. For low concentrations of protein (up to 0.4 g/l) the relation between the protein content of a solution and the observed refractive index is linear ($dn/dc = 0.018$ ml/g [4]). A surface bound affinity reaction will lead to a local enrichment of organic matter (receptor or ligand molecules), which in turn will lead to a localized increase in refractive index. A model calculation (5 ng/mm^2 protein; thickness of the hydrated protein layer 10 nm) indicates a refractive index increase of about 0.1. The actual refractive index of this layer would be about 1.4 to 1.45.

7.3 Probing Bioaffinity Interactions: Techniques and their Applications

In this section we list typical affinity-based applications and then discuss the impact and potential of sensor-based approaches for interaction analysis and screening. Homogeneous phase approaches and heterogeneous phase transducers will be discussed separately.

7.3.1 Historical Aspects

Historically bioaffinity was first employed as a tool in clinical diagnostics [42]. Here, biomolecular interaction was used to improve analytical chemistry. Quantification of binding events was aided by the use of markers (radioactivity, enzyme labels, fluorescent labels, and more recently chemiluminescent and electrochemiluminescent labels). The primary field of application are assays in clinical diagnostics [43], with food and environmental applications coming up [44]. The market for diagnostic use of affinity interactions is dominated by (disposable) assay systems with labeled compounds. The transfer of bioaffinity concepts and techniques in the field of biosensors initially adapted the methods developed in the field of diagnostic use of bioaffinity systems. Therefore research was directed primarily towards the design of tools and probes for a simplified, faster, or easier and more convenient readout or quantification of bioaffinity interactions (for surveys see Scheller [9] or Sutherland [45]). A lot of research has been carried out to link labeled immunoassays to more or less sophisticated transducers. Motivation for these research activities results from different requirements:

- assays with reduced cost;
- automated and semicontinuous assays;
- assays with improved limits of detection;
- faster assays with reduced time to result;
- simplified assays with improved handling by untrained personnel.

While so far no "immunosensor" approach has managed to undercut the costs achieved for diagnostic routine immunoassays, some of the other target requirements have materialized. In particular "affinity sensor" approaches, dispensing with the need for a label or a solid phase or that allow time resolved monitoring of binding processes, open up new routes and deliver a new quality of information. Among the new fields emerging are the study of molecular interaction with sensor-based systems and the above-mentioned approaches in therapeutic drug screening. It is important to note the differences between the well-established use of affinity interactions in analytical chemistry and in the more recent applications. A few characteristics are compiled in the next section in Table 7.7.

7.3.2 Typical Applications and their Specific Requirements

Binding assays are well established in *clinical diagnostics* (*I*). Biomolecular interaction analysis – *BIA* (*II*) – and High Throughput Screening – *HTS* (*III*) – are discussed here as more recent applications. While the underlying molecular interactions are identical for all three fields, the requirements and the restrictions to be observed differ significantly.

Table 7.7. Characteristics of three different applications of affinity analysis.

	(I) Clinical diagnostics	(II) Biomolecular interaction analysis	(III) Binding assays for drug screening
Receptors	1 class (antibodies)	many classes	many classes
Labelling	standard	may cause problems	may cause problems
Immobilization	standard	may cause problems	may cause problems
Optimization of individual assays	standard	acceptable	not wanted
Throughput (no. of assays run per assay system)	very high	low	high
Typical limits of detection	pM	nM	nM

Binding Assays in Clinical Diagnostics (I)

A typical clinical diagnostic immunoassay is well optimized with significant effort, which is based on vast experience from many similar assays. A diagnostic assay is sold in high quantities once it is established. The limits of detection are sometimes very low. The huge market justifies high effort and costs while establishing the system. As the molecular species involved are employed as tools only, modification of their behavior by the assay concept will not matter, as long as the assay itself can be established. This allows one to use all labeling and immobilization strategies which are available and leads to good analytical performance.

Biomolecular Interaction Analysis (II)

Biomolecular interaction analysis aims at a much higher degree of information than quantitative diagnostic assays. Typically a full description of the thermodynamics and eventually the kinetics of the interaction process is desired. If multiple compounds interact, interaction sequences and interfering compounds may be of interest. This puts much more stress on the integrity of the investigated system than in routine immunoassay. The range of molecular targets is much broader than in immunoassays, which means that standard procedures may be less promising. Any impact of the probe or transducer used on the data obtained may compromise the significance of the results. Therefore systems with as little interference as possible are preferred. Under certain aspects radiolabels are ideal, as radiolabeling can be carried out without changing the chemical structure. Also radiotracer assays can often be run without immobilization in the homogeneous phase. However, the considerable cost involved with the preparation and handling of radiotracer materials are a significant drawback which opens the possibilities for competing homogeneous phase techniques or for techniques requiring immobilization of one reactant. Optimization of an assay procedure is acceptable to improve the quality of the data obtained. The concentrations investigated may cover the full range of physiologically relevant concentrations. However, many practical applications today do not go far below the nanomolar level.

Binding Assays for Pharmaceutical High Throughput Screening (III)

Pharmaceutical screening must identify molecular interactions under physiological conditions. As in *BIA*, impact of the test system on the interaction process should be kept to a minimum. While optimization of a screening assay for a certain target is acceptable, optimization of the assay for each individual test substance is precluded by the high number of samples. In primary screening rather a binary information (yes/no) than a quantitative description of molecular interaction (thermodynamics, kinetics) is required to identify possible hits. The test should detect unambiguously binding effects at a preset level. The simplicity of a test is critical for automation and cost effective implementation and is therefore an important parameter. A complex reaction sequence with multiple addition of reagents or with washing steps is less desirable than a wash-free one-pot assay, where compounds are mixed and–after a certain time–a result can be obtained. The amounts of substance required should be as low as possible. Bead-oriented synthesis following the "split-and-combine" approach (one substance synthesized per bead) in combinatorial chemistry limits the amount of substance available for an assay to a few 100 fmole [46], while macrobead-oriented synthesis approaches may yield up to 1 nmole of substance [17]. If substance bases are to be screened substance consumption should also be low, as replenishing substance stocks is often difficult. Typically amounts of up to some 10 ng per test may be acceptable.

Not only the amount of sample, but also the amount and the purity of receptor (protein) required is of importance. In established binding assays a few 10 ng up to a few 100 ng of receptor per assay are common [47, 22]. If crude receptor preparations (e.g. cell extracts) can be used, the cost for preparation of the receptor compound can be significantly reduced.

The throughput of an assay is of high importance in screening. If a throughput of 100 000 substances is targeted per month, which is a common value, the time available per single assay is only 10 to 20 seconds. This requires either a very fast detection system or an effective approach to the parallel detection of binding reactions. These requirements rule out several approaches which may be promising in BIA but require extensive optimization or lack the high throughput required in HTS. Table 7.7 gives an overview of the differing requirements for the three fields of application mentioned.

7.3.3 Homogeneous Phase Detection

The ideal approach to monitor affinity interactions would be a label-free homogeneous phase assay. This ideal is met by techniques using intrinsic properties of one of the reactants. As pointed out above these approaches are not generally applicable. Radioisotope assays require considerable cost and effort in preparation of labeled test compounds.

A wider range of techniques allows monitoring of binding events in the homogeneous phase, but with the use of a label. This eliminates artefacts and influences from immobilization steps. The physicochemical effects used are the change in the diffusional behavior of a small ligand on binding to a comparably big receptor molecule. Therefore, a label is attached preferentially to a low molecular weight reactant. Labels can be electrochemically active groups or fluorophores. The introduction of a label may lead to a considerable change in binding behavior. This problem can be circumvented by the use of a competitive assay format. In this assay format a labeled compound with known binding properties is used to probe the occupancy of the receptor by the structure under investigation. In this format no labeling of the receptor or the target molecule is necessary. Strong interaction between the receptor and the ligand under investigation will lead to replacement or exclusion of the labeled ligand from the binding site. A schematic comparison between direct and competitive assay formats is given in Figure 7.5.

Fluorescence Anisotropy (fluorescence polarization)

The first binding assay with considerable success which employed changes in molecular weight was based on the increase in the rotational relaxation time of a fluorophore attached to a low molecular weight ligand on binding. Fluorophores attached to the low molecular weight compound are used as a probe. The determined parameter is the fluorescence anisotropy. The sample is illuminated with linearly polarized excitation light. Therefore only fluorophores with proper orientation will be excited. If rotational relaxation is slower than the decay of the excited state, the emitted fluorescence light will be polarized to some extent too. This principle called fluorescence polarization immunoassay (FPIA) [32] has been applied to diagnostic

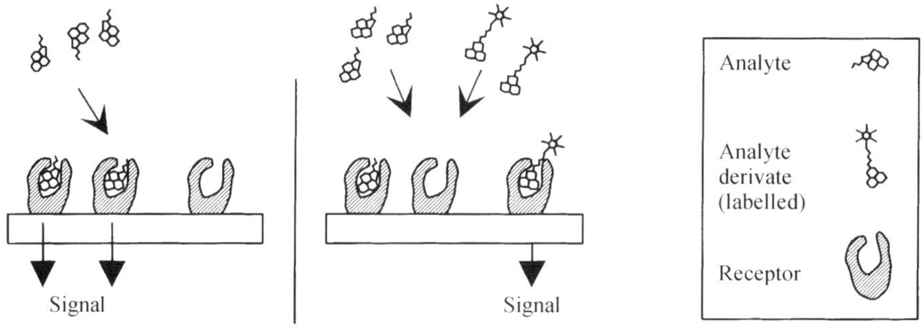

Figure 7.5. Different test formats for affinity-based analytics. Left: direct assay–the binding of the analyte is directly converted to a signal. Right: competitive assay–the analyte competes with a labeled species for the receptor binding site. Only labeled molecules are detected.

assays for various drugs. Binding of the labeled compound to the receptor will increase the fluorescence anisotropy. The measurement of static anisotropy is based on excitation with polarized light and measurement of the orthogonally polarized compounds of the fluorescence light. This is quite straightforward and requires little technical effort. More information can be obtained by dynamic anisotropy measurements with pulsed or modulated excitation and time resolved or phase sensitive detection [48]. Commercial systems are available for basic research, as well as HTS [49].

Fluorescence Correlation Spectroscopy

An elegant technique for determination of translational diffusion coefficients of fluorescence labeled molecules was introduced to affinity applications by Eigen and Rigler [50]. An optical set-up with confocal excitation and detection beam paths is used to focus an excitation laser beam into a liquid sample. By the z-dependence of the confocal optics, the detection of emitted fluorescence light is effectively restricted to the focus of the excitation light beam. The probed volume can be as low as a few fl (Figure 7.6). At concentrations recommended for the technique (nanomolar and

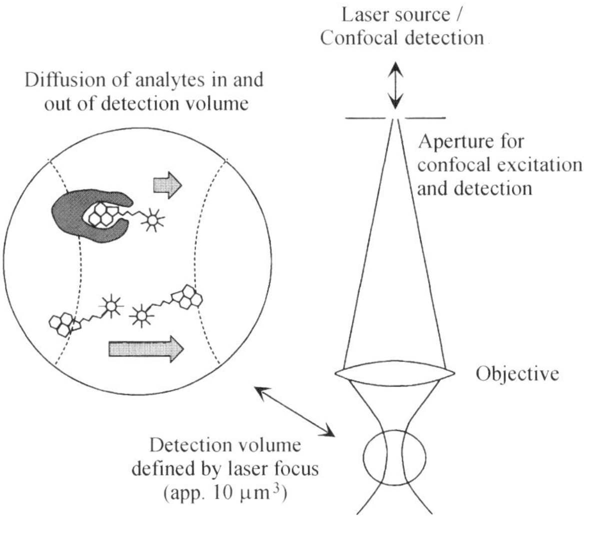

Figure 7.6. Principle of fluorescence correlation spectroscopy (FCS). Fluorescence is excited by a laser beam focused by a microscope objective into the sample. The confocal set-up limits the volume from which fluorescence is detected to the focus of the laser beam (a few μm^3). The fluctuation of the fluorescence signal is correlated with the diffusion of fluorescent molecules in the sample in and out of the detection volume. Autocorrelation analysis of the fluorescence signal allows one to determine the diffusional properties of the fluorescently labeled components. Enlarged section not to scale.

below) only a few analyte molecules – typically less than 10 – are found within the detection volume at a time. The statistical fluctuations in the photon flux over time are correlated to the concentration and the diffusion coefficient of the analyte molecule. Autocorrelation analysis of the time resolved photon flux is used to extract information about binding state and concentrations in the monitored sample. In principle the method may operate with cycle times of one second, the detection volumes are very low and the concentrations required are in the nanomolar range. Commercially available instrumentation allows one to detect changes in the diffusional characteristics if the molecular weight of the complex exceeds the molecular weight of the labeled compound by at least a factor of two.

Electrochemical Detection

Electrochemical techniques may be used to quantify redox-active substances. The electrode current depends on the concentration and the diffusion coefficient of the analyte in the solution. The limiting current i found at a planar electrode in a nonstirred solution is given by the Cottrell equation:

$$i = nFAc\sqrt{\frac{D}{2\pi t}} \tag{7.8}$$

with the number of electrons transferred per molecule n, the Faraday constant F, the electrode area A, and the concentration c. Over the last few years electrochemical methods gained additional attention in modern analytical chemistry as miniaturized electrodes, microdot electrode arrays and interdigitated microelectrode arrays became available. With these devices detection limits in the lower nanomolar range are achievable without preconcentration. Furthermore these miniaturized devices allow investigations in small sample volumes ($< 0.1\ \mu l$).

Such devices were used for applying different techniques like Cyclic Voltammetry and Chronoamperometry. Several pulse methods like Differential Pulse Voltammetry or Square Wave Voltammetry are reported [51, 52]. An overview of electrochemical approaches in pharmaceutical chemistry has been given recently by Kissinger in [52].

An array of microdot electrodes (see Figure 7.7) shows no influence by stirring the samples and higher current densities are a result of the hemispherical diffusion of the analyte to the electrodes. Additionally the charging current is small due to the small electrode area. As a result the signal-to-noise-ratio is enhanced which leads to a decreased detection limit.

Interdigitated microelectrode arrays with typical electrode gaps in the order of 5 μm and below are operated in the redox cycling mode especially with reversible redox active substances. Figure 7.7 shows a typical example of such a transducer consisting of two working electrodes, counter- and reference-electrodes. In the steady state one electrode generates electroactive species (the generator electrode) which diffuse to the neighboring electrode (the collector electrode) where the electrochemical regeneration occurs. This effect leads to an electrochemical feedback. The resulting current is

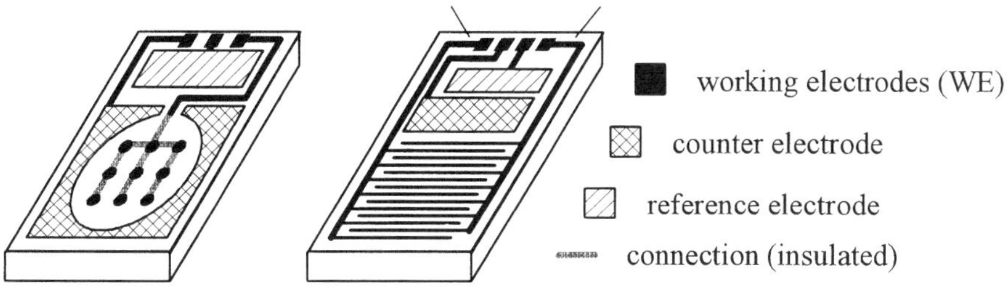

Figure 7.7. Microelectrode geometries. Left: array of microdot electrodes (typical diameter: 10 μm; spacing: 100 μm) [56]. Right: interdigitated microband electrodes for redox cycling (band width $\leq$ 10 μm; band distance $\leq$ 10 μm) [54].

typically amplified by one order of magnitude and therefore the detection limit is improved significantly [53].

The amount of protein needed per assay is dictated by the concentrations of tracer compound required for unambiguous detection of binding events. While this will dictate receptor concentrations up to 100 nanomolar, the potential for microelectrodes requiring only very small assay volumes may lead to small receptor consumption per assay. Furthermore the basic transducers are accessible by low-cost standard microfabrication techniques [54, 55]. Also the required readout circuitry is of limited complexity and cost. First investigations were now performed with parallelized electrochemical devices and the compatibility of miniaturized transducers with the required microtiter format was shown. Electrochemical approaches were shown to be especially suitable when fast sampling is required as in the case of HTS (see Table 7.2).

7.3.3.1 Interaction Analysis

Of the three methods described above, the fluorescence correlation spectroscopy is the most promising for interaction analysis. The high intrinsic time resolution achieved by the autocorrelation of the photon flux signal allows one to resolve binding processes at their intrinsic time scale. The ability of FCS to probe small sample volumes makes this technique unique for *in situ* studies of binding effects, e.g. of the intracellular distribution of receptor occupation.

Electrochemical detection and fluorescence polarization are restricted to slower processes. If binding kinetics are to be investigated, time resolved measurements with controlled timing of the assay can be employed to investigate reaction kinetics. The low limits of detection of the fluorescence techniques allow one under ideal conditions to operate fluorescence polarization studies with tracer levels in the subnanomolar range. Electrochemical techniques are still restricted to higher concentrations. All homogeneous phase approaches are particularly attractive, if procedures are involved, which allow the introduction of a label by routine protocols. This is the case for almost all systems involving the handling of DNA, as labeling is easily integrated with synthesis or isolation procedures like PCR (polymerase chaining reaction).

7.3.3.2 Pharmaceutical Screening

Labeled homogeneous phase assays avoid the influence of immobilization procedures. The label is preferably introduced to the low molecular weight reactand. As the labeling of each test substance is neither feasible nor desirable a competitive test scheme is standard in screening applications. This requires a reference substance which is known to bind to the investigated binding site of the receptor molecule studied. The synthesis of this reference compound must be optimized with respect to acceptable yield, effective purification of the reaction mixture, and low interference of the label with the binding process. As the preparation of a labeled reference compound is done only once per target binding site, this effort is often acceptable. The assay involves mixing of target, labeled reference compound and test substance. Equilibrium binding assays are not time dependent, but a certain incubation time must be ensured to allow establishment of equilibrium. Without any interaction between receptor and test substance, maximum binding of the reference compound to the receptor will occur and low mobility of the label will be detected. With increasing interaction between test substance and binding site replacement of the labeled compound will happen which leads to increased mobility of the label.

High throughput is a major issue in primary screening. Fluorescence polarization binding assays are already used for screening purposes. The equipment for fluorescence polarization assays is of limited complexity. At the moment hardware for microtiterplate assays is available based on static fluorescence anisotropy measurements, which allows detection of assay volumes around 100 μl binding reactants in nanomolar concentrations [49]. The readout time per sample can be reduced to less than one second. Parallel readout of microtiterplates is possible, which ensures sufficient throughput, even for demanding HTS applications. The technique is relatively mature, as considerable experience with clinical fluorescence polarization immunoassays exist.

Fluorescence correlation spectroscopy is especially promising, for three reasons:

(a) The required sample volumes are low; (the volume probed by the focus of the laser beam is a few ten fl), while a practical assay today reaches sample volumes of a few μl.

(b) The measurement technique is potentially very fast. The acquisition of a signal autocorrelation function in the relevant time range can be done within less than a second. If a time below 10 seconds per cycle can be achieved under practical conditions, a single channel system would be appropriate for HTS.

(c) Best performance is expected for concentrations of the labeled compound ranging from picomolar to nanomolar. This meets the requirements of most of the established HTS assay systems.

As the technique is relatively new, only little experience exists so far about complications, sources of error and throughput achievable under practical conditions. FCS technology is explicitly offered for screening purposes. A system (CONFOCOR^{TM}) jointly developed by EVOTEC (FRG) and ZEISS (FRG) adapted to a sample handling system of TECAN (CH) has been sold since 1996. The relatively high price of the equipment (app. 150 000 US$) is not a principal limitation in screening applications, but may delay the quick introduction of the technique to HTS.

Electrochemical assays so far have not been used for screening, but research has been carried out and looks promising [57, 58]. Typically minimum concentrations to be detected are in the ten nanomolar range. As changes in the concentration of the free label are detected, the concentration of target binding sites must also be relatively high. Depending on the readout mode, the technique is considerably fast. Stepped potential techniques (e.g. chronoamperometry) require only a few seconds, while complete potential scans may be acquired within ten seconds. The use of miniaturized electrode geometries, which are available at low cost from microstructuring approaches, can reduce the area of a complete electrode system (working electrode, counter electrode, reference electrode) to less than 1 mm^2 [54, 55, 58]. This makes sample volumes of 1 μl feasible. As the readout circuitry is usually simple, parallel detector arrangements can be established to increase throughput. A set of electrodes can easily be arranged as a comb-like structure. This concept is currently adapted to HTS [57] (Figure 7.16). The comb structure will allow straightforward testing in standard microtiterplates. 96-well and at least 384-well plates should be readily adaptable to this approach. The limits of detection achieved for dopamine and $[Ru(NH_3)_6]^{3+}$ as an electrochemical marker currently are below 50 nM. A further improvement is expected from the use of redox cycling at interdigitated microband electrodes. The use of arrayed microelectrode structures with high density for correspondingly low sample volumes is in the development stage. Manufacturing costs may allow for disposable units, which would facilitate handling and may increase throughput. If a high degree of reproducibility can be ensured, calibration and pretreatment steps for each single electrode can be omitted, which would add to the attraction of the concept.

7.3.4 Heterogeneous Phase Detection

Heterogeneous phase assays are more readily adapted to transducers than homogeneous phase assays. Consequently the number of heterogeneous phase transducers applied to affinity analysis exceeds by far the number of transducer-based approaches in homogeneous phase detection. A lot of experience in heterogeneous phase chemistry or biochemistry and in immobilization strategies exists from conventional solid phase immunoassays and affinity separation techniques [35]. However, the effects of immobilization on the binding behavior of a receptor may be considerable [59, 60]. This must be kept in mind, when designing a heterogeneous phase assay for the study of biomolecular interaction or when extrapolating results from heterogeneous phase studies to homogeneous phase behavior of the same compounds investigated.

Mass Sensitive Transducer

Over the last decade mass-sensitive transduction has become a feasible tool to study interface interactions. These transducers are based on acoustic waves which will propagate in a piezoelectric crystal. The device usually operates as a frequency-determining element in an oscillator circuit, and therefore the frequency shift Δf occurring during an interface change is related to the wave velocity v:

$$\frac{\Delta f}{f_0} = \frac{\Delta v}{v_0}. \tag{7.9}$$

In the late 1950s Sauerbrey described the mass-dependence on quartz crystal microbalances (QCM) [61]:

$$\Delta f \approx \Delta m f_0^2 \tag{7.10}$$

Bulk acoustic wave devices like the QCM (see Figure 7.8) typically operate at resonance frequencies of a few 10 MHz. The measurement effect employed is a frequency decrease of the QCM with increased mass loading. As shown in equation (7.9) the sensitivity of the QCM increases by a factor of four if the resonance frequency is doubled. Higher operating frequencies are therefore necessary to achieve higher sensitivities. This could be realized by using very thin quartz substrates or by evaluating overtones. Typical detection limits achieved with 10 MHz QCMs are about 10 pg/mm^2. This meets the above-mentioned requirements for pharmascreening approaches.

Another type of mass-sensitive transducers is known as surface acoustic wave devices. In this case a surface acoustic wave (SAW) is generated on an elastic transducer surface between interdigital electrodes. In contrast to the QCM the SAW-technique only detects the surface acoustic wave. These devices operate in higher frequency ranges (typically $f_0 > 100\,\mathrm{MHz}$) and lower detection limits compared to the QCM are reported (LOD = 3 pg/mm^2). For application in aqueous solution the SAW technique needs to be modified in order to deal with possible damping. Therefore the use of horizontal polarized shear waves is necessarily combined with mass gratings on the sensor device. The resulting wave type is called a surface transverse wave (STW) and causes only small energy losses in water.

A new approach in bioaffinity studies is given by using the Love wave devices (see Figure 7.8). This method combines the advantages of the bulk acoustic with those of the SAW technique. A SAW-device is typically coated with a thin layer of a thickness in the range of the acoustic wavelength. A suitable layer material is SiO_2 with a thickness of about a few μm. The acoustic wave propagates in the layer with a lower wave velocity than on the piezoelectric substrate. This leads to a higher sensitivity of the device. Preliminary investigations show detection limits below 1 pg/mm^2 [62].

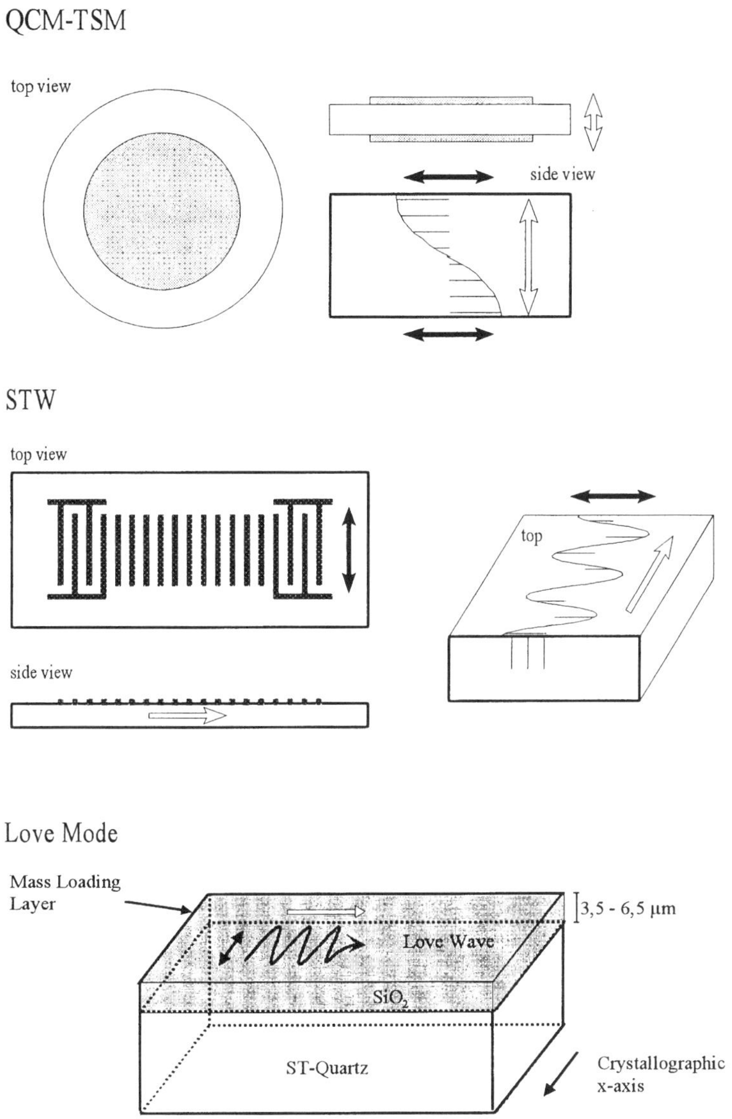

Figure 7.8. Quartz crystal microbalances. Bulk acoustic wave devices (top) induce a thickness shear oscillation in the quartz plate between the electrodes. Surface acoustic wave devices (bottom) use interdigitated transducers to excite and to detect surface waves.

More commonly used acoustic wave devices have surfaces of gold or silica which can be effectively modified by thiol or by silane chemistry. The active surface areas required amount to a few mm^2. Further references to these techniques are given in this series [63] and in other publications [64, 65].

Impedance Spectroscopy

Impedance spectroscopy at electrode surfaces may be carried out with almost any bare electrode surface (the principle is described in [40]). It is recognized as a general tool to study interfaces and molecular properties of surface bound, but also of free molecular species. In studies directed towards affinity analysis very low levels of detection have been realized [39]. The strong nonlinearity in response with increasing coverage and with increasing thickness of a (defect free) adlayer have stimulated research towards layer systems with improved and more controllable performance. Measurements at a fixed frequency are relatively simple to perform. Frequency dependent measurements to study electrodes during surface processes (complex impedance analysis [40]) give very detailed information about capacitive and resistive parts to the overall impedance. This may be correlated with concentrations of free charges and layer capacities. The technique is outstanding in the simplicity of the transducer structure itself. The limits of detection vary considerably and practical devices are under development. The integration of electrodes and drive circuitry in a microstructure will reduce requirements for shielding, referencing, etc., and hence lead to a sensitive practical impedance-based device.

Microrefractometry

As discussed above, the enrichment of organic matter at surfaces will lead to an increase in the refractive index. A variety of optical transducers respond to changes in refractive index close to the device surface. The propagation behavior of a guided wave depends on the dielectric characteristics of the surrounding medium. This effect is mediated by the evanescent field which penetrates from the guiding medium into the environment. As the penetration depth of the evanescent wave is a few hundred nm, all effects mediated by the evanescent field are highly localized. Measurements in aqueous phase are strongly affected by temperature changes, as the refractive index of water changes significantly with temperature ($dn/dT \approx 10^{-4}$/K at 298 K). Transducer elements have been described in detail in previous volumes of this series [66, 67]. Therefore only a brief review will be given here.

Surface plasmons are collective fluctuations of the electron density at the surface of an electrically conductive material–typically gold or silver. Surface plasmon waves have an electrical field distribution with a maximum at the surface and an exponential decay into the metal and the adjacent medium. The penetration depth of the evanescent field into the medium and consequently the region probed is a few 100 nm. A typical set-up for sensor systems based on surface plasmon resonance (SPR) [68] is given in (Figure 7.9).

SPR is detected as a strong attenuation of the reflected light beam. Coupling conditions are sensitive to the surface refractive index and are typically probed by one of two approaches. Either monochromatic light is used and the angle of incidence is varied to find the minimum of reflectance or polychromatic light is used at a fixed angle of incidence and the wavelength of minimum reflectance is determined.

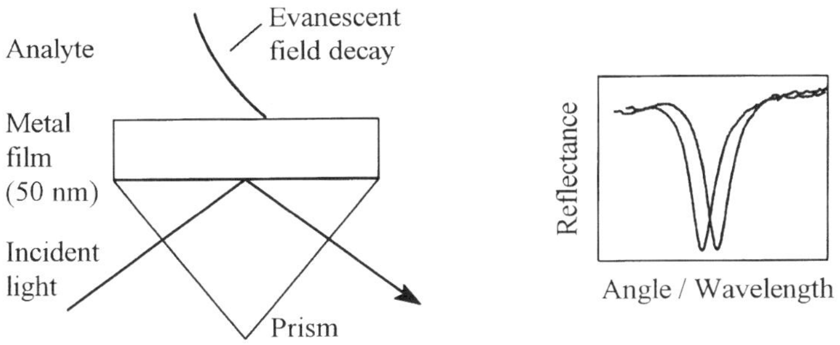

Figure 7.9. Left: Kretschmann set-up for the excitation of surface plasmon waves. Plasmons are excited at the interface between metal and analyte by light totally internally reflected at the base of the coupling prism. Right: the angular or the spectral distribution of reflectance shows a pronounced minimum.

Changes in refractive index below 10^{-6} can be resolved. A variety of optical set-ups have been proposed and in part commercialized. Devices of the angular resolved Kretschmann type [68, 69], spectrally resolved Kretschmann type [70] fiber based [71], or waveguide based [72] with limits of detection below 1 pg protein/mm^2 are reported.

In contrast to plasmon excitation, light propagating in nonabsorbing waveguide structures (and without absorbing cover layers) is not attenuated by changes in the environmental refractive index. The propagation velocity changes, and is reduced by the increasing refractive index in the vicinity of the waveguide. Several set-ups have been proposed and two of them have been commercialized to date.

The first integrated optical device used in bioaffinity analysis was a grating coupler structure embossed in a monomode film waveguide proposed by Lukosz and co-workers [73]. A periodic grating on the surface of a waveguide can be used for in- or outcoupling of radiation from the waveguide. The angle of deflection (coupling angle) depends on the wavelength and the grating period (Figure 7.10). Binding of organic matter on top of the grating alters the coupling angle. The limit of detection is about 10 pg protein/mm^2.

The second coupler device which has found application in a commercial affinity detection system is a prism coupler. In a prism coupler light is coupled into a film waveguide from a prism separated by a small gap (filled with low refractive index material) from the waveguide (Figure 7.11). Light totally internally reflected at the base of the prism can excite a mode in the waveguide. The coupling conditions of a prism coupler are sensitive to the refractive index at the film waveguide. When the coupling conditions are met, light will be coupled into the waveguide, and after some distance back into the prism. This leads to different paths between coupled light and noncoupled light, which can conveniently be detected by forced interference between the p and the s state of incident light [74]. The transducer is just a prism with a thin film structure at its base. A commercial system for interaction analysis is available (see below). The limit of detection is about 10 pg protein/mm^2.

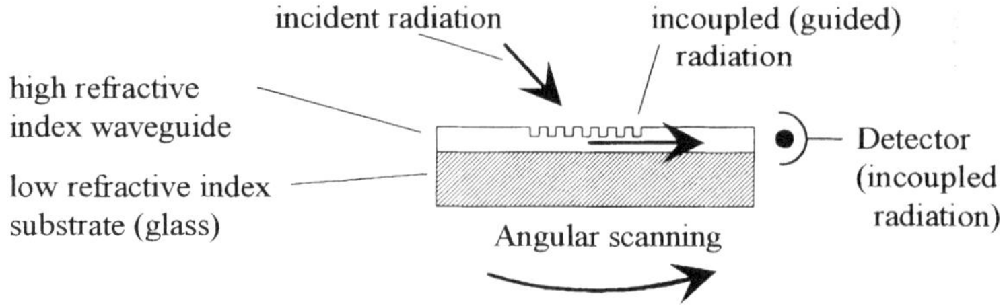

Figure 7.10. Grating coupler structure as surface sensitive refractometer. Typical waveguides are 100–200 nm of TiO_2 or Ta_2O_5. Coupling conditions can be probed by scanning or by imaging techniques.

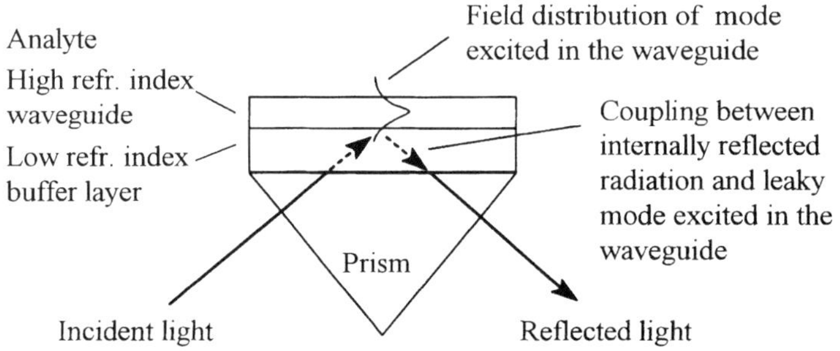

Figure 7.11. Prism coupler structure used as microrefractometer. Light incident at the base of the prism can excite evanescently a leaky mode in the thin film waveguide coated on the low refractive index buffer layer. After some distance the radiation will couple back to the prism. Coupling leads to a phase change which can be detected interferometrically.

Integrated Optical Channel Waveguide Devices

Integrated optics with monomode channel waveguides are routinely used in long-range, high-speed telecommunication. Monomode channel waveguides allow a very straightforward implementation of interferometer structures to be built. A waveguide is split into a measurement and a reference path and recombined after some distance. Any phase difference introduced while the beams were split can be detected by interference effects (Figure 7.12). Binding of organic matter at the surface of the measurement arm leads to a modulation of intensity at the interferometer output. Limits of detection are about 10 pg/mm^2 [75]. Today the costs of manufacturing such devices are prohibitive, but the implementation of IO devices in silicon or in polymers may change this [76, 77]. For example, Bookham Technology (UK) is developing integrated solutions based on its patented ASOC™ silicon technology.

Thin-film Reflectance

Thin films (10 µm and below) can strongly affect the propagation of incident light. This is due to reflections at each of the interfaces of a thin film, which causes multiple

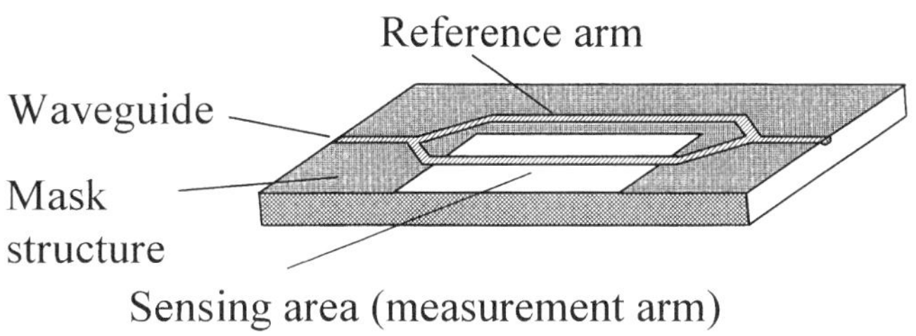

Figure 7.12. Integrated optical Mach–Zehnder interferometer realized by monomode channel waveguides. A mask structure shields most of the waveguide structure. The analyte has access only to one of the interferometer arms.

beams, with small phase shifts. As the phase shifts are small, coherent interaction of these beams occurs. This is used in thin-film interference filters or in dielectric mirror systems [78]. A sensitive technique to monitor thin-film structures is ellipsometry, where the reflectance for p- and s-polarized light is used to determine the properties of a thin-film system [79]. Applications for biomolecular interaction processes have been reported [70]. An imaging (spatially resolving) ellipsometry system for binding studies was described recently by Jin *et al.* [80].

A very simple approach based on thin-film reflectance evaluates the spectral modulation of the reflectance of a thin film under white light illumination. No polarization dependence is evaluated. The spectral characteristics of a thin film are highly dependent on its thickness and therefore any adsorption of organic matter leads to a change in the reflectance spectrum, which can be quantified by reflectometric interference spectroscopy (RIFS) (Figure 7.13) [81]. The transducer is simply a nonstructured thin-film system on a transparent substrate. Limits of detection are about 2 pg/mm^2 [82].

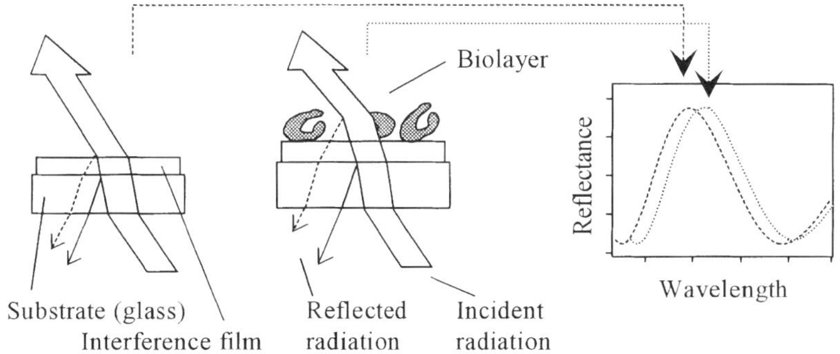

Figure 7.13. Thin film reflectance at thin films shows characteristic spectral modulations due to interference of reflected beams. Differences of reflectance before (———) and after (·····) a binding event allow detection and quantification of binding processes to be carried out.

Atomic Force Microscopy

Atomic force microscopy (AFM) was introduced as a technique for imaging surface topography on the nm-scale. A tip with small radius of curvature is fixed at the end of a cantilever structure. The latter serves as a spring. By scanning this tip towards or away from a surface, attractive and repulsive forces can be detected as a deflection of the cantilever [83]. Recent studies showed the potential of monitoring directly the biomolecular interactions between a tip which was modified with one compound of a ligand/receptor system and a surface modified with the complementary compound [84]. Interaction is monitored by force–distance spectroscopy: the tip is brought in close contact to the surface and then withdrawn again. Specific binding is recognized by a hysteresis in the force/distance curve between the attractive and removal motion. The approach requires immobilization of both interacting components. The required overall surface area involved and the necessary amount of substance can be very low and the technique is relatively fast. At the moment AFM is still an expensive research tool. However, manufacturing by micromachining techniques is possible, and mass applications, e.g. the use of scanning probe microscopes in computer storage devices, are considered. This may open up considerable potential for affinity studies with this approach.

7.3.4.1 Interaction Analysis

Heterogeneous phase affinity analysis was pioneered by Pharmacia Biosensors of Sweden [69], now BIACORE AB. A system based on SPR (angular resolved Kretschmann type device named BIAcore) was introduced to the market in 1990. BIACORE scientists also promoted the term "*Biomolecular Interaction Analysis*". Two further systems are available either based on a grating coupler (BIOS 1; by ASI, CH) [73] or a prism coupler (IASYS; by Affinity Sensors, UK) [74] – an overview is given in Table 7.8. The introduction of further systems has been announced; e.g. Technobiochip (Italy) has announced a multichannel system based on thickness shear mode QCM. The time resolution of the commercial devices is around 1 Hz. So far no system sells at prices below 50 000 US$. Until the end of 1996 about 600 BIAcore systems have been installed world wide, which are used in basic and applied

Table 7.8. Commercially available system for BIA.

Name	Transducer principle, performance	Vendor
BIAcore	SPR, microflow system, up to 4 channels, LOD: <1 pg/mm^2	BIACORE AB, Sweden
BIOS 1	Grating coupler, single channel, LOD: 10 pg/mm^2	ASI, Switzerland
IASys	Prism coupler ("resonant mirror"), up to 2 cuvettes with stirrer, LOD: 10 pg/mm^2	Affinity Sensors, Thermo Group, UK

molecular biology. A significant fraction of the instruments is installed in research laboratories of the pharmaceutical industry.

A LOD for mass coverage of 10 pg/mm^2 is typical for interaction studies with macromolecules (proteins). The latest instrument generation introduced by BIA-CORE in 1995 can detect less than 1 pg/mm^2. This allows operation at very low surface coverages, which improves the quality of kinetic data. High sensitivity also allows the monitoring of binding events with small molecular weight compounds [85]. The latest generation of BIAcore instruments offers the possibility for differential measurements between two (differently modified) channels. This allows one to eliminate nonspecific binding effects which leads to significant improvement at very low signal levels or if high background effects are present.

Transducers are readily available, as SPR and prism couplers operate with non-structured planar film systems, which are easily manufactured. Effective manufacturing techniques for grating couplers by embossing/replicating are under investigation [76]. QCM transducers are available from applications in the electronics industry.

The high sensitivity can only be achieved with a perfect sample handling system. In this context, the BIACore instrument family introduced a tool not only with a sophisticated transducer element, but also with convenient and effective sample handling based on a microfluidic system (Figure 7.14, [86]). Transducers are available with a range of well-developed surface chemistries for rapid implementation of tests; e.g. a carboxydextran matrix for coupling of proteins via free amino residues. The development of the system was facilitated by the vast experience of the Pharmacia group in biochemical instrumentation, biochromatography, and bioprocess technology.

All commercially available systems allow the recording of the surface coverage Γ as a function of time by assuming a constant calibration factor. The following principal parameters can be assessed, typically as a function of analyte concentration:

- equilibrium surface coverage;
- binding kinetics (nonequilibrium conditions);
- mass transport and/or diffusional characteristics (nonequilibrium conditions).

The steps involved in the detection of biomolecular interaction with immobilized compounds are rather simple, due to the use of a heterogeneous phase system. For the immobilization of one reactant at the transducer an excess of reagents can be used. Separation of unreacted compounds is achieved by simple washing steps. Partial inactivation of immobilized molecules will not affect the results obtained, as long as a sufficient concentration of active sites is achieved at the transducer surface. Also limited purity of the mobile compound will not interfere with the assay, as long as non-specific binding can be kept to minute values.

Applications of BIA started in the field of immunoassays. The first instrument introduced in this field by Pharmacia Biosensors met no existing market, as no competing technology was available. The potential usefulness of the device was first recognized in fields where the concepts of affinity interaction were already present. Early applications were characterization and optimization of immunoreagents. Other

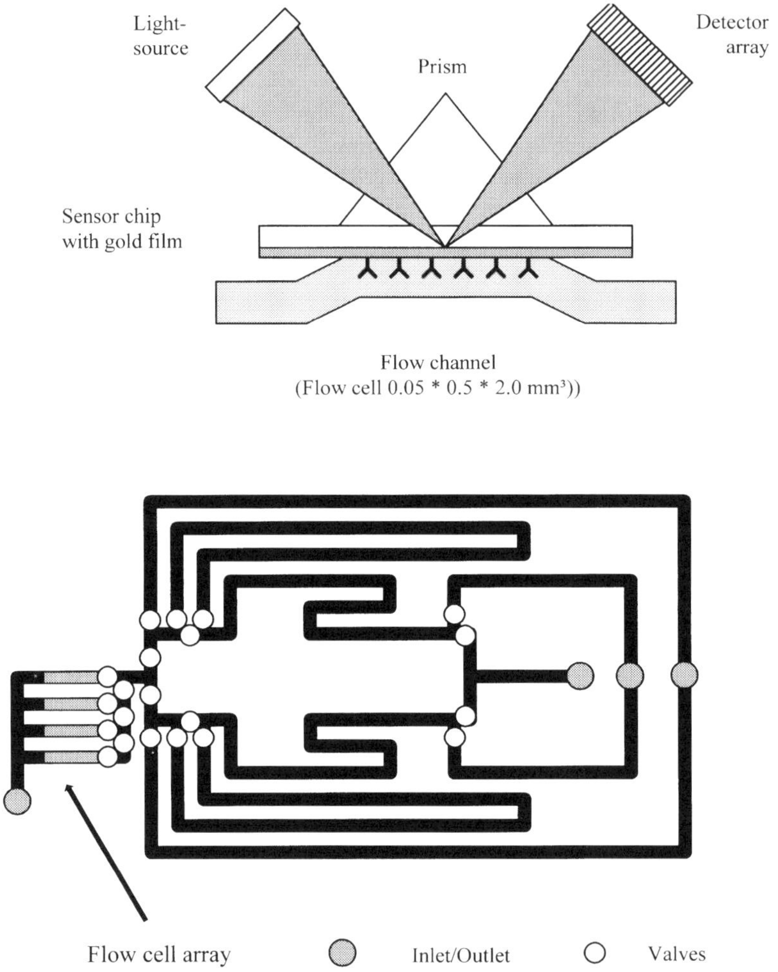

Figure 7.14. BIAcore system details (not to scale). Top: a gold film coated on a glass slide serves as transducer. SPR is interrogated by a wedge of monochromatic light. SPR is detected as minimum in reflectance. Bottom: four parallel flow cells can be addressed by a microconduit system comprising sample loops of different sizes and pneumatic valves. The size of the individual flow cells is $0.5 \times 2 \times 0.05\,\text{mm}^3$.

fields have emerged since then and more than 400 papers related to BIA has been published by 1996. Studies range from simple yes/no binding assays [87], to the assembly of multiprotein complexes [88], quantitative assays [89, 90], and investigations on interaction kinetics [91]. Models for the evaluation of surface binding kinetics are derived from homogeneous phase kinetics. The calculation of kinetic constants from the time resolved monitoring of molecular interaction was heavily advocated by BIACORE scientists. However, there is an ongoing controversy about the interpretation of results from heterogeneous phase BIA and the applicability of results to the understanding of homogeneous phase reactions [59, 60]. Nonetheless the success of the technique is impressive and the range of applications will certainly grow further.

Novel transducers may enter the market for BIA-systems either by competitive pricing or by challenging the performance of the established systems. The current price level leaves room for systems with LOD levels appropriate for investigations in macromolecular interaction (app. 10 pg/mm^2), but significantly reduced prices. Further reduced limits of detection would certainly be desirable. At a surface coverage of 1 pg/mm^2 the surface density of an average protein is about 10^7 molecules/mm^2. Therefore, in principle LODs in the 1–10 fg/mm^2 range may be possible. Theoretical performance values for some transducers may reach this level, but it is not yet clear whether working solutions will emerge or not.

7.3.4.2 Pharmaceutical Screening

Binding assays are a well-established tool in HTS. The migration of established BIA-techniques to the HTS market should be possible. The relative simplicity of BIA-techniques could make them an attractive alternative to established assay systems. The potential of heterogeneous phase affinity biosensor techniques has been recognized [24], but not yet exploited. A brief comparison between prime requirements in HTS (Table 7.2) and current performance of BIA techniques allows one to identify the major problems of the existing approaches in BIA so far:

- *Simplicity*: Assay protocols and transducer preparation are typically simple.
- *Applicability*: Heterogeneous phase BIA-techniques are not restricted to a particular calls of targets.
- *Substance consumption*: Depending on the assay type and the sample handling concept, the volumes (up to a few 100 μl) and concentrations required may be at the upper limit tolerable in HTS.
- *Cost effectiveness*: The costs of a screening approach is determined by consumables for each assay, by development costs for novel targets, and to a lesser extent by hardware costs. Relatively high system costs should be no major drawback for application of BIA-techniques in HTS.
- *High throughput*: This is the most critical point. Heterogeneous phase BIA involves diffusion of analytes to the transducer surface. As concentrations should be low, a finite amount of time is required to achieve a detectable signal level. The per cycle times are in the range of minutes, while HTS demands 1–10 seconds.

To meet throughput requirements a certain reduction of per cycle times may be possible, but values in the range of a few seconds at low concentrations may be out of reach. Parallel operation of several transducers or development of transducer arrays may be a possible solution. Transducers that allow for parallel multichannel systems are best suited for HTS. Particularly attractive may be arrays, which match the size of the common microtiterplate format. HTS requires simple solutions. Therefore it seems at least doubtful whether transducers array can be matched by (an array of) dedicated flow through sample handling systems. Low limits of detection are attractive as these will allow either to reduce consumption of test substances or reduction of assay times.

Assay Formats

Not all assay formats that are useful in BIA will be suitable for HTS. The use of a heterogeneous phase affinity sensor makes the immobilization of one test compound necessary. Two basic approaches are possible (Figure 7.15):

(1) Direct detection: immobilization of the receptor compound and direct administration of test substances to this surface. Binding is detected without the use of a further reference compound.

(2) Binding inhibition assay: immobilization of a (reference) compound to the transducer surface with known affinity to the receptor binding site under investigation. The test protocol involves preincubation of receptor compound and test substance until equilibrium binding is achieved. After delivery of the sample to the transducer surface, unoccupied receptor molecules will bind to the

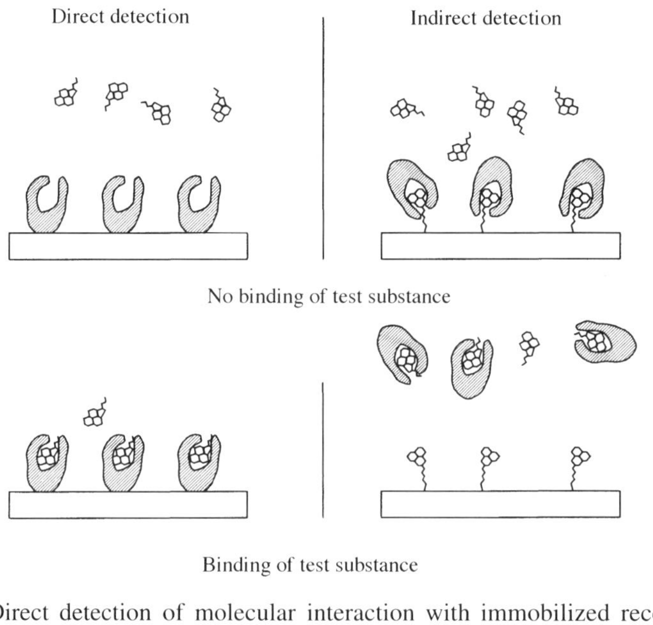

Figure 7.15. Direct detection of molecular interaction with immobilized receptor (left), or binding inhibition format (right).

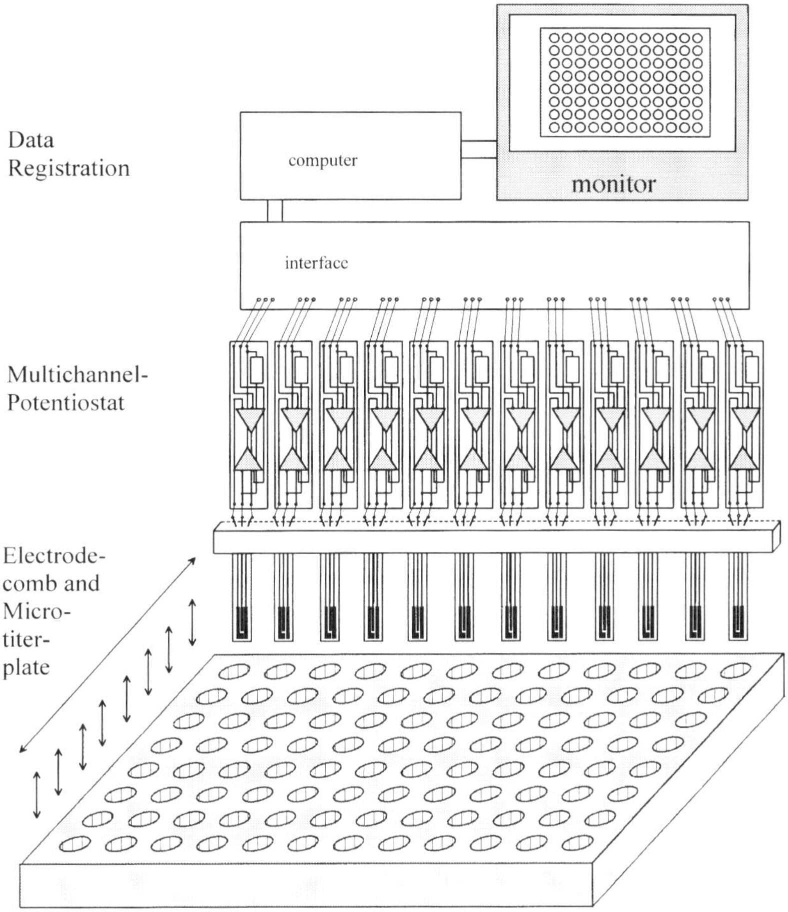

Figure 7.16. Parallel testing in a microtiterplate with electrochemical microelectrode transducers. Transducers and potentiostat circuitry can be adapted to the standard microtiterplate grid of 9 mm.

immobilized reference compound. Binding may be quantified after a fixed time or dynamically. The binding rate is inversely related to the binding strength of the test substance.

The first approach is attractive due to its simplicity, but is limited to high sensitivities of the transducers, as the overall effects will be minute. Assuming a ratio of molecular weights between ligand and receptor of at least 100, the changes in mass coverage to be expected will rarely exceed 100 pg/mm^2. A key requirement is the effective immobilization of receptor molecules in high density. This requires purified preparations of receptor material. Under practical conditions (limited effectivity of immobilization, use of complex material like cell membrane preparations with limited receptor density) the maximum signal levels to be expected may drop to values around 1 pg/mm^2. Depending on the binding kinetics, the dissociation of a ligand bound to a receptor will require considerable time. Physicochemical agents (low pH, chaotropic salts) may accelerate dissociation, but may also affect the integrity of the immobilized receptor. Depending on the immobilization procedure, single use of a given layer may be possible, but this might compromise the advantages of a sensor-type approach with a fixed detection element.

The second approach allows regeneration and repeated use of the transducer surface. Up to a few 100 cycles have been reached in comparable applications [92]. In this approach the concentration of the immobilized compound can be much higher than the receptor density achievable in the first approach. The receptor compound is sacrificed after each test. This contributes to the costs of the assay, but removes any problems with stability of the receptor compound. As long as nonspecific binding can be maintained at a sufficiently low level, crude target preparations can be used. Concentrations are dictated by assay time and by transducer resolution. Generally the maximum binding signal over the test time should be at least ten times the baseline noise plus drift. For a device capable of detecting 1‰ of a protein monolayer, a blank value (binding of receptor without any inhibition of a test compound) of at least 1% of a monolayer would be desirable.

Application of QCM

Individual QCM transducers are available from high-volume applications. The readout circuitry is of limited complexity and imposes – other than in optics – no constraints for the geometry of the readout system. This makes the development of a transducer array from individual transducers attractive [99–101]. In a feasibility study (carried out at Tuebingen University) [57] commercially available 8.5 mm thickness shear mode QCM were directly mounted as the bottom of wells (6 mm diameter) spaced in the 9 mm grid of a standard microtiterplate (Figure 7.17). A readout system was developed, which allows addressing of QCM transducers by rows of eight individual elements. Binding reactions proceed out of an unstirred sample volume, to avoid a complex multichannel flow system. A binding inhibition assay with thrombin (serin protease from blood clotting cascade) and thrombin inhibitors was established as a model system to test the performance of the QCM-transducer array. Aminodextran was linked via a thiol monolayer to the transducer surface and subsequently covalently modified with a reference thrombin inhibitor. Figure 7.18 (left) shows binding curves of thrombin binding to the immobilized reference inhibitor. A fixed concentration of thrombin was incubated with increasing concentrations of a soluble thrombin inhibitor. As the samples are simply added to the cavities, the binding curve is governed by the square root law predicted by Ficks second law for substance diffusion to a planar surface. Figure 7.18 (right) shows the effect of increasing concentrations of two different inhibitors at a fixed thrombin concentration (titration curves). Interaction of the inhibitor and thrombin leads to a significant decrease in binding to the transducer surface. A per cycle time of 10

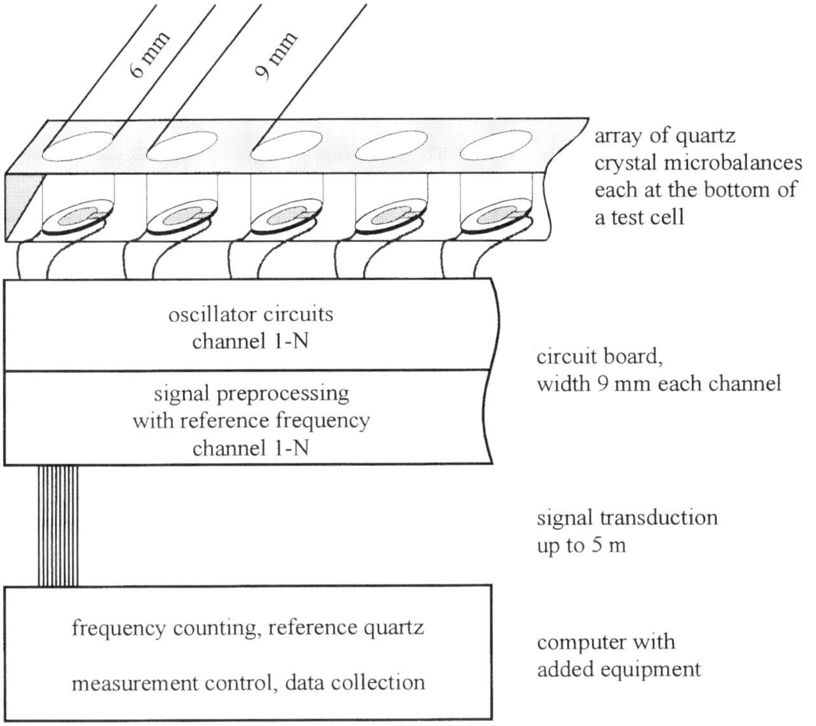

Figure 7.17. Integration of QCM transducers in the bottom of microtiterplate cavities. SMD driver circuitry allows mounting in 9 mm microtiterplate grid.

minutes could be reached. The active surface area is about 20 mm^2 and the sample volume was 50 µl per assay. Binding curves can be recorded with a time resolution of 1 Hz.

As mentioned above for the established optical BIA systems some increase in throughput seems possible by laterally adding interaction sites. The maximum number of spots addressable by this approach may be limited. Also a miniaturized linear transducer array may require dedicated sample handling approaches, which are not fully compatible with the microplate oriented standards in automated sample handling.

Concepts where transducer and readout system form separate entities – which is the case with all optical systems – allow one in principle to use removable transducers, which are interrogated only once or twice per assay. So far little experience with such systems exists and it is not yet clear how far up the accuracy can be tuned (at least to the high level of performance which at present is achieved by continuous readings from a transducer fixed in the readout system).

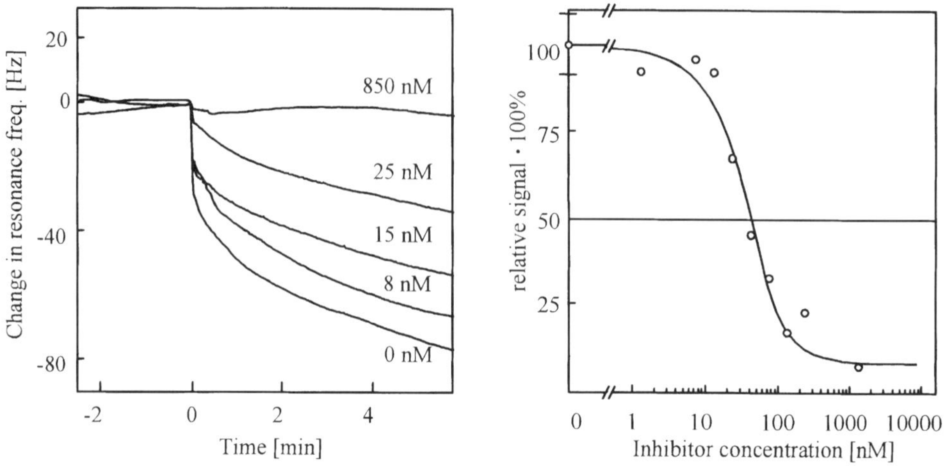

Figure 7.18. Binding inhibition assay for a thrombin inhibitor. Detection by QCM transducer covalently modified with a reference inhibitor. Left: binding curves. Right: resulting calibration plot.

Application of RIFS

In a feasibility study (carried out jointly by Tuebingen University and Carl Zeiss, Germany) the potential for two-dimensional optical transducer arrays in HTS was investigated. The approach is adaptable to a range of transducers. A first demonstrator was based on thin-film interference [81], as this transducer shows only little temperature sensitivity. The entire transducer surface is interrogated in reflection

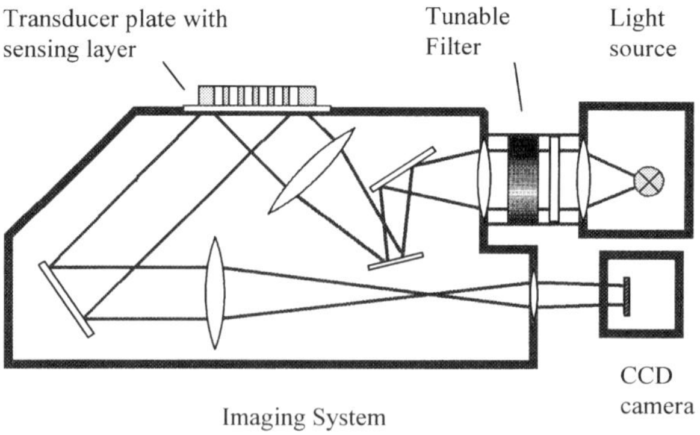

Figure 7.19. Parallel interrogation of planar optical transducers by a spectroscopic imaging system. Reflectance spectra are acquired wavelength sequential. The setup is format-free and can be adapted to a wide range of transducer areas ($< 1\,cm^2$ to $> 100\,cm^2$).

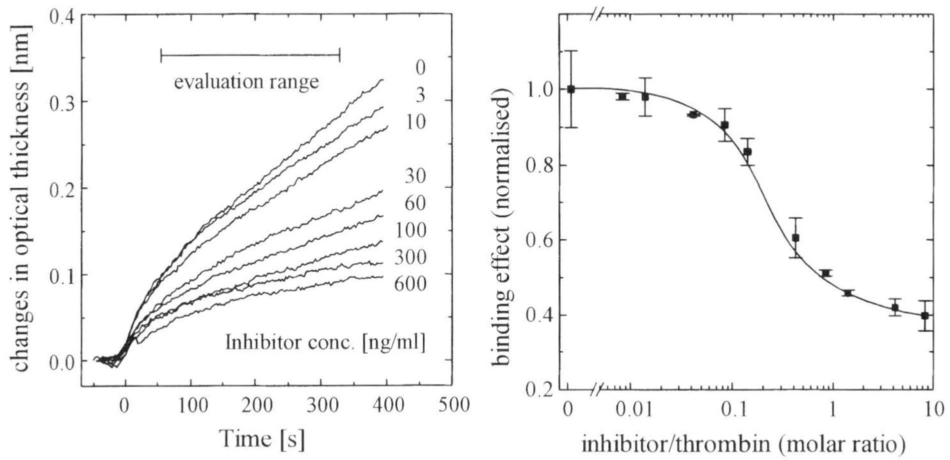

Figure 7.20. Binding inhibition assay for a thrombin inhibitor. Detection by reflectometric transducer (RIFS) covalently modified with a reference inhibitor. Left: binding curves. Right: resulting calibration plot.

mode. The reflectance profile is recorded as a function of wavelength. A fast tunable filter device is used to sequentially illuminate the transducer with a set of wavelengths. A CCD camera is used to acquire images of the transducer at each selected wavelength (Figure 7.19). From a sequence of images acquired during a wavelength scan, the reflectance spectrum of any part of the transducer surface can be derived. Binding of proteins to the device surface leads to a change of the reflectance spectrum. By repetitive scanning binding curves can be obtained for randomly selected areas of the transducer surface. Initially a system was set up addressing 96 sample positions in a 4.5 mm grid. A system addressing 96 locations in a 9 mm grid is under development. Initial trials were conducted with the same model system (thrombin/thrombin-inhibitor) as used with the QCM transducer. Binding curves could be recorded with a time resolution of 0.1 Hz. An assay time of less than 10 minutes was possible. A sample volume of 30 μl was used. Increasing concentrations of an inhibitor lead to a significant decrease of the binding effect observed (Figure 7.20). It should be noted that the parallel optical approach is not restricted to a given sample geometry. Also the active area per sample cavity can be reduced to less than 1 mm^2, which opens a route to further increased throughput or drastically reduced sample volumes.

Both feasibility studies show that high throughput–the major requirement in HTS– can be achieved by existing transducer technologies. Further studies with a broader range of screening targets are required to underpin the potential of affinity sensors in HTS.

7.4 Summary and Outlook

Sensor-based approaches are established to study biomolecular interaction in various applications. Prospects in three main fields differ significantly:

- The clinical diagnostics market by far exceeds the research market addressed by BIA technology. It is only natural that the hope to enter this market with concepts from BIA sensor approaches is of constant and high attraction. However, it seems noteworthy that more than five years after the introduction of a commercial platform for BIA measurements, BIA techniques have not entered the clinical diagnostics market. Heterogeneous phase BIA as it was introduced, has clear and distinct advantages over competing techniques in the field of biomolecular research. The lack of competing techniques has contributed to the success of the instruments introduced up to now. The migration of BIA techniques to markets and applications, where strong competing techniques exist, and where the strengths of BIA do not count fully, will certainly put an accent on the weaknesses and limitations of BIA concepts and prove to be at least slower than the introduction to the research market.

- The near future of BIA will be dominated by heterogeneous phase techniques. At the moment instrumentation has reached a high level of performance at relatively high prices. A further improvement in device sensitivity may be possible, but is not required for most applications. The high-end price market may be relatively small. In contrast, low-cost approaches with limited performance could match the requirements of many applications. Resolutions around 10 pg/mm^2 are sufficient for most studies involving macromolecular compounds. Also sample consumption is in many cases not highly critical. This may open a market for acoustic wave devices, which can be operated with readout circuitry of low complexity and cost. If this market is served, BIA instrumentation will become a standard tool in the routine biochemical laboratory within the next five to ten years.

  The range of applications of BIA in molecular biology is constantly broadening. More and more DNA and RNA interactions are finding interest. However, DNA sequence based assays are dominated by labeled techniques. In this field real time information is not of prime interest and the introduction of labels is straightforward during well-established assay protocols. Heterogeneous phase BIA may therefore be restricted to investigations on the interaction between DNA/RNA and other molecular species, primarily proteins. Fluorescence-based approaches will maintain a strong position, where labeling of reactants is routine. The potential of fluorescence correlation spectroscopy for spatially resolved *in situ* binding studies (e.g. on living cells) is unchallenged. The potential of AFM-based techniques is unclear yet, but dramatic improvements may be driven by emerging routine applications of AFM-systems in other fields.

  The particular characteristics of sensor-based BIA approaches have allowed dedicated equipment for label-free heterogeneous phase detection to open up within limited time a new market, which will grow further. Low-cost equipment as well as high-end devices will find customers in this market. The range of transducers proposed for BIA applications is considerably broader than what has reached the market yet. The transducer structures commercialized so far are relatively simple (planar devices without lateral structuring). In the field of very low limits of detection (below 1 pg/mm^2) new and maybe more complex devices may enter the market.

- High throughput screening is one aspect of a change in paradigms in pharmaceutical and agrochemical research. The budgets involved and the future markets are of considerable size. So far BIA techniques do not play an important role in this emerging field. However, their potential is high. Feasibility studies on two transducers (QCM and thin-film interference) demonstrated that throughput requirements and concentration levels can be met. Homogeneous phase assays are also attractive for HTS. The effort for the preparation of a labelled reference compound is less critical than in research applications, as in HTS typically several 10 000 assays are run per established target.

  BIA techniques are only one route to HTS. Strong competitors exists in the field of established immunoassay techniques. Therefore, ultimately versatility and cost of a technique will determine its success in that market. High throughput can be achieved by rapid detection techniques and/or parallelization. The costs are primarily determined by the costs of implementation of a target for screening and the costs of the screening run itself. The costs of the equipment typically only make up for a fraction of the total costs. The use of BIA techniques from characterization of a potential protein target, for the implementation of a screening assay and finally for the screening run itself will speed up testing and reduce costs.

  Approaches that allow low-volume detection are promising with respect to the cost of consumables. Homogeneous phase detection (fluorescence or electrochemical label-based) as well as optical heterogeneous phase techniques should allow operation from one µl sample volume or even less. Today the sample handling technique for such volumes is not yet mature. However the ability to perform low-volume detection may become the key to the success of the technology.

  One approach in studying molecular diversity are spatially resolved surface immobilized libraries [93, 94]. These libraries are synthesized *in situ* and at the moment focused to DNA-based diagnostics. Spot sizes below 100 µm are possible and 10 µm seem to be within reach. Transducer techniques that allow one to interrogate directly a spatially resolved immobilized library will become of interest within the next few years. High potential can be foreseen for optical imaging techniques and for scanning probe microscopes.

Sensors for bioaffinity analysis have only recently entered the market. Their particular capabilities make them attractive also for other emerging fields. The potential for growth is high for both already established and commercialized approaches, but also for entirely new systems.

7.5 References

[1] Goffeau, A. et al., *Science.* **274** (1996), 563–567.
[2] Stryer, L., *Biochemistry*; New York: Freeman, 1995.
[3] Göpel, G., Hesse, J., Zemel, N., (eds.), *Sensors*, volume 2: *Chemical and Biochemical Sensors Part I*; Weinheim: Verlag Chemie, 1991.
[4] Göpel, G., Hesse, J., Zemel, N., (eds.), *Sensors*, volume 3: *Chemical and Biochemical Sensors Part II*; Weinheim: Verlag Chemie, 1992.
[5] Price, Ch. P., Newman, D. J. (eds.), *Principles and Practice of Immunoassay*; New York: Stockton Press, 1991.
[6] Morgan, C. L., Newman, D. J., Price, C. P., *Clin.–Chem.* **42** (1996), 193–209.
[7] Suzuki, H., *J. Biochem. Tokyo* **115** (1994), 623–628.
[8] *Skerra-A, Curr-Opin-Immunol.* **5** (1993), 256–262.
[9] Scheller, F., Schubert, F., *Biosensors*; Amsterdam: Elsevier, 1992.
[10] Lodish, H., Baltimore, D., Berk, A., Zipursky, S. L., Matsudaira, P., Darnell, J., *Molecular Cell Biology*; New York: Scientific American Books, 1995.
[11] Gilman, A. G., *Biosci. Rep.* **15**(2), (1995), 65–97.
[12] Ortells, M. O., Lunt, G. G., *Trends Neurosci.* **18** (1995), 121–127.
[13] Roitt, I. M., *Essential Immunology*; Oxford: Blackwell Scientific Publications, 1994.
[14] Tijssen, P., *Practice and Theory of Enzyme Immunoassays*; Amsterdam: Elsevier, 1985.
[15] Schmidt, H. L., Schuhmann, W., Scheller, F. W., Schubert, F., in: [4], pp. 717–818.
[16] Merrifield, R. B., *Adv. Enzymol. Relat. Areas Mol. Biol.* **32** (1969), 221–296.
[17] Jung, G. (ed.), *Combinatorial Peptide and Nonpeptide Libraries*; Weinheim: Verlag Chemie, 1996.
[18] Hassan, M., Bielawski, J. P., Hempel, J. C., Waldmann, M., *Mol. Diversity* **2** (1996), 64–74.
[19] Society of Biomolecular Screening, *Proc. of the Second Annual Conference*, Basel, 1996.
[20] McConnell, H. M., Owicki, J. C., Parce, J. W., Miller, D. L., Baxter, G. T., Wada, H. G., Pitchford, S., *Science* **257** (1992), 1906–1912.
[21] Lo, C. M., Keese, C. R., Giaever, I., *Biophys. J.* **69** (1995), 2800–2807.
[22] Schullek, J. R., Butler, J. H., Ni, Z. J., Chen, D., Yuan, Z. Y., *Anal. Biochem.* **246** (1997), 20–29.
[23] Bosworth, N., Towers, P., *Nature* **341** (1989), 167.
[24] Taremi, S. S., Prosise, W., Durkin, J., Zhang, R., Windsor, W. T., Le, H. V., *BIA Journal* **3**(1) (1996), 21.
[25] Bruham, N. A., Colton, R. J., Force Microscopy in Scanning Tunneling Microscopy and Spectroscopy (Theory, Techniques and Applications), edited by Bornell, D. A., VCH Publishers, Inc (1993), New York.
[26] Atkins, P. W., *Physical Chemistry*, Oxford: Oxford University Press, 1990.
[27] Takeda et al., *Proc. Nat. Acad. Sci. USA* **89** (1992), 8180–8184.
[28] Berger, R., Gerber, Ch., Gimzewski, J. K., *Anal. Meth. Instr.*, Special Issue MicroTAS (1996) 74–77.
[29] Jelesarov, I., Leder, L., Bosshard, H. R., *Methods* (Orlando) **9** (1996), 533–541.
[30] Zukin, R. S., *Biochemistry* **18** (1979), 2139–2145.
[31] Sohanpal, K., Watsuji, T., Zhou, L. Q., Cass, A. E. G., *Sens. and Act. B* **11** (1993), 547–552.

[32] Dandliker, W. B., de Saussure, V. A., *Immunochemistry* **7** (1970), 799.
[33] Guilbault, G. G., *Analytical Uses of Immobilized Enzymes*; New York: Dekker, 1984.
[34] Johnsson, B., Löfas, S., Lindquist, G., *Anal. Biochem.* **198** (1991), 268–277.
[35] Hermanson, G. T., Mallia, A. K., Smith, P. K., *Immobilized Affinity Ligand Techniques*; San Diego: Academic Press, 1992.
[36] Arwin, H., Aspnes, D. E., *Thin Solid Films* **138** (1986), 195.
[37] Brash, J. L., Horbett, T. A. (eds.), *Proteins at Interfaces–Physicochemical and Biochemical Studies*, Washington, DC: American Chemical Society, 1987.
[38] Rickert, J., Goepel, W., Beck, W., Jung, G., Heiduschka, P., *Bios. and Bioelect.* **11** (1996), 757–768.
[39] Maupas, H., Saby, C., Martelet, C., Jaffrezic-Renault, N., Soldatkin, A. P., Charles, M. H., Delair, T., Mandrand, B., *J. Electroanal. Chem.* **406** (1996), 53–58.
[40] Bone, S., Zaba, B., *Bioelectronics*; Chichester: John Wiley, 1992.
[41] De Feijter, J. A., Benjamins, J., Veer, F. A., *Biopolymers* **17** (1978), 1795.
[42] Yalow, R. S., *Les Prix Nobel 1977*; Nobel Foundation Stockholm, 1978, pp. 237–241.
[43] Owen, V. M., in: *Sensors 8*, Göpel, W., Hesse, J., Zemel, J. N. (eds.); Weinheim: Verlag Chemie (1995), pp. 431–450.
[44] Jones, K., in: *Sensors 8*, Göpel, W., Hesse, J., Zemel, J. N. (eds.); Weinheim: Verlag Chemie, 1995, pp. 451–490.
[45] Place, J. F., Sutherland, R. M., Dahne, C., *Biosensors* **1** (1985), 321–353.
[46] Lebl, M., Krchnak, V., Sepetov, N. F., Seligmann, B., Strop, P., Felder, S., Lam, K. S., *Biopolymers* **37** (1995), 177–198.
[47] Harris, S. R., Garlick, R. K., Miller, J. J., Harney, H. N., Monroe, P. J., *J. Recept. Res.* **11** (1991), 115–128.
[48] Maliwal, B. P., Lakowicz, J. R., Kupryszewski, G., Rekowski, P., *Biochemistry* **32** (1993), 12337–12345.
[49] Jolley, M. E., *J. Anal. Toxicol.* **5** (1981), 236 and Jolley, M. E., *J. Biomol. Screen.* **1** (1996), 33–38.
[50] Eigen, M., Rigler, R., *Proc. Natl. Acad. Sci. USA* **91** (1994), 5740–5747.
[51] Bard, A. J., Faulkner, L. R., *Electrochemical Methods, Fundamentals and Applications*; New York: John Wiley. 1980.
[52] Kissinger, P. T., Heineman, W. R. (eds.), *Laboratory Techniques in Electroanalytical Chemistry*; New York: Marcel Dekker, 1996.
[53] Niwa, O., Xu, Y., Halsall, H. B., Heineman, W. R., *Anal. Chem.* **65** (1993), 1559–1563.
[54] Fiaccabrino, G. C. et al., *Anal. Chim. Acta* **326** (1996), 155–161.
[55] Hintsche, R., Paeschke, M., Wollenberger, U., Schnakenberg, U., Wagner, B., Lisec, T., *Bios. Bioelect.* **9** (1994), 697–705.
[56] Ross, B., Cammann, K., Mokwa, W., Rospert, M., *Sens. Act. B* **7** (1992), 758–762.
[57] Brecht, A., Burckardt, R., Rickert, J., Stemmler, I., Schuetz, A., Fischer, S., Friedrich, Th., Gauglitz, G., Goepel, W., *J. Biomol. Screen.* **1** (1996), 191–201.
[58] Bratten, C. D. T., Cobbold, Ph., Cooper, J. M., *Anal. Chem.* **69** (1997), 253–258.
[59] O'Shannessy, D. J., *Curr. Opin. Biotechnol.* **5** (1994), 65–71.
[60] Nieba, L., Krebber, A., Plückthun, A., *Anal. Biochem.* **234** (1996), 155–165.
[61] Sauerbrey, G., *Z. Phys.* **155** (1959), 206.
[62] Du, J., Harding, G. L., Ogilvy, J. A., Dencher, P. R., Lake, M., *Sens. Act. A* **56** (1996), 211–219.
[63] Nieuwenhuizen, M. S., Venema, A., in: [3], pp. 647 –680.
[64] Suleiman, A. A., Guilbault, G. G., *Anal. Lett.* **24** (1991), 1283.

[65] Ballantine, D. S., Martin, S. J., Ricco, A. J., Frye, G. C., Wohltjen, H., White, R. M., Zellers, E. T., *Acoustic Wave Sensors–Theory, Design and Physico-Chemical Applications*; San Diego: Academic Press. 1997.
[66] Wolfbeis, O., Boisde, G. E., Gauglitz, G., in: [4], pp. 573–646.
[67] Gauglitz, G., in: *Sensors Update 1*, Baltes, H., Göpel, W., Hesse, J. (eds.); New York: Verlag Chemie, 1996, pp. 1– 48.
[68] Kretschmann, E., Raether, H., *Z. Naturf.* **23a** (1968), 2135.
[69] Joensson, U., Faegerstam, L., Ivarsson, B., Johnsson, B., Karlsson, R., Lundh, K., Loefaas, S., Persson, B., Roos, H. et al., *Bio Techniques* **11**(5), (1991), 620.
[70] Striebel, C., Brecht, A., Gauglitz, G., *Biosens. Bioelectron.* **9** (1994) 139.
[71] Jorgenson, R. C., Yee, S. S., *Sens. Actuators B* **12**(3), (1993), 213–220.
[72] Lavers, C. R., Wilkinson, J. S., *Sens. Actuators B* **22** (1994), 75–81.
[73] Nellen, Ph., Lukosz, W., *Biosens. Bioelectron.* **8** (1993), 129–147.
[74] Edwards, P. R., Gill, A., Pollard-Knight, D. V., Hoare, M., Buckle, P. E., Lowe, P. A., Leatherbarrow, R. J., *Anal. Biochem.* **231** (1995), 210–217.
[75] Gauglitz, G., Ingenhoff, J., Fres, J., *Anal. Chem.* **349** (1994), 355–359.
[76] Gale Michael, T., Baraldi, Luca G., Kunz Rino, E., *Proc. SPIE-Int. Soc. Opt. Eng.* **2213** (1994), 2–10.
[77] Lai, Q., Bachmann, M., Hunziker, W., Besse, P. A., Melchior, H., *Elect. Letters* **32** (1996), 1576–1577.
[78] Hecht, E., *Optics*; Reading, MA: Addison-Wesley, 1987.
[79] Azzam, R. M. A., Bashara, N. M., *Ellipsometry and Polarized Light*; New York: North Holland, 1988.
[80] Jin, G., Tengvall, P., Lundstrom, I., Arwin, H., *Anal. Biochem.* **232** (1995), 69–72.
[81] Gauglitz, G., Brecht, A., Kraus, G., Nahm, W., *Sens. Actuators B* **11** (1993), 21.
[82] Piehler, J., Brecht, A., Gauglitz, G., *Anal. Chem.* **68** (1996), 139.
[83] Lindsay, S. M., Biological Applications of the Scanning Probe Microscope in Scanning Tunneling Microscopy and Spectroscopy (Theory, Techniques and Applications), edited by Bonnell, D. A., VCH Publishers, Inc. (1993), New York.
[84] Florin, E. L., Moy, V. T., Gaub, H. E., *Science* **264** (1994), 415–417.
[85] Karlsson, R., Stahlberg, R., *Anal. Biochem.* **228** (1995), 274–280.
[86] Sjölander, S., Urbaniczky, C., *Anal. Chem.* **63** (1991), 2338–2345.
[87] Saunal, H., Van-Regenmortel, M. H. V., *J. Imm. Meth.* **183** (1995), 33–41.
[88] Persson, E., Ezban, M., Shymko, R. M., *Biochemistry* **34** (1995), 12775–12781.
[89] Sternesjo, A., Mellgren, C., Bjorck, L., *Anal. Biochem.* **226** (1995), 175–181.
[90] Minnuni, M., Mascini, M., *Anal. Lett.* **26** (1993), 1441–1460.
[91] Johne, B., Gadnell, M., Hansen, K., *J. Imm. Meth.* **160** (1993), 191–198.
[92] Piehler, J., Brecht, A., Geckeler, K. E., Gauglitz, G., *Bios. and Bioelect.* **11** (1996), 579–590.
[93] Fodor, S. P., Read, J. L., Pirrung, M. C., Stryer, L., Lu, A. T., Solas, D., *Science* **251** (1991), 767–773, and Chee, M., Yang, R., Hubbell, E., Berno, A., Huang, X. C., Stern, D., Winkler, J., Lockhart, D. J., Morris, M. S., Fodor, S. P., *Science* **274** (1996), 610–614.
[94] Blanchard, A. P., Kaiser, R. J., Hood, L. E., *Bios. Bioelect.* **11** (1996), 687–690.
[95] Roelant, C. H., Burns, D. A., Scheirer, W., *Biotechniques* **20** (1996), 914–917.
[96] Lebsanft, J., McMahon, J. B., Steinmann, G. G., Shoemaker, R. H., *Biochem. Pharmacol.* **38** (1989), 4477–4483.
[97] Rickert, J., Brecht, A., Goepel, W., *Anal. Chem.* **69** (1997), 1441–1448.
[98] Tom-Moy, M., Baer, R. L., Spira-Solomon, D., Doherty, T. P., *Anal. Chem.* **67** (1995), 1510.

[99] Rickert, J., Weiß, T., Göpel, W., *Proc. EMRS Spring Meeting*, May 1995, Straßburg (F) and *Sensors and Actuators B* **31** (1996), 45–50.

[100] Rickert, J., Weiß, T., Kraas, W., Jung, G., Göpel, W., *Proc. Eurosensors IX*, June 1995, Stockholm (S) and *Biosensors and Bioelectronics* **11** (1996), 591–598.

[101] Rickert, J., Brecht, A., Göpel, W., *Proc. Biosensors 96*, May 1996, Bangkok and *Biosensors and Bioelectronics*, in press.

8 Sensors in Ophthalmology

T. Hellmuth, Fachhochschule Aalen, Germany

Contents

8.1 Introduction

In this chapter we focus on sensor applications in ophthalmology. This field has a very long tradition of quantitative diagnosis based on physical measurement techniques. Refractometry as a way to measure the refractive power of the human eye in order to select the correct spectacle lens is only one example of methods which have been in use for generations.

Due to this long history the art of "measuring" parameters characterizing the eye has become very specialized. As a consequence a unique discipline known as "optometry" became established especially in the Anglo Saxon countries. However, in the following sections we don't distinguish between sensors in opthalmology and optometry because the same instruments are used in both fields.

Methods for the measurement of parameters such as refractive power, intraocular pressure, shape of the eye, etc., have been extensively studied in the nineteenth century and famous names such as, for example, Helmholtz are associated with these techniques. Methods were mainly based on mechanical, optical and geometric approaches. The next consequent step was to extend the potential of these methods by introducing electronics. Sensors such as devices converting nonelectrical measurable quantities into electrical signals have become slowly established in ophthalmology since the 1950s. However, it was not much more convenient to read the result from the pointer position of an electronic device than from a mechanical meter. The situation changed drastically with the availability of low-priced computers in the 1970s. First, it has become possible to register the results easily and almost with unlimited memory capacity. Computers allowed more complex sensor systems to be designed with the advantage of better repeatability and reliability of data. The instruments have become more user-friendly and consequently more economic because less skilled personnel can now operate the devices.

In the following sections we want to give an overview of the most important sensors and sensor systems used in ophthalmic and optometric practice. We focus our attention primarily on the technical aspects of the instruments and discuss their medical applications only as far as they are necessary to understand the product features and functions.

Before we describe the various sensor techniques the following section shall give a brief introduction to the anatomy of the eye and its structures.

8.2 Anatomy of the Eye

Figure 8.1 shows a schematic cross-section of the human eye. The outer coat comprises the white sclera and the transparent cornea on the front. The cornea and the eye lens constitute the dioptric system of the eye. The iris functions as the aperture diaphragm and determines the amount of light entering by changing the diameter of the entrance pupil. The segment of the eye bounded by the cornea and the front surface of the iris is referred to as the anterior chamber. The much smaller posterior chamber is bounded by the lens and the rear surface of the iris. The lens is contained by the lens capsule peripherally fixated by the lens zonules at the ciliary body with the ciliary muscle. The refractive power of the lens can be accomodated by changing the lens shape with the ciliary muscle. Aqueous humor is secreted by the ciliary body into the posterior chamber, passing into the anterior chamber, and leaving through the trabecular meshwork. The secretion and drainage process of the aqueous humor determines the intraocular pressure.

The third chamber of the eye, the vitreous cavity, is filled with transparent vitreous and is bounded anteriorly by the lens and posteriorly by the retina. The purpose of the retina is light detection. It consists of the nerve fibers connected with the receptor layer via transmitter cells (ganglion and bipolar cells) and modulator cells (horizontal and amacrine cells). The nerve fiber layer is located between vitreous and receptors. In contrast to technical detector systems where the wiring is disposed behind the detector array we have here the remarkable situation that the transparent "wires" are in front of the "detector array". The light detecting receptor cells can be divided into rod shaped cells and cone shaped cells. The rod cells are the most sensitive receptors

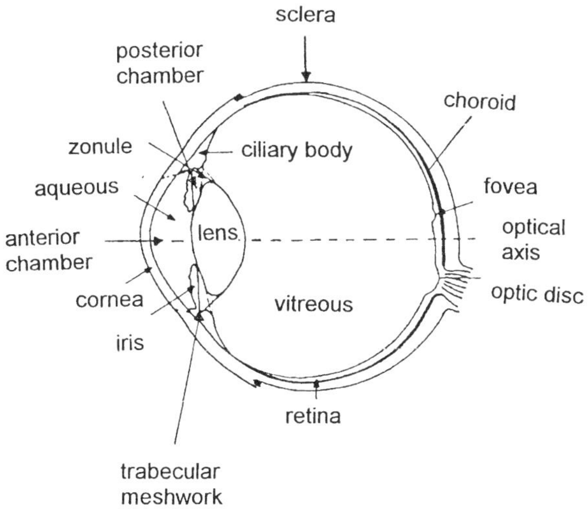

Figure 8.1. Anatomy of the eye.

with the ability to detect single photons and are optimized for dim light illumination (known as scotopic vision). The cone cells are much less sensitive (photopic vision). Being the receptors responsible for color vision there are three types of cone cells differing in their spectral sensitivity. One distinguishes between S-cells (short wavelength), M-cells (medium wavelength) and L-cells (long wavelength). The boundary of the retina is the pigment epithelium dumping light which is not absorbed by the receptor cells. The choroid is a highly circulated vascular tissue supplying the retina with blood and is located behind the pigment epithelium. The retina and the choroid together constitute the fundus of the eye.

The retina can be divided into various regions. Important for our context is the fovea and the optical disk. The fovea with an angular diameter of about 3° is the area with the highest density of receptor cells (cone cells only) with consequently the highest visual acuity. The fovea is slightly displaced in the nasal direction from the optical axis of the eye. The optical axis is defined as the axis connecting the vertices of the cornea and the lens. The optical disc is the location where the nerve fibers converge to leave the eye as the optical nerve. This area corresponds to the blind spot.

8.3 Refractometry

Ametropia can be described as the optical condition of the eye where the refractive power of the lens and the eye length do not correspond. As a result the image of an object is focussed either in front of (myopia) or behind (hyperopia) the fundus. This refractive error is usually referred to as "sphere" error. It is caused by paraxial defocussing and has nothing to do with spherical aberration. In addition to sphere astigmatism causes a refractive error referred to as "cylinder" error.

The result in both cases is decreased visual acuity. In practice it can be well corrected with spectacle lenses to achieve the ideal situation of emmetropia where the image of an object at infinity is focussed on the fundus.

Refractometry is a very important technique in optometry. There are subjective and objective methods to assess the refractive error. The classical objective "manual" method is retinoscopy (or skiaskopy) constituting the standard which automatic refractometers are compared to.

The principle is shown in Figure 8.2a. A light source is reflected here by a semitransparent mirror into the pupil of the patient's eye generating a patch of light on the retina referred to as the secondary light source in the following. The examiner observes the patient's pupil. He holds the mirror and tilts it up and down causing an up and down movement of the secondary light source on the patient's retina. For reason of simplicity we assume a point-like secondary light source. If the examined eye is emmetropic the patient accomodates to the examiner's iris if a test lens with a focal length corresponding to the distance between patient and examiner is introduced. The examiner accomodates to the iris of the patient. The secondary light source on the patient's fundus is imaged into the examiner's pupil plane creating

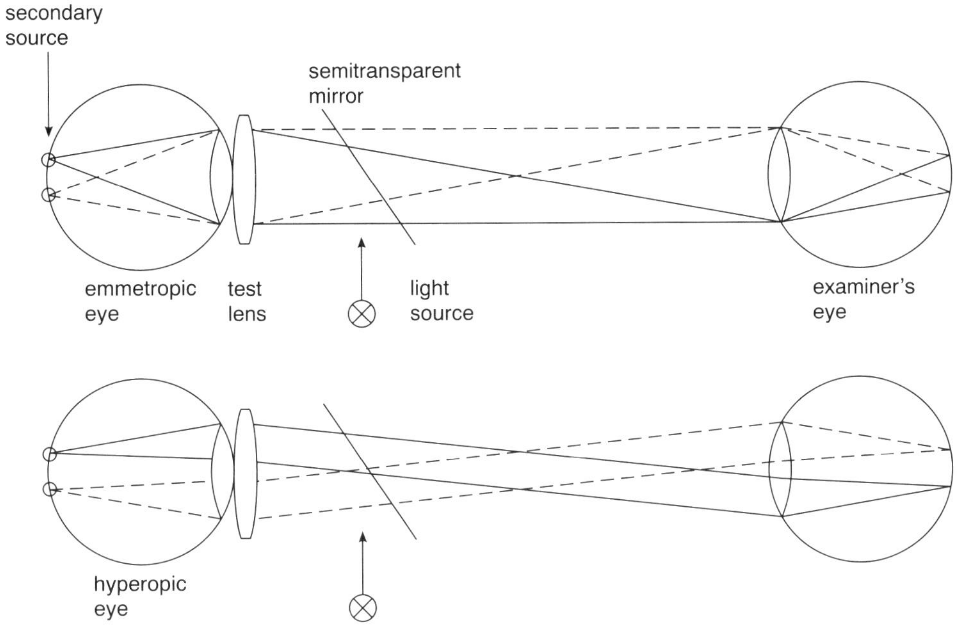

Figure 8.2. Principle of retinoscopy. The secondary source is moved across the fundus of the examined eye by tilting the semitransparent mirror. In the case of emmetropia (a) the secondary light source generates a resting light patch on the examiner's fundus. In the case of a hyperopic eye the image of the secondary source creates a moving light patch on the examiner's fundus (b).

a defocussed light patch on the examiner's fundus. As a result the examiner sees the pupil of the patient as a homogeneously illuminated disk. As mentioned above the secondary light source is moving across the patient's fundus such that its image is moving across the examiner's pupil. However, the defocussed light patch on the examiner's fundus is resting. The illumination of the patient's pupil appears to be stationary to the examiner.

Figure 8.2b shows the case of a hyperopic eye. The patient is not able to accomodate to the examiner's iris. For reasons of simplicity we have assumed that the virtual image of the secondary light source relating to the hyperopic eye lens is located at the focal plane of the test lens such that a collimated beam cmanates from the test lens and is focussed on the fundus of the examiner who sees a small moving reflex in the patient's eye.

The examiner tries to compensate the refractive error of the patient with further test lenses until he sees a stationarily illuminated pupil indicating emmetropia. This technique is not only applicable to determine sphere but also cylinder by varying the orientation of the tilt angle and compensating the refractive error with cylinder lenses.

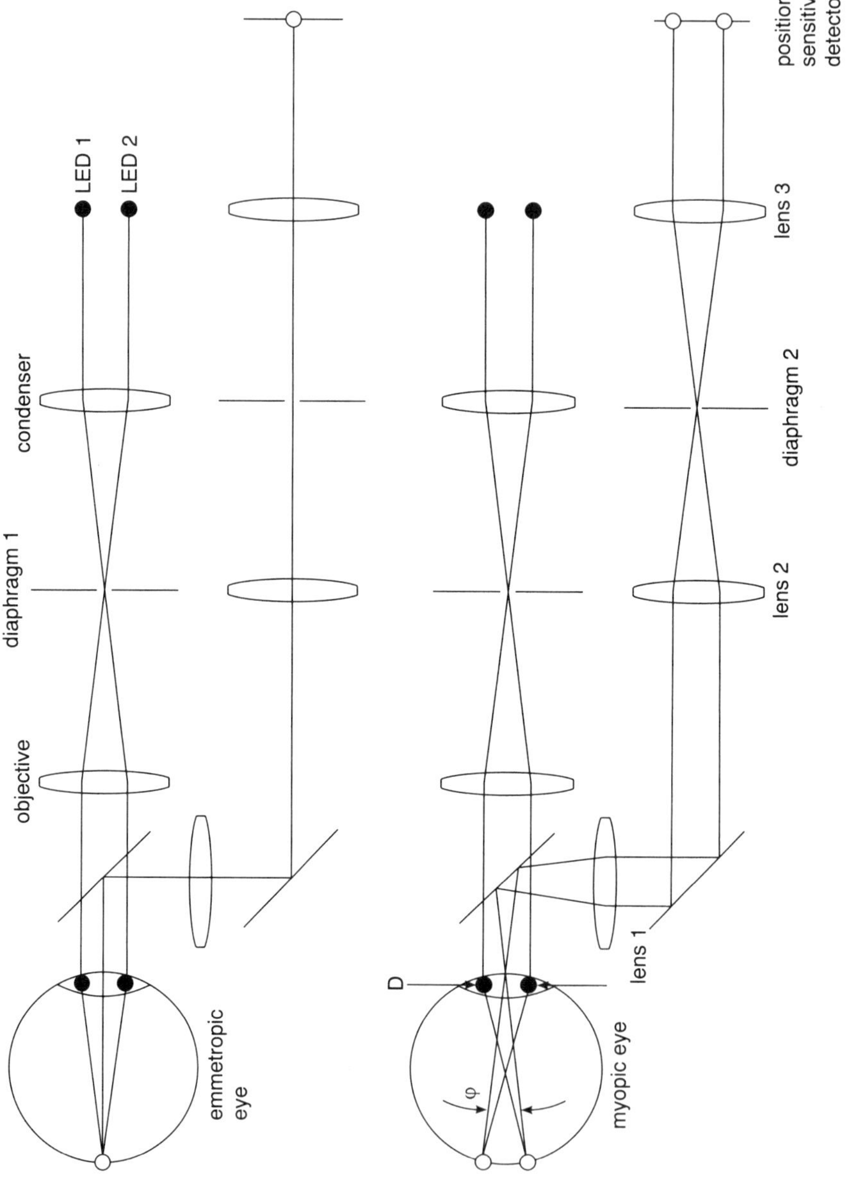

Figure 8.3. Scheiner principle. LED 1 and 2 are imaged into the pupil of the examined eye. In the case of emmetropia (a) a single light patch is created on the fundus of the examined eye which is finally imaged onto the position-sensitive detector. In the case of myopia (b) two light patches are generated on the fundus also imaged onto the position-sensitive detector. The distance between the two patches is proportional to the refractive error.

Although with this technique an accuracy of ± 0.25 diopters can be achieved the examiner must be very skilled. It is also a rather time consuming procedure because it takes several iterations with test lenses to compensate the refractive error. For that reason efforts have been undertaken very early to automate the measurement. The first automated refractometer was designed in 1937 by Collins [1] with many features typical also of modern refractometers. Meanwhile there are piles of patents describing a confusing zoo of refractometer principles. Nevertheless all can be sorted into two classes characterized by the following two approaches.

One approach dates back to Scheiner (1573–1650) the inventor of the double pinhole method to determine the focal length of a lens. Figure 8.3 shows the setup principle [2]. Two LEDs emitting in the near infrared wavelength regime are imaged with the condenser lens and the objective lens into the eye pupil of the patient. The diaphragm between objective and condenser lens obliquely illuminated by the two LEDs is imaged on to the fundus by the objective lens and the dioptric system of the eye. In the case of emmetropia a single light patch is located on the retina as the image of the diaphragm as shown in Figure 8.3a. In the case of ametropia the image of the diaphragm is defocussed and split into two patches. The distance of the light patches on the fundus is proportional to the refractive error. However it is not possible to measure the displacement on the fundus directly but only the angular displacement φ. On the other hand this parameter is sufficient to determine the refractive error given by

$$\Delta = \frac{1}{l} - \frac{1}{f}$$

with the unknown eye length l and the unknown focal length f. With the known distance D of the two LED images at the eye pupil one gets with Figure 8.3b and the theorem of intersecting lines

$$\frac{l-f}{\varphi \cdot l} = \frac{f}{D}$$

and we finally get

$$\Delta = \frac{\varphi}{D}$$

The angle φ can be measured as shown in Figure 8.3b by imaging the pupil center through lenses 1 and 2 on diaphragm 2. With real refractive errors the displacement of the light patches on the fundus is smaller than the patch diameters. Thus a differential diode is sufficient to measure the displacement of the images of the LED light patches. In order to distinguish the light from the respective LED in the diode signal the LEDs are modulated with different frequencies. This Scheiner principle which can also be described as a retinal image analysis technique is being used in the refractometer of Nidek and in a modified version in the Topcon refractometer.

The second class of approaches can be described as techniques for the analysis of the directions of beams at the pupil emanating from a secondary point source on the retina. In terms of optical testing one would say these techniques analyze the OPD

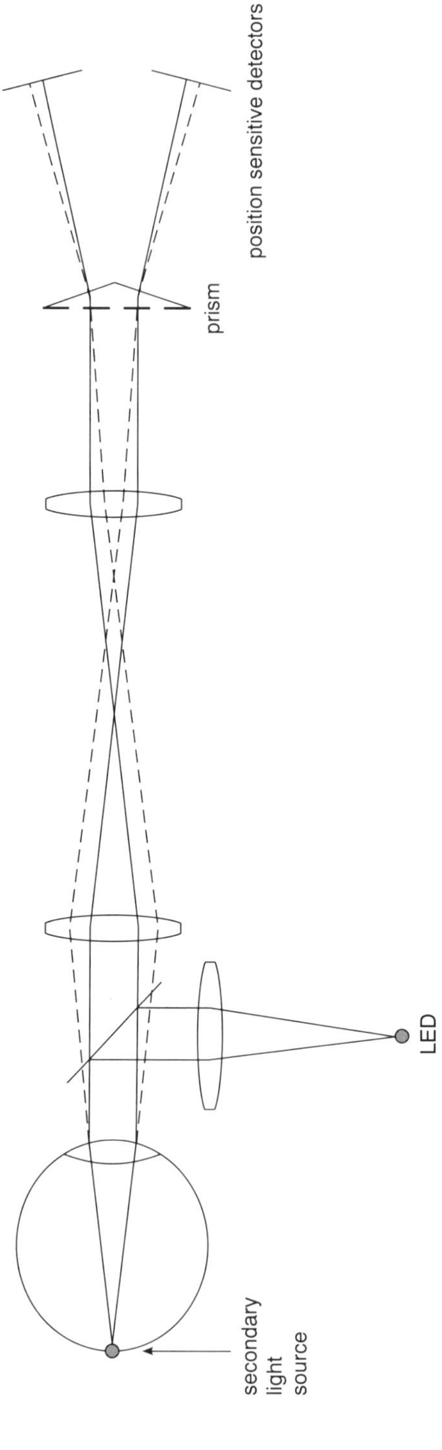

Figure 8.4. Analysis of beam directions. The secondary light source generates two beams hitting position-sensitive detectors. The positions on the detectors depend on the beam deflection angles at the eye corresponding to the refractive error.

(optical path deviation) of the eye's dioptric system. The most straightforward realization of this approach is shown in Figure 8.4 (see also [3]). A single LED is imaged on to the retina generating a secondary light source. Two lenses image the pupil into a plane where a prism with a double pinhole diaphragm is located. The prism generates two light beams hitting two position-resolving detectors. These can be differential diodes or better CCDs. Ametropia causes a divergent (hyperopia) or convergent (myopia) light bundle emanating from the pupil. The corresponding beams at the detectors are deflected according to the refractive error. The beam angles and the intersection positions of the beams at the pupil can be calculated with the optical data of the two-lens system and the position of the detectors. The dashed beams in Figure 8.4 constitute the boundary rays of the divergent beams corresponding to hyperopia. The angles in the x- and y-direction θ_x and θ_y and the positions of the two beams are sufficient to determine sphere and cylinder. The OPD of a lens can be expanded into the three Zernike polynomials for defocussing and astigmatism [4]:

$$\mathrm{OPD}\,(\rho,\,\varphi) = \alpha Z_{\mathrm{sph}}\,(\rho,\,\varphi) + \beta Z_{\mathrm{cyl1}}(\rho,\varphi) + \gamma Z_{\mathrm{cly2}}(\rho,\,\varphi)$$

with the "spherical" Zernike polynomial

$$Z_{\mathrm{sph}}(\rho,\,\varphi) = 2\rho^2 - 1$$

and the two "cylindrical" Zernike polynomials

$$Z_{\mathrm{cyl1}}(\rho,\varphi) = \rho^2 \cos(2\varphi)$$
$$Z_{\mathrm{cyl2}}(\rho,\varphi) = \rho^2 \sin(2\varphi).$$

ρ, φ polar coordinates at the pupil.

The angles of the rays at the pupil are equal to the orientation of the gradient of the OPD. With the knowledge of the angles θ_{xi} and θ_{yi} of two rays at two points at the pupil with known coordinates ρ_i, φ_i $(i = 1,\,2)$ the expansion coefficients α, β, γ can be calculated by solving the following linear equation system $(i = 1,\,2)$:

$$\frac{\nabla_x \mathrm{OPD}(\rho_i,\,\varphi_i)}{|\nabla \mathrm{OPD}(\rho_i,\,\varphi_i)|} = \cos\theta_{xi}$$
$$\frac{\nabla_y \mathrm{OPD}(\rho_i,\,\varphi_i)}{|\nabla \mathrm{OPD}(\rho_i,\,\varphi_i)|} = \cos\theta_{yi}$$

providing the three constants α, β, γ. The parameters to describe the spherical refractive error (S) and the cylindrical refractive error (C) and the angular orientation of the cylindrical refractive power Φ can be easily calculated with the parameters

$$C = \sqrt{\beta^2 + \gamma^2}$$
$$\Phi = \tfrac{1}{2}\arctan\left(\tfrac{\gamma}{\beta}\right) - 45^\circ$$
$$S = \alpha - \frac{C}{2}.$$

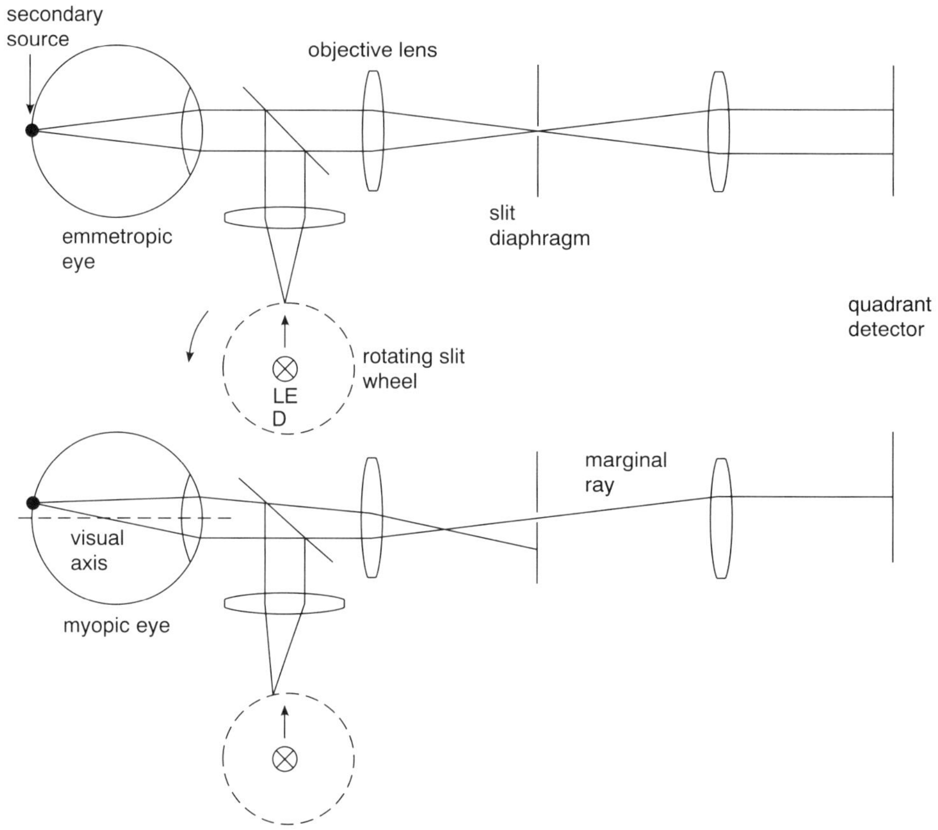

Figure 8.5. Retinoscopic principle. The illuminated rotating slit wheel is imaged on to the fundus of the examined eye generating a moving secondary light source. Due to the diaphragm a symmetric light distribution is generated on the quadrant detector in the case of emmetropia (a). The light distribution is asymmetric in the case of ametropia (b).

Usually more than two rays are used to achieve a higher redundancy in respect of measurement errors and to determine the two tilt angles of the OPD usually referred to as prism error not taken into account in the analysis above.

Another representative of this second class of refractors is the automatic refractometer principle equivalent to the manual retinoscopic method shown in Figure 8.5a [5]. An LED illuminates a rotating slit wheel which is imaged on to the retina generating a moving secondary source. The retina is optically conjugate to the slit diaphragm in the focal plane of the objective lens. A quadrant detector is located behind the slit and is conjugate to the pupil plane of the eye. In the case of an emmetropic eye the secondary source is imaged into the slit plane. Light hits the quadrant detector symmetrically when the image of the moving secondary light source passes the slit aperture. In the case of a myopic eye the beams leaving the pupil of the eye are convergent and the image of the secondary light source is in front of the slit plane as shown in Figure 8.5b. In that situation the marginal rays of the light pencil hitting the slit plane pass the slit before the center of the defocussed secondary source image patch reaches the slit. These marginal rays hit the quadrant detector nonsymmetrically. In Figure 8.5b the upper marginal ray hits the upper detector first when the image of the secondary light source is moving from bottom to top. The time difference between the detector signals is a measure for the ametropia. The sequence of the detector signals indicates whether the examined eye is hyperopic or myopic.

8.4 Ultrasonometry

The eye is the ideal object for optical diagnosis methods because of its transparency. Optical methods fail when the eye lens is turbid as it is the case with cataracts. Optical diagnosis also fails in detecting pathological deviations behind the iris or the sclera. Under these circumstances ultrasound techniques can provide diagnostic information. Two types of ultrasonography are used in ophthalmology. The A-scan uses a stationary transducer primarily to measure the distances of the various structures differing in their index of refraction for ultrasound. The most important application is the measurement of the eye length necessary to determine the power of an intraocular lens. The B-scan uses a linearily scanned transducer providing a tomographic cross-section of the eye to image all sorts of abnormalities.

8.4.1 Physical Basics of Ultrasonometry

Ultrasound for diagnostic purposes is generated piezoelectrically. An oscillating voltage is applied to a piezoelectric disk referred to as the transducer. The voltage frequency is chosen to be the fundamental axial resonance frequency of the disk. The disk is brought into contact with the object. In order to avoid reflection losses at the surfaces a contact jelly is necessary. The ultrasound wave diverges due to diffraction. One has to distinguish between the near field and the far field. The near field is defined to be within the distance $x_0 = D^2/(4\lambda)$ to the transducer with the disk diameter D and the resonance wavelength λ [6]. The asymptotic divergence angle α obeys the simple rule known also from optics

$$\sin \alpha = 1.22 \frac{\lambda}{D}.$$

A typical ultrasound frequency used in ophthalmology is 10 MHz. The corresponding wavelength is in the order of 150 μm for ocular media. The disk diameter of the transducer is usually more than 3 mm. The corresponding length x_0 of the near field is then 1.5 m. Therefore the divergence of the ultrasound beam does not play a significant role.

Ultrasonography is an echographic technique. The transducer is driven by short pulses consisting of only a few cycles. The transducer must be mechanically damped. This can be achieved by contacting the transducer with a damping material on the rear side. Also piezoelectric polymers are used in practice with high intrinsic damping coefficients. In addition the polymer transducer thickness can be made very small; this is necessary to achieve high frequencies.

The pulses are reflected at the tissue boundaries of the eye. Each reflex hits the transducer causing a mechanical deformation which generates a piezoelectric voltage signal. The signal is registered as a function of time and can be displayed on a monitor. This echograph, also referred to as an A-scan, reflects the tissue distribution in the eye and can be utilized for the measurement of distances.

8.4.2 Biometry

Measurement of axial eye length with ultrasound A-scan is also known as biometry. The A-scan probe is typically operated at a frequency of 10 MHz with a resolution of 0.12 mm. The scan rate normally corresponds to the power source frequency of 60/50 Hz. The low signals of the reflexes from the front and back surface of the eye are measured with a logarithmic amplifier with a dynamic range of 80 dB. In many instruments the A-scan mode is implemented in a B-scan system with a monitor to display a two-dimensional cross-section of the eye (Figure 8.6). The B-scan is normally not used for measurement purposes but to display morphologic details of the eye.

8.4.3 Pachometry

A pachometer is an A-scan device optimized to measure the corneal thickness in the range between 0.3 and 1 mm. It is used primarily in the field of refractive surgery but also as a diagnostic tool to examine pathologic changes of the corneal thickness such as corneal swelling. The resolution of a pachometer is in the order of $\pm 2.5\,\mu m$ which is considerably better than the resolution of the standard A-scan being 0.12 mm although the transducer frequency of a pachometer is only slightly higher, namely at 12 MHz. The reason for this improvement is based on the simple structure of the cornea which generates only two dominant reflexes from the front and the end surface of the cornea. The timing of these two reflex pulses can be determined with a much higher precision than it is achievable by measuring the time difference between corneal reflex and fundus reflex of the A-scan because of the complex layer structure

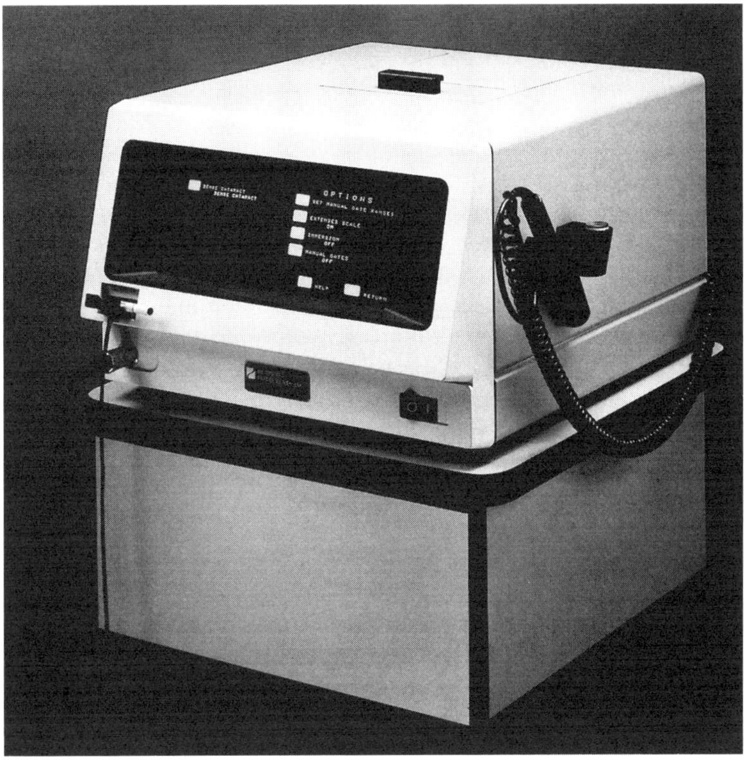

Figure 8.6. Ultrasonic biometer (courtesy Humphrey Instruments).

of the fundus. Pachometers are usually dedicated instruments without a monitor but with a liquid crystal display to show the corneal thickness in microns.

8.5 Keratometry and Corneal Mapping

The surface topography of the cornea plays an important role in contact lens fitting, intraocular lens fitting and refractive surgery.

8.5.1 Contact Lens Fitting

When a patient needs a contact lens it is important that the inner surface of the contact lens fits the surface of the cornea. The contact lens floating on a tear film is expected to be centered relative to the apex of the cornea to achieve the proper refractive correction. If the inner surface of the contact lens is too flat the lens is drifting on the cornea losing centration. If the surface curvature is too high the rim of the lens irritates the cornea causing inflammation and pain.

8.5.2 Intraocular Lens Fitting

Ultrasound A-scan is used to measure the eye length in order to determine the focal length of the right intraocular lens. However, the dominant part of the refractive power of the eye is determined by the refractive power of the cornea. In addition to the index of refraction of the corneal tissue and the tear film the radius of curvature of the cornea determines its refractive power. Therefore this radius and the eye length need to be known to select the right intraocular lens.

8.5.3 Refractive Surgery

The traditional way to correct the refractive error of the human eye is to prescribe spectacles or contact lenses. More recently refractive surgery has become an attractive alternative whereby the cornea is either shaped by radial incisions at the corneal periphery or by laser ablation of the corneal tissue. In both cases it is necessary to measure the corneal shape before and after operation.

In all three cases keratometry and corneal mapping are the standard techniques in practice. Whereas keratometry provides only the radii of curvature, corneal mapping instruments generate a distribution map of the radii of curvature over the whole corneal surface.

8.6 Keratometry

The cornea is in general an aspheric, toroidal surface. The concept of keratometry is based on the idea of approximating the corneal surface with an ellipsoidal surface which can be characterized by the two radii of curvature of the principal meridians (spherical meridians) and their angular orientation at one of the poles of the ellipsoid [7].

Figure 8.7 shows the principal set-up of an automatic keratometer. For reasons of simplicity we restrict the discussion to the measurement of the radius of curvature for a single meridian shown in Figure 8.7 as a circle with radius R. Two collimated light sources referred to as mires are imaged by the convex reflecting surface of the cornea and the objective lens of the keratometer on to a CCD target. The CCD image is recorded by a frame grabber and processed by an image analysis routine calculating

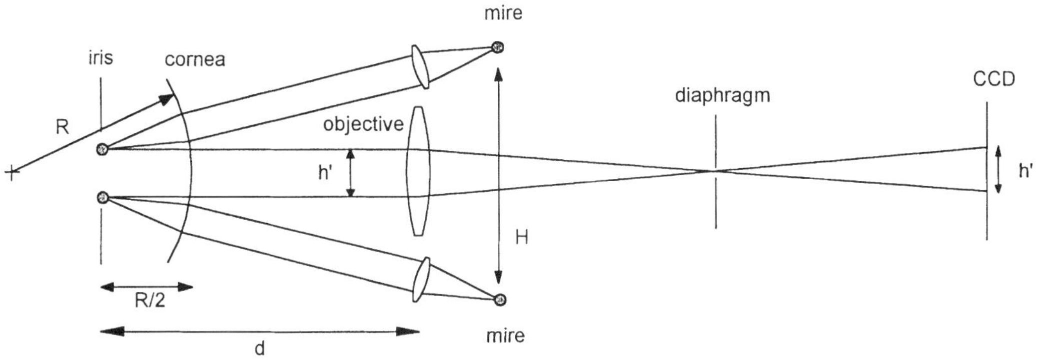

Figure 8.7. Keratometer. The cornea generates virtual images of the mires in the plane of the iris which are imaged on to the CCD target by the objective. The image hight h' is inversely proportional to the radius of the cornea.

the distance h' between the mire images on the CCD target. A diaphragm is located at the image focal plane of the keratometer objective restricting the principal rays from the two mire objects to be parallel on the object side of the keratometer objective (telecentricity). The distance between the keratometer and the eye is manually aligned so that the eye pupil image is sharp on the CCD monitor. The cornea acting as a convex spherical mirror generates a virtual image of the mires at a distance of $R/2$ behind the cornea. The radius of curvature of the normal eye is $R = 7.7\,\text{mm}$. The distance of the iris to the apex of the cornea is 3.6 mm which corresponds approximately to the position of the virtual mire image. The distance d between this image plane and the objective is therefore equal to the known object distance of the keratometer objective which is focussed to the iris with fixed image distance behind the objective lens of $2f$. With the known mire distance H, the image distance h' and the object distance d the radius of curvature R can be calculated with the keratometer formula which is a straightforward consequence of the geometry in Figure 8.7 ($R/2$ small compared to d):

$$\frac{R/2}{d} = \frac{h'}{H} \Rightarrow R = \frac{2\,dh'}{H}$$

It is obvious that with more than two mires the other radius of curvature and the orientation of the ellipsoid can also be determined.

8.7 Corneal Mapping

A corneal mapper is in a sense an extension of the keratometer which measures the apical radius of curvature. For refractive surgery the apical radii are not sufficient to

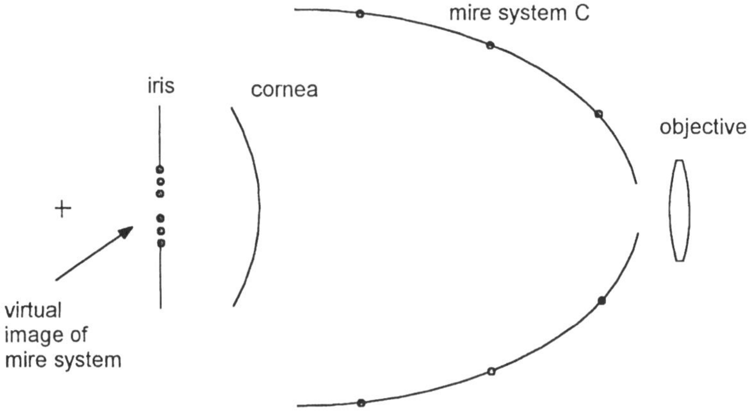

Figure 8.8. Corneal mapper. The mire system C is shaped such that its corneal image is disposed in the iris plane. The objective transfers the virtual image to the CCD target not shown in the figure.

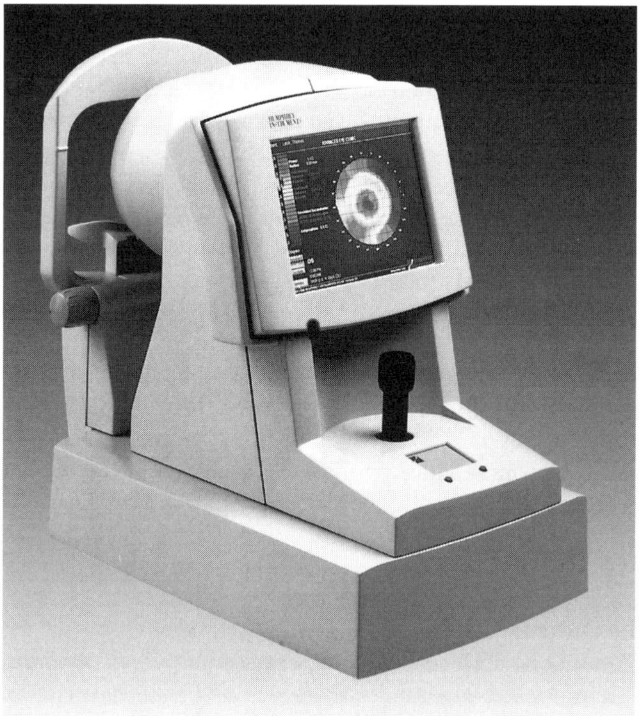

Figure 8.9. Corneal mapper (courtesy Humphrey Instruments).

characterize the shape of the cornea. It is necessary to map the surface of the entire cornea across a diameter of about 12 mm. This can be achieved with the optical set-up of the keratometer but with an increased number of mires located at much higher angular distances compared to the keratometer [8]. As a consequence the field curvature of the corneal image surface has to be taken into consideration. Because the virtual image needs to be conjugate to the flat CCD target the mires have to be arranged on a curved surface such that its virtual image surface is plane. This can be achieved with a surface shown as the curve C in Figure 8.8. A cone shaped mire body can be used as a sufficient approximation. A corneal topographer is shown in Figure 8.9.

8.8 Perimetry

Perimeters constitute an extraordinary sensor principle because the objects of this measurement method are sensors themselves and even more extraordinary is the fact that these sensors are of a biological nature. We are talking about sensor devices for the test of the visual sensoric system.

In many ocular diseases the sensitivity of the retina decreases slowly or vanishes completely. The patient cannot perceive these changes because the defects occur mainly in the periphery and reach the foveal area very late. One reason for the late perception is the fact that the visual system has the ability to mask and compensate defects in the periphery. A good example is the blind spot with the optical disk. Although the sensitivity there is zero it is difficult to realize the blind spot.

8.8.1 Visual Field

Perimetry is used in practice to measure the sensitivity distribution of the visual field, defined as the angular space where a fixating eye can perceive an optical stimulus. Its maximum size is about 60° upward, 75° downward, 100° temporally and 60° nasally. Further boundaries are imposed by the eyelids and the nose varying individually. The sensitivity of the retina is under photopic conditions (daylight vision) maximum at the central area (fovea). It decreases towards the periphery. Under scotopic conditions (night vision) the sensitivity distribution has a trough at the fovea, reaches a maximum at about 10° and finally decreases again towards the periphery. The reason for this difference in the sensitivity distribution between scotopic and photopic conditions is the different distribution of rods and cones on the retina. The fovea contains exclusively cone cells responsible for photopic vision whereas the density of the rods which are more sensitive under scotopic conditions has its maximum at about 20° in the visual field. Another important landmark of the fundus is the blind spot, the area of the optical nerve. There are no receptors at all, such that the sensitivity is zero within a region of about 5° to 8° in diameter. The blind spot is found temporally at about 15° off the visual field center.

8.8.2 Instruments

Most automatic perimeters currently in the market are variants of bowl perimeters which can be derived from the Goldmann perimeter developed in the 1940s [9]. Its essential part is a hemispheric bowl with a matt white inner surface and a radius of 30 cm. The patient's head is positioned on a head and chin rest such that the eye to be examined is located at the center of the bowl. The patient gazes at a fixation target usually an LED opposite to the patients eye on the bowl surface. The bowl is illuminated homogeneously with white light with a luminance of 10 cd/m^2 constituting photopic light conditions.

During the examination light stimuli with a duration of about 200 ms are sequentially generated on the bowl surface at different locations. The stimulus duration is not critical for the determination of the threshold sensitivity of the eye if the period is longer than 100 ms. (Below 100 ms Bloch's law is valid.) The brightness of the stimuli is varied between successive stimulus presentations. For the standard examination with the Goldmann perimeter stimulus luminance values are used with factors of 31.8, 10, 3.18 and 1.0017 larger than the luminance of the bowl illumination of 10 cd/m^2. The brightest stimulus referred to as the 0 dB stimulus has a luminance of 318 cd/m^2. The faintest stimulus of the Goldmann standard corresponds to the luminance of the background. Thus, the dynamic range of the relative stimulus luminance is 42.7 dB. With automated perimeters dynamic ranges of 50 dB are possible. The angular diameter of the circular light patches can be varied in steps between 0.05° and 1.72°. In contrast to the stimulus duration which does not influence the sensitivity threshold if it is above 100 ms the stimulus diameter needs to be taken into account [10]. For stimulus diameters below 10$'$ Ricco's law describes the relation between stimulus area and threshold luminance. The product of threshold luminance I and stimulus area A is constant. Above a stimulus diameter of 10$'$ Piper's law has to be considered which implies $\sqrt{A} \cdot I = \text{const}$. The constants depend on the location within the visual field. The knowledge of the angular distribution is necessary to compare perimetry charts acquired with different stimulus sizes.

The original Goldmann perimeter is a manually operated instrument. The stimuli are projected on the bowl with a projector moved by the operator, who has also to handle the filters for the stimulus attenuation and the registration of the patient's response. One can imagine that this procedure is very time consuming and can only be successfully carried out with cooperative patients. For that reason automated perimeters are preferred in todays practices. The positioning of the stimulus and the filter wheels is controlled by a computer integrated in the instrument. The patient is instructed to push a hand button when he perceives a stimulus. The patient's response and the position of filter wheel and stimulus is registered by the computer. The data

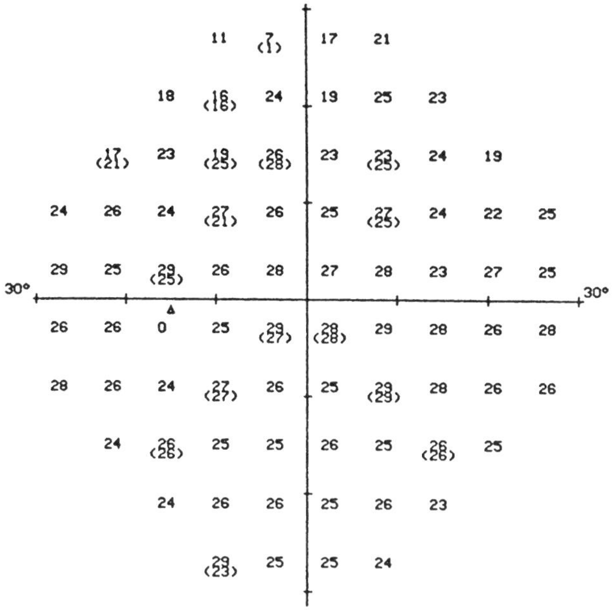

Figure 8.10. Numerical data chart of Humphrey Field Analyzer HFA (courtesy Humphrey Instruments).

are processed and statistically evaluated by the computer program. The result of the procedure is a sensitivity chart of the visual field of the patient (Figure 8.10).

8.8.3 Automatic Perimeters

One can divide automatic perimeters into projection perimeters where stimuli are projected on the bowl as in the Goldmann perimeter and instruments with stimulus sources at fixed positions on the bowl.

Figure 8.11 shows the set-up principle of a projection perimeter. It consists of an incandescent lamp with a condenser lens illuminating an aperture wheel which is imaged with the projector lens via mirrors on to the perimeter bowl. The aperture sizes cover the Goldmann stimulus sizes. Figure 8.12 shows the Humphrey perimeter.

There are also two filter wheels. The attenuation filter wheel consists of a circular filter wedge. The color filter wheel carries various color filters including an open aperture for the standard white stimuli. All filter wheels can be rotated by stepper motors.

The stimulus is positioned in the vertical direction by tilting the projector mirror and in the horizontal direction by rotating the turret on which the projector mirror is mounted. Both mirror and turret are driven by stepper motors via gear belts. In order to achieve a sharp image of the aperture on the bowl surface for all field positions the

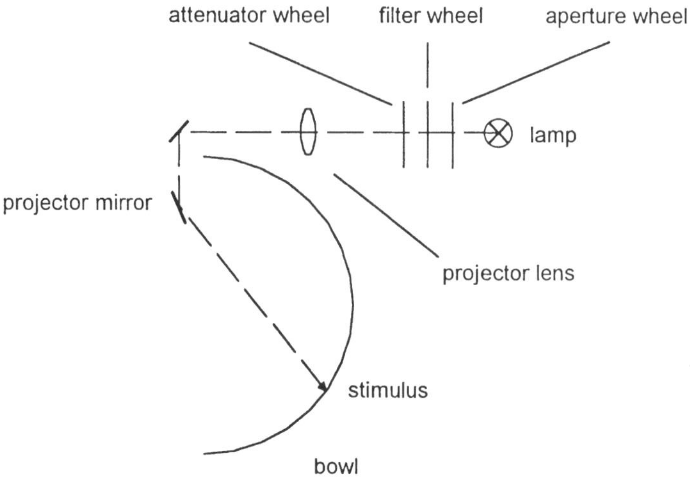

Figure 8.11. Set-up principle of a projection perimeter. The illuminated aperture wheel is imaged via the projector lens and the projector mirror on to the perimeter bowl. The filter wheel and the attenuator wheel allow change of color and stimulus brightness.

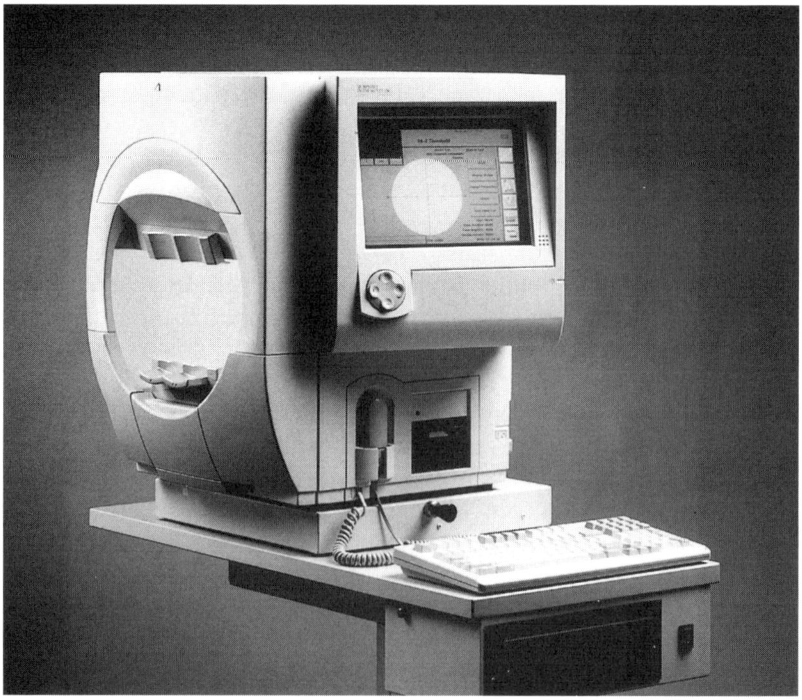

Figure 8.12. Automatic Perimeter (courtesy Humphrey Instruments).

mirror would be ideally located at the bowl center. However it must not interfere with the patient's head. For that reason the projector lens needs to be focussed depending on the stimulus position.

All stepper motors are controlled by a computer. The sequence of stimulus position and filter wheel position is determined automatically by the computer program.

8.8.4 Fixed Stimulus Perimeters

The advantage of projection perimeters is their flexibility concerning stimulus sizes and stimulus color. Another beneficious feature is the large accessible field of vision. The price for this high flexibility is the mechanical complexity entailing increasing manufacturing costs and risk of wear. Alternatives are fixed stimulus perimeters with light sources located at fixed positions of the perimeter bowl. Some manufacturers use yellow-green LEDs installed behind the semitransparent scattering bowl surface. The dynamic range of the stimulus luminance is between zero and more than 30 dB which is less than the Goldmann standard . Other manufacturers use fiber optic illumination. A single incandescent light source is coupled successively via an opto-mechanical switch device into the fiber bundles. The stimulus diameter is only $10'$. This approach allows different colors and a higher dynamic range of up to 40 dB. The disadvantage are the necessary moving parts.

Both types of fixed stimulus perimeters use up to 250 fixed stimulus positions within a field diameter of about 30° which corresponds to a stimulus distance of about 2°. This stimulus density is sufficient in the periphery. In the foveal region higher densities are necessary. This can be achieved by moving the fixation target fractions of the angular raster spacing of 2°. A disadvantage of using multiple light sources (as opposed to projection systems or fiber bundle systems) is that each source has to be calibrated for its light intensity output separately.

The advantage of the fixed stimulus perimeters is their mechanical and optical simplicity. The limited field and the limited flexibility concerning color, stimulus size and dynamic range are drawbacks.

8.8.5 Strategies

All automatic perimeters provide perimetry charts showing the distribution of the sensitivity threshold within the visual field of the examined eye. The stimuli are presented according to examination programs run by the computer. All manufacturers offer their own perimetry software with their specific strategies. The optimization criteria for these strategies are acquistion time, accuracy of the threshold value and its angular resolution. The acquisition time is limited by the ability of the patient to concentrate on the fixation target. The typical acquisition time per eye is about ten to fifteen minutes.

One can distinguish non-adaptive and adaptive strategies. Non-adaptive strategies use fixed coordinates where the stimuli appear. At all test points the procedure for determining the sensitivity level is the same whereas adaptive strategies choose the stimulus location and the method for the threshold determination depending on the already acquired data of the current test series.

For the determination of the threshold sensitivity in most cases an iterative procedure is used, known as the staircase method. The most important one is the 4/2 dB method developed by Fankhauser [11]. At particular points of the perimeter bowl (test points) stimuli are presented in a random sequence. Each test point is addressed several times with stimuli of different luminance. This sequence of stimuli is started with a stimulus luminance derived from age corrected standard values or from threshold values of previous examinations. Depending on the patient's response, the stimulus luminance is either reduced (patient has seen the stimulus) or increased (patient has not seen the stimulus) by a fixed step size until the patient's response changes. Then the step size is divided by two and the change of the stimulus luminance is reversed. This staircase method is repeated until the threshold can be estimated with the required accuracy.

The reaction of the patient is of course not always correct. The patient might push the button although no stimulus was seen. Another possibility is that a stimulus above threshold has been presented and the patient does not respond. This frequency of "false positive" and "false negative" responses can be determined by presenting randomly "dark" stimuli and stimuli above threshold and register the responses.

Another source of error is loss of fixation. The patient gazes in different directions during the test. Proper fixation can be probed by directing stimuli at the position corresponding to the blind spot several times during the test. Another method is to observe the eye, e.g. with a CCD camera.

The disadvantage of these non-adaptive strategies is a relatively long acquisition time based on the high number of stimulus probes necessary to achieve the required accuracy. However, the diagnostic value does not necessarily depend on the threshold accuracy at each point of the visual field. More relevant is the clinical validity of the results. Adaptive strategies are in that sense as precise as non-adaptive strategies in respect of validity with considerably fewer test steps leading to reduced acquisition time.

Manufacturers use a great variety of adaptive strategies in their programs. A few examples shall illustrate some concepts.

At field locations with deep defects, the achievable accuracy of the determination of the threshold is much lower than with normal thresholds. The step sizes of the staircase method can therefore be chosen to be much larger. Another approach is to determine the start values of the stimulus luminance at each field position based on the threshold values determined at the neighboring test points. Also the density of test point patterns can be adapted such that detected defect areas are sampled with higher density. Other strategies use maximum likelihood methods to estimate the field chart based on some reference points measured with high accuracy and points with less accuracy [12].

8.8.6 Other Methods

So far we have described perimetry with stimuli of varying luminance at fixed test points. This method is also known as static white light perimetry which is the most common perimetry technique used in practice. Another examination method more frequently used before the advent of automatic perimeters is kinetic perimetry where a stimulus with constant luminance is moved radially from the periphery with a constant angular speed of about 5° per second. The patient pushes the butten as soon as he perceives the simulus. This is done for various angles. The result is a line of equal thresholds (isopter). Repeating this procedure with different stimulus luminances provides an isopter map of the visual field.

Flicker perimetry is a rather new approach determining the angular distribution of the flicker sensitivity threshold defined as the flicker fusion frequency [13]. Here, the stimulus is modulated with a variable frequency which is changed between successive stimulus presentations whereas the luminance is kept constant. The advantage of this technique is based on the fact that the fusion frequency (i.e. the modulation frequency which can be still perceived) increases from the fovea towards the periphery. The perception of moving objects is more sensitive in the periphery than at the fovea, a peculiarity which has proven to be very advantageous for prey and predator during evolution. Flicker perimetry promises to be a sensitive diagnostic method for detecting scotomas in the periphery, a typical symptom of early glaucoma.

Blue yellow perimetry measures the sensitivity threshold distribution for blue stimuli against a yellow bowl illumination [14]. The yellow background saturates the red and green color reception channels of the visual system without irritating the blue channel. The blue stimulus luminance is varied within a staircase routine determining the thresholds for the blue light. The advantage of this technique is that changes of the blue light threshold appear in an earlier stage of glaucoma than changes with white light stimuli. The concept is based on the theory of color vision according to which the S-cones (blue sensitive) are less interconnected than L- and M-cones sensitive for yellow light. Lower interconnection means less redundancy against defects.

8.8.7 Fixation Control

As mentioned above, a central role for the reliability of the test results of all perimetry techniques requires the cooperation of the patient. Of particular importance is fixation. Staring at one point for several minutes needs much concentration which especially old patients are not always able to bring up. This hurdle can be removed with a gaze tracking device [15]. A CCD camera monitors the eye. The images are acquired by a frame grabber. The eye is illuminated by several near infrared LEDs appearing on the CCD as virtual images (Purkinji images) generated by the reflection at the cornea. Lateral movements of the head can be calculated by registering the displacement of the Purkinji images on the CCD target. However, changes of the gaze direction of the examined eye are not only caused by lateral head movements but even more by rotations of the eye causing a lateral displacement of the pupil on the CCD image. Therefore the pupil edge coordinates are determined with standard edge detection algorithms. The coordinates of the center of gravity of the pupil area can then be calculated. These coordinates together with the coordinates of the Purkinji images can be used to calculate the gaze direction and therefore the visual axis of the eye and its changes between successive video frames. In addition this information can be used to rescale the respective test point coordinates of the test series considerably reducing the acquisition time because of the reduced rejection rate of false patient responses.

8.9 Tonometry

The intraocular pressure (IOP) plays an important role in the development of glaucoma. This widespread disease causes a progressing decrease of the visual field. It is related to the degeneration of the nerve fiber layer and the optic nerve head. The glaucoma syndrome is classsified into low tension and high tension glaucoma. Both syndromes show a decreasing visual field but high tension glaucoma seems to be related to an increased intraocular pressure. Therapeutic measures reducing the intraocular pressure slow down the progression of the disease in many cases. Measurement of the intraocular pressure is therefore a routine examination method in medical eye care.

The IOP can be measured only indirectly. There are basically two methods used in practice. The indentation tonometer dates back to Schioetz. It measures the indentation of the anaesthetized cornea caused by a movable plunger of known weight. The movement of the plunger is a measure of the IOP. A high IOP causes a hard eye ball such that the indentation is small. However there are several inherent problems associated with indentation tonometry. The hardness of the eye measured with the indentation tonometer is not only determined by the IOP but also by the rigidity of the cornea and the viscosity of the aqueous displaced by the deformation of the anterior chamber through the indentation. In addition the adhesional forces between the indentation plunger and the tear film are acting against the weight force of the indentation plunger. Further errors are caused by the friction generated when the plunger squeezes the tear liquid out of the space between plunger and cornea.

These disadvantages can be avoided with another class of tonometers based on the applanation principle. The current standard of IOP measurement is the Goldmann applanation tonometer introduced in 1954 [16]. The principle is shown in Figure 8.13. It consists of a hollow cylinder with a flat front face which is brought into contact with the anaesthetized cornea of the patient's eye. A weight causes a force on the cylinder against the cornea transduced through a lever system not shown in Figure 8.13. The applanated cornea can be viewed through the cylinder with a biomicroscope. The applanated area is a circular disk. Two prisms split the field of view into two halves

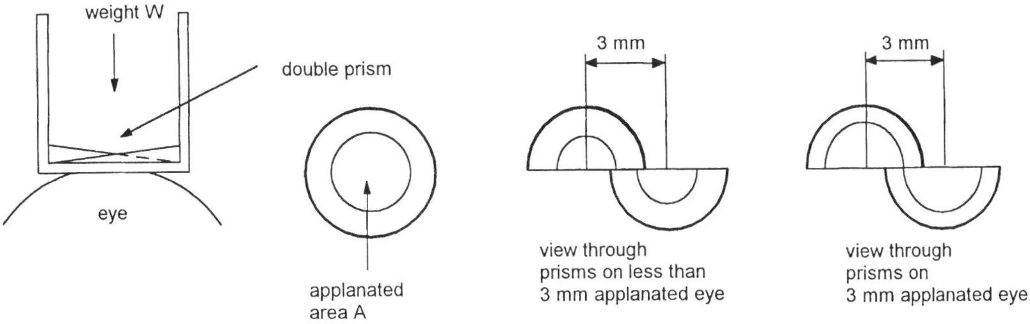

Figure 8.13. Applanation tonometer. The tonometer is pressed against the cornea with a weight W. The applanated cornea is viewed through the tonometer cylinder with a biomicroscope. Without the prisms the applanated area would be visible as a circular disk. The prisms are dimensioned such that the circular disk is split into two halves laterally shifted by three millimeters.

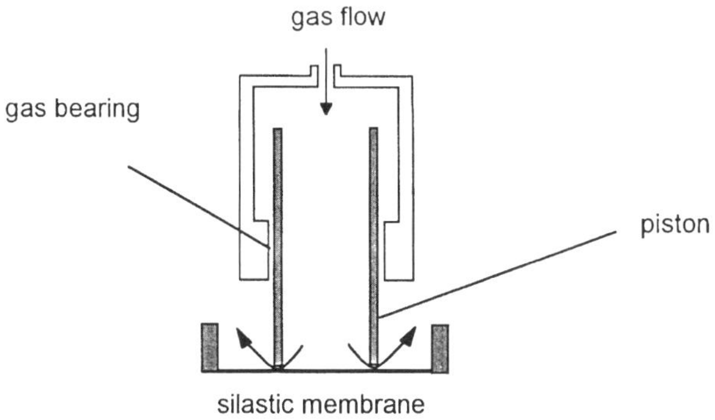

Figure 8.14. Principle of Digilab noncontact tonometer. The gas flow escapes through the gap between the membrane and the piston as soon as the pressure within the piston exceeds the IOP.

and shift it relative to each other by 3 mm. If the tonometer weight is chosen by the practitioner such that the applanated surface on the cornea is exactly 3 mm the inner edges of the lower and upper half disks are in contact as shown in Figure 8.13. In that case the IOP equals W/A with the weight W and the area of the disk with diameter $D = 3\,\text{mm}$. This particular diameter has been found by Goldmann to be optimal because in that case the attractive force caused by the tear film adhesion is balanced by the corneal rigidity and the frictional forces of the squeezed tear film layer so that the counterforce of the tonometer weight is only caused by the IOP. Another advantage of the applanation tonometer is that the displaced aqueous volume is negligible.

For these reasons all electronic tonometers are based on the applanation principle. They can be divided into contact and non-contact tonometers. The principle of an example of a contact tonometer shows Figure 8.14. The Digilab floating tonometer is a pneumatic applanation tonometer [17]. It consists of a silastic membrane which is

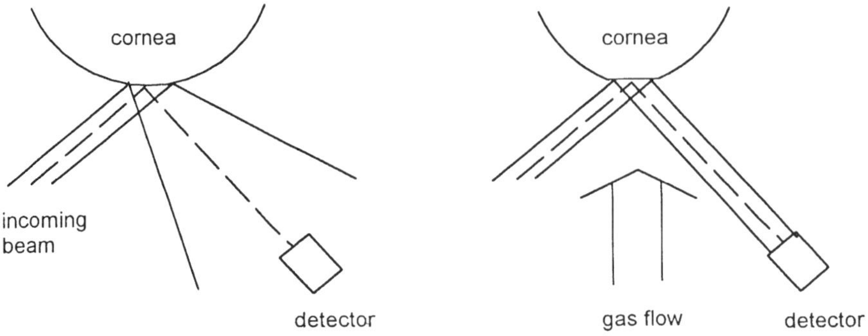

Figure 8.15. Principle of American Optical noncontact tonometer. The collimated beam hitting the conea diverges without gas flow causing a low detector signal. With the proper gas flow pressure the cornea is flattened reflecting the light beam collimated to the detector generating a maximum signal.

brought into contact with the cornea. The membrane seals the cylindrical chamber volume of a piston into which a gas flow is streaming. The pressure in the cylinder increases until it exceeds the IOP. At that moment the membrane is minimally displaced allowing gas to escape until the chamber pressure equals the IOP. Again the diameter of the piston defining the area on the cornea for the IOP measurement is chosen according to the Goldmann standard.

The main advantage of this electronic tonometer is the fact that it measures the IOP continuously such that variations of the IOP because of the cardiac cycle can be registered. The floating piston is mounted on an air bearing overcoming the effect of movements of the patient or the examiner.

The inherent disadvantage of contact tonometers is the fact that the patient's cornea has to be anaesthetized. It is time consuming and needs skilled examiners. Non-contact tonometers do not mechanically contact the cornea. The applanation of 3 mm in diameter is achieved with a collimated pulse of air. The air flow is increased until it reaches its peak after about 12 ms. An optical device detects when the cornea is flattened across 3 mm. As an example Figure 8.15 shows the setup of the American Optical non-contact tonometer [18]. A collimated light beam hits the cornea. Without the air flow the cornea is spherical such that the reflected ray bundle is divergent and only a small fraction of the reflected light hits the detector. As soon as the cornea is applanated the reflected beam is also collimated and focussed on the detector generating a signal which triggers the reading of the pressure meter.

8.10 Electro-ophthalmology

Vision can be described physiologically as a sequence of electrochemical events. The resting potential in the receptor cells is changed when photons are absorbed. This change of the potential is transmitted via the bipolar and ganglion cells to the nerve fibers leaving the eye through the optic nerve to the cortex of the brain. Due to disease this signal chain can be interrupted or disturbed at various stages generating disorders of visual perception without any visible changes on the fundus image.

On the other hand the electrochemical nature of vision can be used to localize disorders by measuring changes of the potentials associated with these electrochemical processes [19] [20].

There are three types of electro-ophthalmic examinations used in practice: EOG (electrooculography), ERG (electroretinography) and VECP (visually evoked cortical potential).

8.10.1 EOG

There is a resting potential between cornea and retina of the normal eye brought about by the receptor layer and the pigment epithelium. The cornea is positive and the potential is in the order of several millivolts. Disorders of the pigment epithelium reduce this resting potential. It cannot be directly measured but it can be indirectly measured through electro-oculography.

Electrodes are applied to the skin on both sides of the eye. The electrostatic field of the positive charged cornea induces negative charges on the electrodes. When the eye is turned towards one electrode the charge increases there and decreases on the other electrode. A difference of potential between the two electrodes can be measured. When the eye is turned in the other direction the same potential difference can be recorded but with opposite sign.

The signal depends of course on a variety of parameters which have to be standardized to get comparable results. The standardized registration procedure is as follows. Within a dark period of twelve minutes a steady state of the potential is achieved. Then the visual field is illuminated with a luminance of 1800 lux. The patient is instructed with a frequency of 15/minute to look sequentially at two fixation light targets separated by an angular distance of 30°. The electrodes also used in electrocardiography and electroencephalography are applied to the left and right of the eye. The signal amplitudes are in the range of up to 2 mV.

8.10.2 ERG

Whereas the EOG measures the resting potential between cornea and retina electroretinography (ERG) measures the reaction potential to a flash of light led off with a contact lens electrode placed on the cornea. The other electrode is usually attached to the ear lobe.

The standard corneal electrode is designed to allow the patient to look through it to see the stimulus flash. Silver wires are wrapped around this artificial pupil kept at a

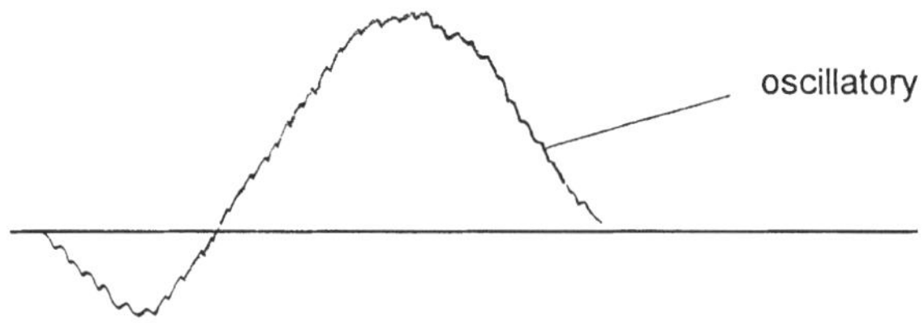

Figure 8.16. ERG signal. It consists of the negative a-wave, the large positive b-wave and the superimposed oscillatory potential.

distance from the corneal surface. A saline solution fills the gap between cornea and electrode. This contact lens electrode is not very convenient for the patient. Alternatives are electrodes attached to the lid, but with the disadvantage of reduced sensitivity.

The registration procedure for the ERG is the following. After some minutes of dark adaptation a xenon lamp flashing with a frequency of several cycles per second illuminates a hemisphere comparable to the perimeter bowl generating a homogeneous illumination of the visual field. The repetitive ERG signal is recorded and averaged by a computer. A schematic ERG signal is shown in Figure 8.16. It consists of the negative a-wave, the larger positive b-wave and the superimposed oscillatory potential. The duration of the whole event is about 200 ms. The maximum amplitude is in the order of 200 μV.

The a-wave is due to the activity of cone cells whereas the b-wave reflects activity of bipolar cells. Because cone cells are responsible for photopic vision the a-wave does not appear under scotopic light conditions where rod cells are involved. For that reason the background illumination determines whether a scotopic ERG or a photopic ERG is recorded. A photopic ERG is obtained with a background illumination corresponding to an illuminance of 2000 lux at the cornea, effectively suppressing the rod activity.

The ERG signal is not affected by defects or disorders of ganglion cells, nerve fibers or the nerve head. For that reason ERG is a diagnostic tool to identify disorders of the retinal receptor layer.

8.10.3 VECP

The visually evoked cortical potential can be measured between electrodes attached to the occipital pole (on the backside of the head) and to the ear lobe. The VECP can be regarded as an electroencephalographic signal correlated to a light stimulus. The VECP amplitude is very low – in the range of a few microvolts. The latency period between stimulus and signal is about 100 ms. Repetetive flash stimuli are necessary to separate the stimulus correlated VECP signal from the uncorrelated EEG signals by averaging procedures. At least 10 flashes are necessary to distinguish the VECP signal. Normally the entire retina contributes to the VECP signal. One can restrict the signal generation to the foveal area if one projects flashed pattern stimuli (e.g. a checker-board) on the retina. The enhanced VECP signal can only be related to the foveal area because the cone density is high enough to resolve the projected pattern.

The VECP reflects the potential change of the cerebral cortex as a reaction to a light stimulus. Wheras ERG is used for the diagnosis of disorders of the receptor layer VECP can be applied for the diagnosis of optical nerve defects (optical neuritis).

8.11 Sensors in Ophthalmologic Research

The sensor principles described above are well established and standardized in clinical practice. There are of course many more principles, devices and methods described in the literature which have not yet fully found their way into routine diagnosis and therapy, although instruments might be manufactured and distributed.

Several conditions have to be fulfilled before a new method is successful. One condition is that the clinical efficacy has to be proven in respect of diagnosis or therapy of frequently occurring diseases. That means diseases have to be diagnosed earlier than with conventional techniques such that therapeutic measures can be taken to avoid or impede the progress of the disease in an early stage. In addition an ordinary practitioner must be able to afford and handle the instrument with a reasonable amount of ease. Not all criteria are yet proven for the sensor principles to be discussed in the following sections. In that sense they are still in a research state concerning their potential applications.

It is necessarily arbitrary to choose a representative selection of ophthalmologic measurement techniques out of a huge variety of methods published in the literature. The methods and devices discussed in the following have in common that they attract considerable attention in ophthalmologic research because they address the diagnosis of the most relevant eye diseases in the developed world.

The most important causes leading to blindness in developed countries are diseases of the fundus. These are diabetes related retinopathies, age related macular degeneration and glaucoma. The mechanism of their generation is barely understood. One believes that they are related to disorders of blood circulation and metabolism in the fundus. Therefore research activities focus on diagnostic measurement devices which can differentiate the various vascular systems of the fundus in respect of morphology and perfusion.

The nerve fiber layer and the optical nerve head are structures of the fundus playing an important role for the diagnosis of glaucoma. Traditionally glaucoma is diagnosed with perimetry identifying regions of the visional field already lost. There are indications that the nerve fiber layer is thinning before field losses are apparent. Also the nerve head shows structural changes which seem to be related to the progression of glaucoma. Early glaucoma diagnosis might be tied to the detection of these changes of the nerve fiber layer and the optical nerve head. The techniques described in the following address these diagnostic needs in several ways.

8.11.1 Doppler Flowmetry

The retina is nourished by two separate vascular systems. The retinal circulation supplies the nerve fiber layer and the ganglion cells. The choroidal circulation supplies the photoreceptors (rods and cones). Malfunctions of these systems are supposed to have an essential influence on the etiology of the above-mentioned diseases. Doppler flowmetry is a technique which measures both the flow and the blood volume [21].

A Doppler flowmeter comprises a laser preferably with a wavelength in the near infrared region to achieve better penetration into the fundus delivered into the eye with a standard fundus camera. The laser beam is focused by the eye lens on to the fundus. The light is multiply scattered by the fundus tissue hitting red blood cells moving in capillaries in different directions. Due to the scattering process in the tissue the red blood cells are irradiated from all sides. The light scattered by the moving blood cells is Doppler shifted with respect to the irradiating light scattered by the stationary tissue. The total light emerging from the illuminated fundus spot is composed of tissue scattered and Doppler shifted light superimposed at the detector. Their beat causes a modulated photocurrent. Because the velocity directions of the red blood cells and the directions of the k-vectors of the irradiating light waves are arbitrary all sorts of Doppler shift frequencies can be found in the Fourier transform of the photocurrent. The average velocity of the red blood cells is proportional to the width of the photocurrent power spectrum (more precisely its first moment). As the strength of the Doppler signal depends on the number of scattering red blood cells the integral of the power spectrum is proportional to the moving blood volume. With this technique it is possible to determine locally the bloodflow although it is hard to distinguish between the retinal and choroidal circulation.

8.11.2 Laser Scan Ophthalmoscope

Laser beam scanning is applied in many fields. However, particular conditions have to be considered for the design of a laser scan ophthalmoscope. The fundus is a very bad reflector. Only 10^{-4} of the incident laser light is typically reflected back. In addition the optical quality of the eye lens is not very good. Only about 2 mm of the diameter of the pupil can be utilized to achieve good images. On the other hand high apertures are required to gather enough light reflected from the fundus. Furthermore, wide field angles (60 degrees) have to be covered by the scanning laser.

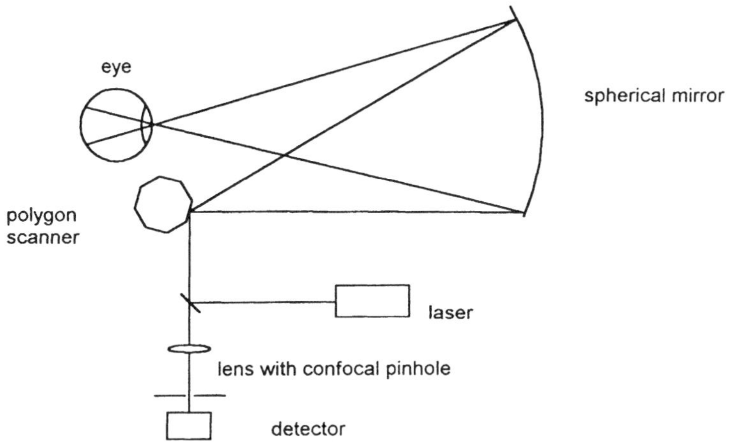

Figure 8.17. Schematic of confocal scanning opthalmoscope.

All three requirements can be addressed by the setup shown in Figure 8.17 [22], [23]. The laser beam is scanned laterally with a polygonal mirror and vertically with a galvanometer driven mirror (not shown). Both mirrors are imaged with a spherical mirror into the pupil of the eye. The laser beam diameter is about 1 to 2 mm to avoid spherical aberration of the eye lens. The pupil is typically dilated to a diameter of about 9 mm to achieve a high aperture for light detection. The reflected light is imaged via a beamsplitter to a detector. The image quality is not very important for the detection path. Therefore the spherical aberration associated with the 9 mm pupil diameter is tolerable.

There are several ways to utilize such a device for measurement purposes. One is confocal imaging. The laser spot on the fundus is imaged to a pinhole in front of the detector. Only light from the focal plane of the eye lens is detected. This allows one to measure the topography of the nerve head since a correlation seems to exist between the excavation of the nerve head and glaucoma.

Another approach measures with a polarization analyzer as detector device the elliptic polarization state of the light scattered back from the fundus [24]. It has passed the transparent but birefringent nerve fiber layer twice causing elliptic polarization. The thickness of the layer is proportional to the ellipticity of the detected light. Because reduced NFL thickness is supposed to be an indicator for early glaucoma this method might have some diagnostic value.

8.11.3 Optical Coherence Tomography

Optical coherence tomography (OCT) is a scanning technique combining the advantages of confocal imaging and short coherence interferometry for three-dimensional imaging of objects [25, 26]. OCT demonstrates its strength most spectacularly in ophthalmology where *in vivo* cross-sectional images of the fundus are not accessible by any other technique with a comparable resolution.

One advantage of OCT is the high resolution in the axial direction. This axial resolution is determined by the coherence length of the light source whereas the lateral resolution is determined by the numerical aperture of the eye lens.

There are two reasons why OCT is being successfully applied in the field of ophthalmology. First is the fact that the fundus of the human eye has to be imaged through the ocular lens system. Because of spherical aberration the useful numerical aperture of the human eye is only in the order of NA = 0.05. As a consequence the diffraction limited depth of field of this imaging system is in the order of 150 micron, limiting the depth resolution of confocal or stereoscopic techniques. The histologically important structures of the fundus (for example the nerve fiber layer, the receptor layer and the choroid) have a thickness below 100 micron.

The second advantage of OCT is its inherently high signal-to-noise ratio which allows fast image acquisition which is necessary since patients cannot fixate the eye for long periods.

Figure 8.18 shows the OCT-interferometer which is a Michelson interferometer. The light beam from a superluminescent diode is divided into a reference beam and a probe beam hitting the patient's eye and being focussed onto the fundus by the eye lens. The reference mirror is scanned along the beam axis and the detector measures a time dependent interferogram. The position of the interferogram depends on the relative position of the object point to the beam splitter. Because of the short coherence length of the light source the interferogram appears only when the length of the reference arm equals the length of the probe arm. The signal strength depends on the scattering amplitude of the respective object point.

The probe beam is scanned in the lateral direction by a scanning mirror as shown in Figure 8.19 [27]. The fundus is scanned along a line. The mirror is imaged into the eye pupil.

The interferometer can be realized either with bulk optics or as a single-mode fiber-optic interferometer. The advantage of the second approach is to avoid air turbulence and improve mechanical stability. Another useful side effect is that a single-mode

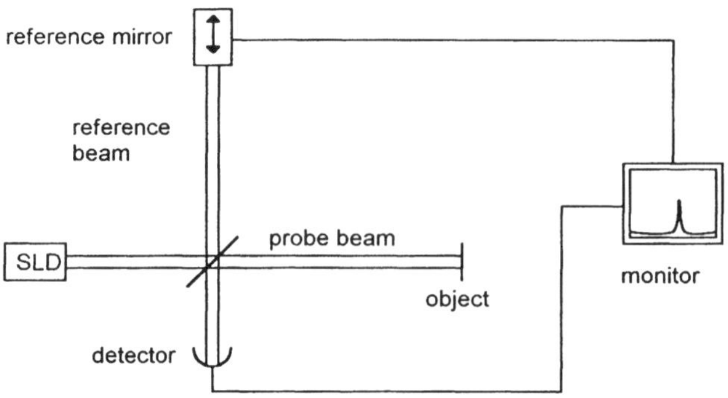

Figure 8.18. OCT interferometer.

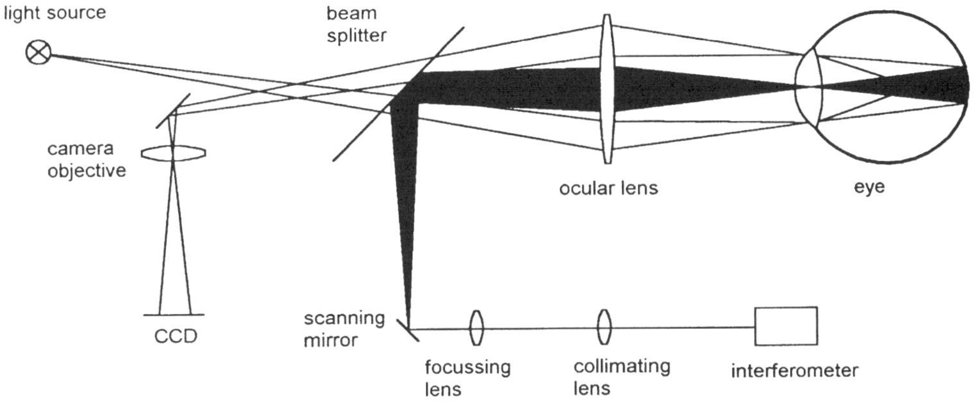

Figure 8.19. Optical system of optical coherence tomograph.

fiber-optic interferometer acts like a spatial filter. The divergent fiber output is collimated by the collimating lens. The focussing lens focusses the collimated probe beam on the back focal plane of the ocular lens generating a collimated beam hitting the eye and being focussed on to the fundus. The light scattered back from the various fundus layers goes back the same way. But only light within the depth of focus of the eye is received by the interferometer fiber. This confocal imaging system suppresses the reflex from the cornea. However, the depth resolution of the confocal system alone is not sufficient to resolve single fundus layers. The back scattered light generates an OCT echo signal displayed on the screen as shown in Figure 8.20. Each pixel corresponds to a point of a cross-sectional view of the fundus. The cross-section

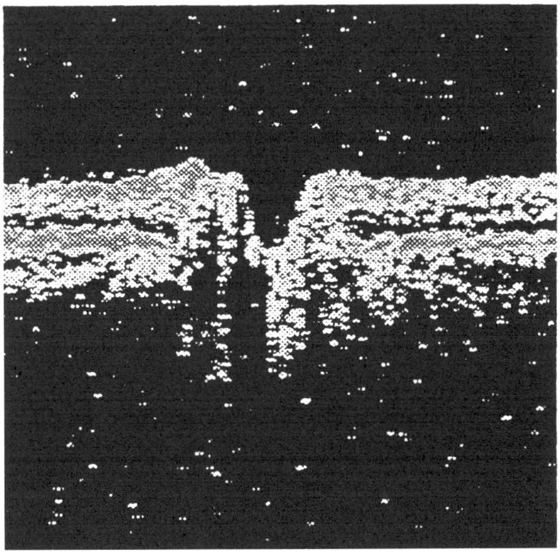

Figure 8.20. Cross-sectional view of the retina acquired with an optical coherence tomograph. The thickness of the structure is in the order of 300 micron (courtesy Humphrey Instruments).

consists of 128 lateral scan positions. For alignment and documentation the fundus is simultaneously imaged via a CCD camera (Figure 8.19). A light source with a NIR transmitting filter is imaged with the ocular lens into the pupil of the eye and illuminates an area on the fundus. The same field is imaged via eye lens, ocular lens and camera objective on to the CCD target and finally displayed on a monitor.

The basic problem with all ophthalmic instruments is the corneal reflex of the illuminating light. Here we have two light sources the probe beam and the fundus illumination. The reflex problem is avoided by separating geometrically the pupils of the illumination optics, the scanning optics and the camera optics at the eye pupil as shown in Figure 8.19.

With this setup the nerve fiber layer thickness and the shape of the optical nerve head can be measured with unsurpassed resolution. But also thickness measurements of the cornea and eye length measurements are possible. OCT is a non-contact method such that routine diagnosis is much easier and faster compared to the corresponding ultrasound techniques of pachometry and biometry.

8.12 Conclusion

The trend in the development of new sensors in ophthalmology is determined by the immense cost pressure in health care in all developed countries. New techniques have a chance to get introduced into routine practice only if they reduce costs either directly or indirectly.

Only one example shall illustrate what sort of new technologies are necessary to address this new situation. Sensors that register and document data almost automatically produce huge volumes of data. A very important step in sensor development is therefore the development of techniques to compress, store and transfer these data streams. A straightforward step in this direction is to link individual instruments within single practices through local networks to avoid paperwork. These paperless practices are further linked with each other through telephone communication networks. Artificial intelligence will be used to interpret and evaluate the data. This ongoing trend is further nourished by the consolidation process in health care also driven by exploding costs.

8.13 References

[1] Collins, G., *Br. J. Physiol. Opt.* **11** (1937), 30–42.
[2] Kamiya, M., *et al.*, US Patent 4,591,247, 1986.
[3] Sekiguchi, K., US Patent 4,878,750, 1989.
[4] Born, M., Wolf, E., *Principles of Optics*, 6th edn; Oxford: Pergamon Press, 1980, p. 464.

[5] Nohda, M., US Patent 4,364,646, 1982.
[6] Wells, P. N. T., *Biomedical Ultrasonics*, London: Academic Press, 1977.
[7] Helmholtz, H. v., *Handbuch der physiologischen Optik*; Leipzig: 1856.
[8] Placido, A., *Periodico d'oftalmologica practica*; Lisbon: 1880, pp. 27–30.
[9] Goldmann, H., *Ophthalmologica* **109** (1945), 71.
[10] Lachenmayer, B. J., Vivell, P. M. O., *Perimetrie*; Stuttgart: Thieme, 1992, p. 5.
[11] Fankhauser, F., *et al.*, *Perimeter Digest*; Schlieren: Interzeag AG, 1983.
[12] Rootzen, H., *et al.*, US Patent 5,381,195, 1995.
[13] Lachenmayer, B., *Habil.-Schr.*; Med. Fakultaet, Munich, 1988.
[14] Sample, P. A., *et al.*, *Invest. Ophthalmol. Vis. Sci.* **31** (1990), 1869.
[15] Lehmer, D. E., *et al.*, US Patent 5,491,757, 1996.
[16] Goldmann, H., *Bull. Soc. Fr. Ophthalmol.* **67** (1955), 474–478.
[17] Langham, M. E., McCarthy, E., *Arch. Ophthalmol.* **79** (1968), 389.
[18] Forbes, M., *et al.*, *Arch. Ophthalmol.*, **91** (1974), 134.
[19] Ermers, H. J. M., van Lith, G. H. M., in: *Photography, Electro-Ophthalmology and Echo-Ophthalmology in Ophthalmic Practice*, Henkes, H. E. ed; The Hague: Dr Junk, W., B. V. Publishers, 1973, pp. 245–255.
[20] McClure, E., in: *Optometry*, Edwards, K., Llewellyn, R., eds., London: Butterworths, 1988, Chapter 22.
[21] Riva, C. E., et al., *Optical Engineering* **34** (1995), 746–752.
[22] Sabban, J. C., et al., *J. Optics* **15** (1984), 425–430.
[23] Marshall, J., *et al.*, *Br. J. of Ophthalmology* **76** (1992), 470–474.
[24] Dreher, A. W., *et al.*, *Clin. Vision Sci.* **7** (1992), 481.
[25] Fujimoto, J. G., *et al.*, *Science* **254** (1991), 1178–1181.
[26] Fercher, A. F., *et al.*, *Opt. Lett.* **13** (1988), 186–189.
[27] Hellmuth, T., *SPIE Proc.* **2926** (1996), 228–237.

PART 3

Sensor Market

9 Market Overview: Sensors for Scanning Force Microscopy

U. GRUNEWALD, S. MÜLLER-PFEIFFER, Carl Zeiss, Jena, Germany

Contents

9.1 Brief Outline

Since the introduction of scanning tunneling microscopy by Binning and Rohrer [1] in 1981, the techniques combined under the term scanning probe microscopy have undergone a breath-taking development. The 1980s were marked by intensive research and development. The early 1990s have seen a dramatic rise in the commercial utilization and technical improvement of these microscopes. Instruments allowing a wide variety of measuring techniques became available. As in light microscopy, these are also called contrasting techniques.

In the early 1990s the vast majority of scanning probe microscopy was used in basic research in physics and chemistry. Now these microscopes are increasingly making inroads into the field of applied research in biology and medicine and into quality assurance, for example in the semiconductor industry.

The increasing interest in the investigation of natural or artificially produced microscopic objects has greatly inspired development in microscopy. However, time and time again, the demand for higher resolution has increased both expenditure and costs. The scanning probe microscope has halted this trend to a certain extent and has thus considerably contributed to its widespread utilization.

In analogy to the various contrasting techniques in light microscopy such as fluorescence or polarization contrast scanning probe microscopy has developed operation modes such as topographic imaging, frictional force microscopy or potential contrast. In contrast to conventional light microscopy the contrasting techniques (i.e. operation modes) of a SPM are not diffraction limited since no wave optics is involved.

Scanning probe microscopy has developed various techniques to overcome this seemingly natural restriction. From the viewpoint of the concept [2] which it implements, the so-called near-field optical microscope is the oldest scanning probe microscope, and the newest as far as its technical implementation is concerned [3]. The currently most promising principle involves a light-emitting aperture with a diameter which is smaller than the wavelength (typically a few 10 nm, called a nanoaperture) and is guided over the sample at a distance which is also smaller than the diameter of the aperture. One major problem in near-field optical microscopy is the fact that the loss of energy at this nanoaperture totals several orders of magnitude. An increase in energy is not possible because of the resultant thermal destruction of the nanoaperture. If we consider the further reduction of detectable energy in fluorescence or even Raman spectroscopy, the practical implementation of a near-field optical microscope can primarily be seen as a problem of energy. Scientists throughout the world are attempting to solve this problem.

In the field of atomic resolution, mechanically or acoustically induced vibrations of the probe or the sample in the range of a few nanometers can make instrument operation impossible. Even very slight thermal drifts may endanger the measurements. If we compare nm-vibrations to the usual vibration of buildings, it becomes evident that the mechanical design of a scanning probe microscope must absorb vibrations in several orders of magnitude. One of the most ingenious principles of implementing this absorption is the self-compensating, so-called Beetle principle, which has made history in scanning probe microscopy [4]. This principle incorporates all the mechanical and design principles required for the vibration inert and completely thermally compensated arrangement of a scanner and a probe. All of the mechanical eigenfrequencies are considerably above 10 kHz due to the dimensioning of the instruments, and an extremely short path for the transfer of force between the probe and the sample is guaranteed.

Almost every day, the still very novel methods of scanning probe microscopy are revealing new, important details in many areas of research and industry and are opening up new applications especially due to the invention of new probes.

To design and built those microscopes is however the easier problem. Without any doubt, the key to success lies in the ability to produce a big variety of probes reproducible at low cost, since their quality determines directly the image quality and probes will always wear after a certain time of use.

9.2 SFM Explained in a Nutshell

9.2.1 Basic Experimental Set-up

9.2.1.1 Scanner Set-up

The basic principle (as can be seen from Figure 9.1) of scanning probe microscopy consists in the scanning of the sample surface with a probe featuring a tip which is as sharp as possible by utilizing a wide range of near-field effects, such as the tunneling current, interatomic or intermolecular forces or magnetic field. Scanning probe microscopy is therefore also called near-field microscopy. Its advantage over classical light microscope is that a lateral resolution into the atomic range can be attained. The advantage over various types of electron microscopes is that scanning probe microscopes are considerably less expensive and in some cases offer a superior resolution.

As far as the lateral resolution is concerned, the near-field microscopy resolution – unlike far-field microscopy – is not restricted by the diffraction of electromagnetic radiation. The currently most important scanning probe microscopes are the atomic force microscope (AFM) and the scanning tunneling microscope (STM).

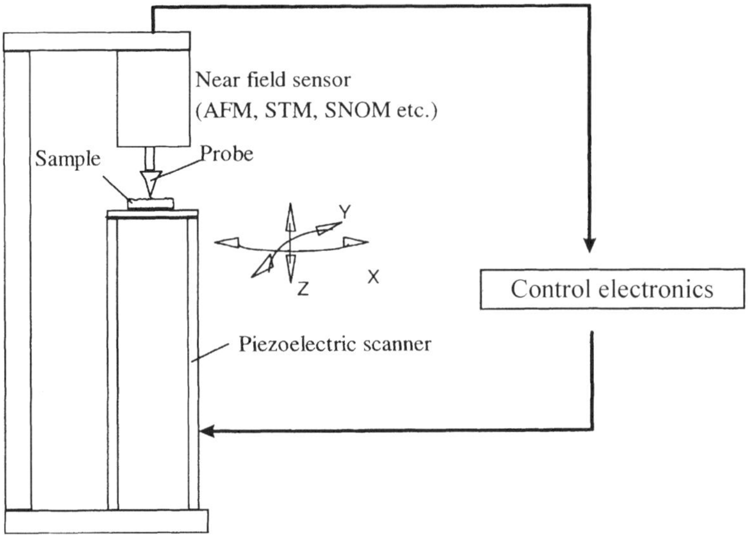

Figure 9.1. The principle of a Scanning Probe Microscope.

Despite the fact that near-field microscopy is a sequential technique, there is some reason to believe that scanning probe techniques will be applied in the future to produce structures in the nanometer range. So-called near-field structuring up to the atomic range has already been verified in research laboratories [5]. The development of scanning probe microscopy could therefore act as a catalyst for the development of nanolithography, which will probably require new probes with specific features such as hardness or chemical resistance.

9.2.1.2 Common Detection Principles

Beam Deflection

As can be seen from Figure 9.2, the most common setup to raster-scan the tip over the surface consists of the tip mounted on to a so-called cantilever and a beam deflection sensor, which indicates the smallest deflections of the cantilever when the tip is moved into close proximity of the surface. This beam deflection method was first described by Meyer and Amer [6]. The actual sensor (or probe) is the–often microfabricated–combination of the tip and the cantilever and not the beam deflection sensor.

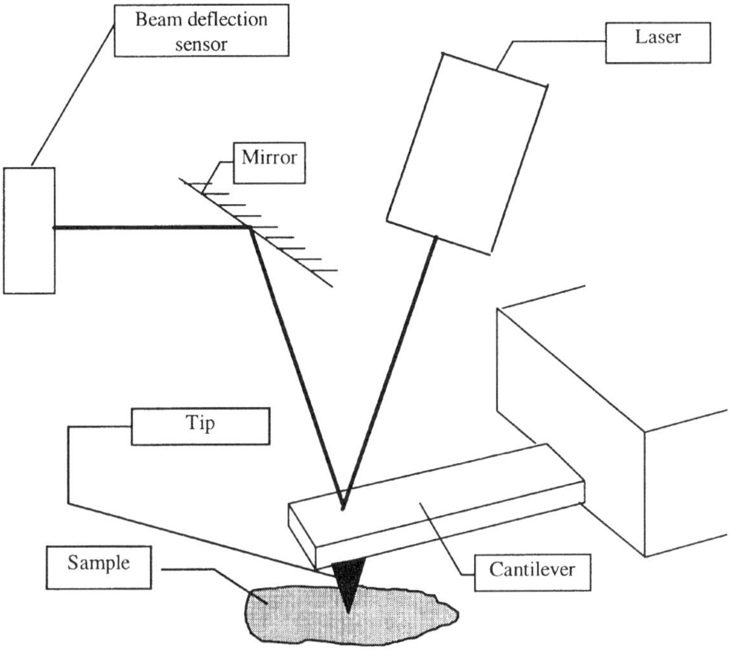

Figure 9.2. The beam deflection method.

Fiber-Optical Interferometer

Another widely used approach to detect the cantilever deflection is the fiber-optical interferometer first applied to scanning probe microscopy by Rugar [7]. The central element of this setup is a fiber-optical directional coupler (i.e. x-coupler.). One arm of the coupler is coupled to a diode laser and the other arm at the same side of the coupler goes to a detector, usually a silicon photo diode.

The cleaved end of one arm at the opposite end of the coupler and the back of the cantilever now form the interferometer. At the cleaved end approximately 4% of the incoming wave inside the optical fiber will be reflected. The remaining 96% of the intensity will be emitted. Some of this intensity is reflected from the back of the cantilever into the fiber. The wave reflected from the cleaved face of the fiber and the wave reflected from the cantilever now interfere. The interference signal will be detected by the photo detector.

This detection scheme works with virtually any kind of cantilevers. The advantage of this setup is that it can be integrated into vacuum chambers or cyrostats.

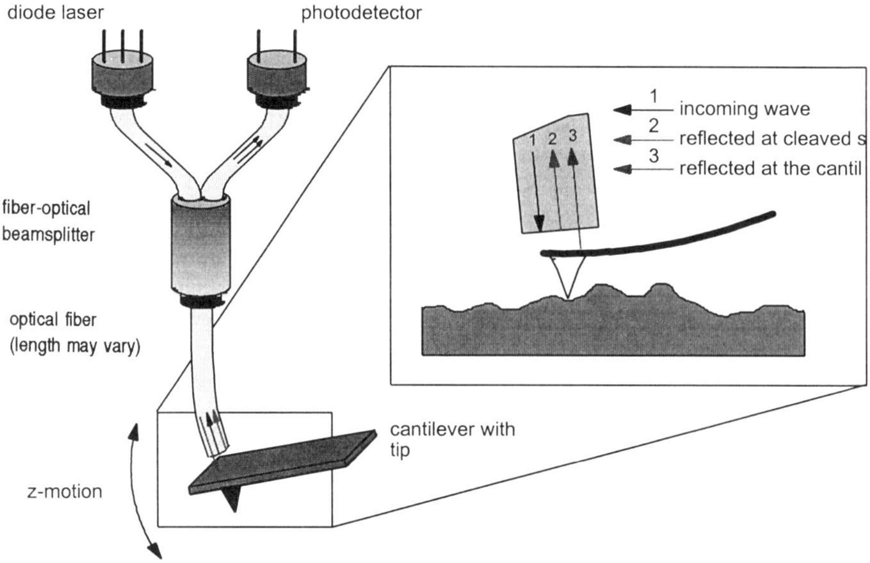

Figure 9.3. Working principle of the fiber-optical interferometer.

9.2.1.3 Less Common Detection Principles, Mostly of Historical Interest

Integrated Piezoresistive Element

Replacing the optical setup for the detection of cantilever deflection by integrated piezoresistive strips as described in [8, 9] has two main advantages. Changing the cantilevers would be easier because no optics adjustment is needed during this procedure. Additionally the scanner design is less complicated because no optics at all have to be included. This is especially important if the scanner should be placed into a vacuum chamber or other "hostile" environment.

The main disadvantage of this approach is the smaller sensitivity of cantilevers with integrated piezoresistive element compared to the use of beam deflection or the fiber-optical interferometer with any kind of cantilevers.

Macroscopic Interferometers

Various authors have described more or less complicated macroscopic interferometers to detect cantilever deflection [10–12]. These set-ups are expensive, cumbersome and difficult to use. They have been abandoned by most people and replaced by more compact set-ups.

Macroscopic interferometers work with virtually any kind of cantilevers including bent wires. However, if one uses micromachined sensors longer cantilevers are preferred. At least one vendor (DI) has special cantilevers in his list for use with interferometers which are longer than usual.

Capacitive Detection

An alternative way of detecting the cantilever deflection uses the cantilever and some counter electrode as a capacitor. This technique was demonstrated for example by Göddenhenrich [13]. The deflection of the cantilever will alter the capacitance which can be detected using an ac (i.e. rf) bridge scheme. Since the capacitance to be measured is rather small (some pF), the whole set-up is sensitive to interference. Moreover, the set-up is less sensitive then the ubiquitous beam deflection arrangement.

All conductive cantilevers can be used with this detection scheme.

9.2.2 Basic Ways of Running an SFM

For the understanding of Atomic Force Microscopy (AFM) it is essential to know the near-field force interactions between the probe and surface to be scanned. As can be seen from Figure 9.4 there are two regions of the surface force potential, the repulsive and the attractive forces. According to where the probe is operated, we distinguish between contact and noncontact AFM.

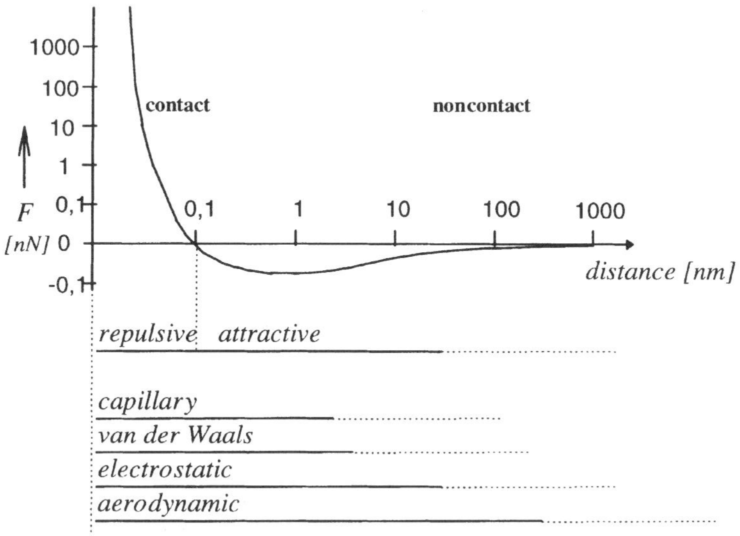

Figure 9.4. Force potential of a solid state surface.

According to the region in which the probe is operated, we distinguish two major modes: the static and the dynamic mode.

In the static mode the probe physically touches the surface, therefore the mode is normally called the contact mode (AFM cm). One uses very soft cantilevers with spring constants between 0.02 and 1 N/m not to stress the surface too much and to avoid wear of the tips. The resonance frequencies of these cantilevers are in the order of 6–20 kHz.

Advantages of the contact mode are very high lateral resolution down 0.01 nm and the possibility of measuring under liquids. It is disadvantageous that the tip is moved within the contamination layer, which causes sticking forces, contaminates the tip and always has the risk of damaging the surface and the tip as well.

To make use of the much weaker attractive (mainly van der Waals) forces, one has to excite the cantilever to oscillate close to its resonance frequency. This allows one sensitively to detect shifts in the resonance frequency, phase or amplitude, to control the tip over the surface without touching it. This mode is called the noncontact (AFM ncm) or dynamic mode. The advantage is that this mode is less prone to damaging the surface and to wear the tip. However the lateral resolution achievable is in the range of a few nm and is therefore much lower than for the contact mode. For this mode the cantilevers' spring constants range from 10 to 100 Nm^{-1}, their resonance frequencies from 30 to 400 kHz.

There are also various semicontact/noncontact modes known, by which the oscillating probe is scanned over the surface in such way, that more or less strong contacts between the tip and surface/contamination layer occur while the tip oscillates. The advantages and disadvantages are somewhere in between both static and dynamic modes.

Which cantilever for which purpose?

As a rule of thumb, for normal AFM contact imaging there should be chosen soft cantilevers ($k < 1$ N/m) while for noncontact topographic imaging stiff cantilevers ($k > 10$ N/m) with a high resonant frequency are recommended. Sensors for noncontact force modulation techniques have low spring constants and high resonance frequencies.

9.3 Non Batch Produced SFM-Sensors (Hand-crafted SFM Sensors)

9.3.1 Bent Wires

Bent wires have been the oldest form of SFM cantilevers. They were derived from (electro-) chemically etched STM tips by bending the last 100–200 µm between sharp knives, usually razor blades.

While the manufacturing of these tips is "low tech" and can be performed in virtually any laboratory, it is relatively time consuming and the results greatly depend upon the skills and the ability of the person involved. Because metals are softer than Si or Si_3N_4 bent wire tips wear faster then microfabricated ones. Additionally, the use of the widely used beam deflection scheme with this cantilever is difficult because the small wire does not give a specular reflex as sharp as the microfabricated spring does.

Since the wide and affordable availability of microfabricated SFM cantilevers most people have abandoned the manufacturing of SFM cantilevers from wires.

9.3.2 Bent Capillaries and Bent and Tapered Optical Fibers

Bent capillaries and bent and tapered optical fibres have been introduced into scanning near-field optical microscopy by Lewis [14]. Concerning their properties as pure SFM sensors they have nearly the same quality microfabricated cantilevers have. Compared to bent wires the surface is much smoother and has a good mirror finish. This allows them to be used in beam deflection set-ups. Because of their relatively high price compared to Si or Si_3N_4 cantilevers it is uncommon to use them as pure SFM sensors. Their main use is that they allow a conventional cantilever based SFM set-up to be used as a scanning optical near-field microscope.

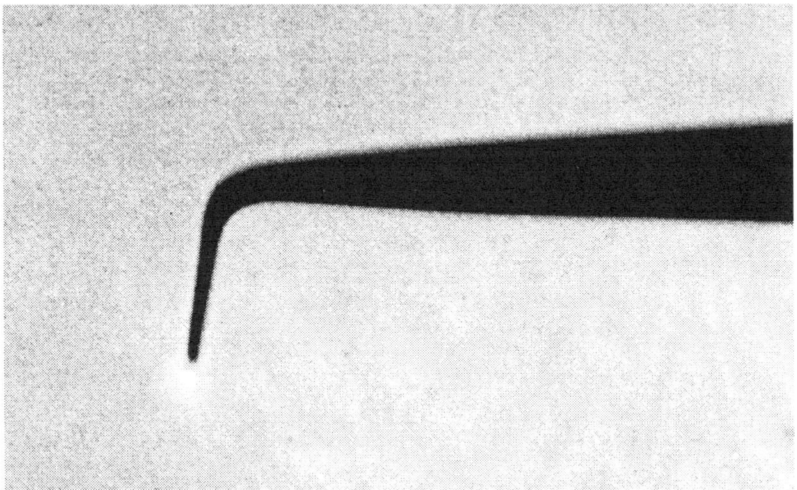

Figure 9.5. Micropipette, picture courtesy of Nanonics Ltd., Jerusalem.

9.4 Microfabricated SFM-Sensors

9.4.1 Basic Forms

9.4.1.1 Microfabricated Cantilevers

Most cantilevers currently used in scanning force microscopy are produced by microlithography well known from the manufacturing of integrated circuits or micromechanic devices. The first microfabricated cantilever was described by Akmine et al. [15].

As in the above cases microfabrication of cantilevers offers a couple of advantages which result from the batch fabrication process: cantilevers produced this way are inexpensive, exist in a wide range of different geometries and have relatively reproducible properties.

Three basic tip geometries are offered by the manufacturers of SFM cantilevers: pyramidal, tetrahedral and conical. Conical tips can be made sharp with high aspect ratio and a small curvature radius (as small as 5 nm) at the end of the tip. Pyramidal tips instead have a lower aspect ratio and a much blunter end (some 10 nm) but tend to be more durable.

SFM tips can be made from silicon or silicon nitride. The fabrication process is different for the two materials and influences the properties of the resulting tips as well as the properties of both materials do.

Conical tips are made from monocrystalline silicon. They are made by etching into the silicon around a silicon dioxide cap. The conical tips are rather slim which makes them suitable for imaging deep narrow features. The disadvantage of this geometry is that these tips tend to break more easily than pyramidal tips. Silicon has the advantage that it can be doped and thus the tips can be made electrically conductive. This can be useful for applying a bias between the tip and the sample or for preventing unwanted charge buildup at the tip.

Cantilevers made from silicon have additionally the advantage over silicon nitride cantilevers that they have a lower intrinsic mechanical damping than silicone nitride

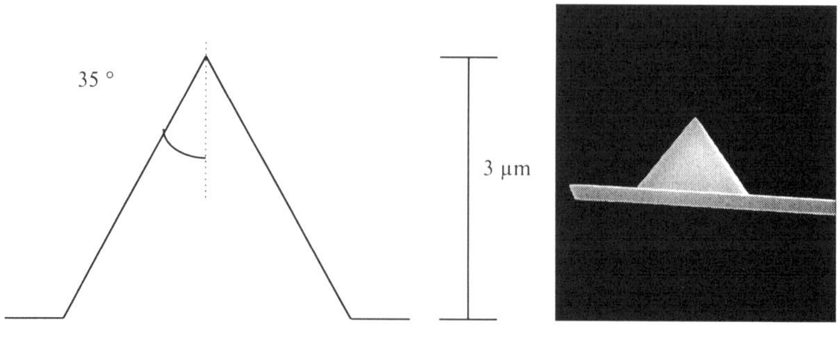

Figure 9.6. Typical shape of a standard SiNi tip. Radius of curvature is 10–50 nm (picture courtesy: Park Scientific Instruments, Sunnyvale, CA).

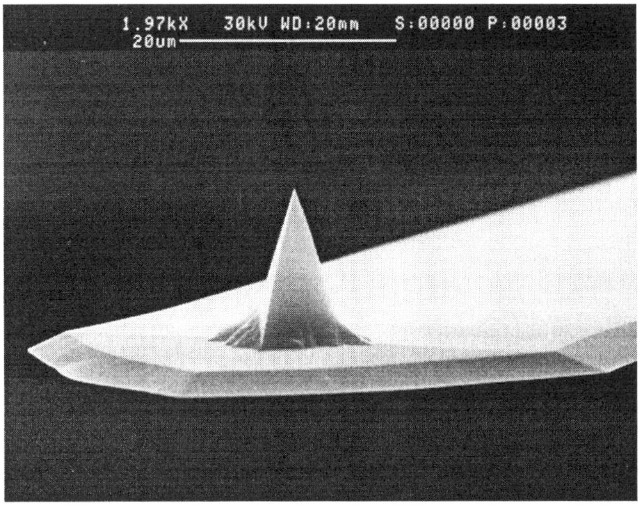

Figure 9.7. Typical shape of a conical tip etched from monocrystaline silicon (picture courtesy: Nanosensors GmbH, Aidlingen).

cantilevers. This results from the low dislocation density in silicon monocrystals manufactured for the semiconductor industry. This property can be used if a silicon cantilever is used in the dynamic force modulation mode in vacuum. If operated in ambient air the hydrodynamic damping limits the maximum sensitivity. In vacuum the intrinsic damping is the limiting factor for the sensitivity of the setup.

Silicon nitride tips are made by depositing a layer of silicon nitride over an etched pit in a crystalline silicon surface. This process produces tips of tetrahedral or pyramidal form. Since the pits to be filled are made by an anisotropic etch process, the geometry of the tips depends on the cyrstallographic structure of the etch pit material, silicon. The silicon nitride tips are broader and blunter than conical silicon tips which makes them less suitable for imaging deep narrow structures. On the other hand, silicon nitride is much harder than silicon which makes the tips less susceptible to wear.

The silicon nitride films normally made by a standard chemical vapor deposition process contain stresses which cause film deformation when the film thickness increases. This is the reason why thick cantilevers (for high spring constants or high resonant frequencies) cannot be made from silicon nitride. They have to be made from monocrystalline silicon instead.

Both silicon and silicon nitride cantilevers come in different form: straight beams and V-shaped. Both types of cantilevers are mounted on a small silicon chip, at the end, which allows the cantilevers to be handled easily and to be mounted into the scanning force microscope itself.

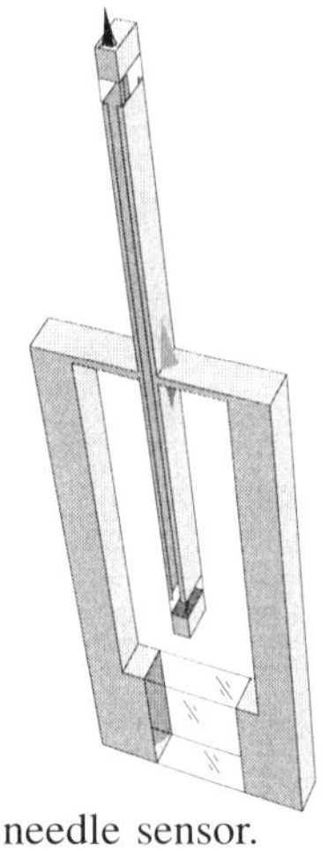

Figure 9.8. The needle sensor.

9.4.1.2 Special Solution: Needle Sensor

A completely different way of running a SFM is using the "needle sensor". The needle sensor is not like a cantilever. The needle sensor is a microfabricated small longitudinal quartz oscillator with a SFM tip at one end. It is possible to mount tips of almost all forms discussed below to the needle sensor. It is operated as follows. The needle sensor is excited near its resonance frequency (1 MHz). When the tip during an oscillation cycle touches the surface the resonance frequency of the quartz is shifted a little bit. This frequency shift can be detected as a phase shift between the response signal of the quartz and the excitation signal.

Compared to common cantilevers the needle sensor is extremely stiff ($k = 10^6$ N/m) which renders it impossible to use it for force modulation microscopy and makes the imaging of soft matter difficult.

The big advantage of the needle sensor is that it does not need any separate deflection detection scheme. This makes it easy to operate a needle sensor based SFM in a UHV chamber or a cryostat. It has the big advantage over the cantilever with piezoresistive detection of being much more sensitive. Atomic resolution can be achieved with this set-up.

9.4.2 Sensors with Modified Tip Geometries

9.4.2.1 Artifacts Caused by Tip Geometry

The most important role for high lateral and vertical resolution is played by the geometrical shape of the tip. One can assume that the lateral resolution on pretty flat surfaces is approximately twice the tip radius.

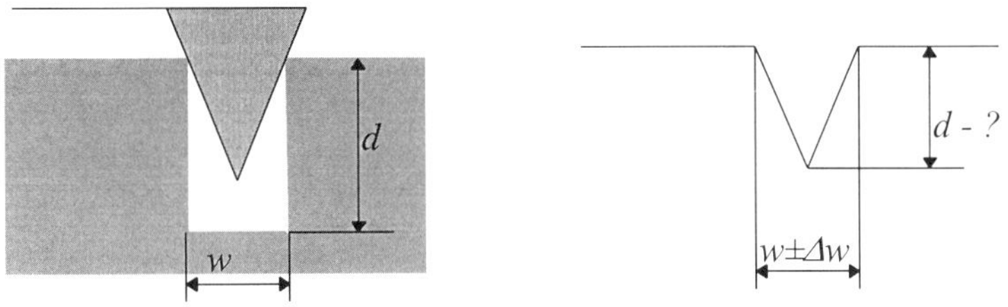

Figure 9.9. Artifacts caused by tip shape.

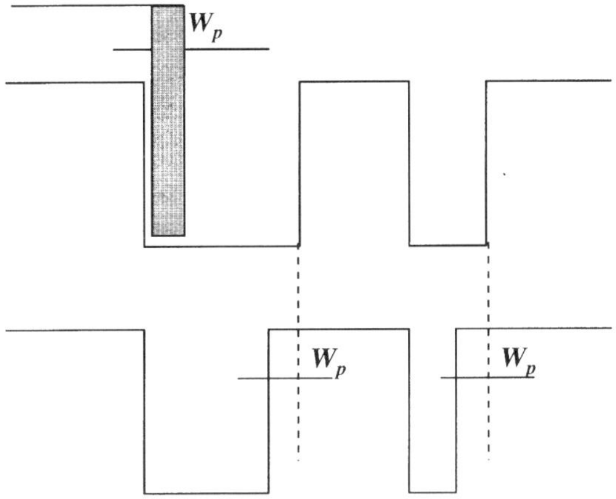

Figure 9.10. Measurement errors caused by tip dimension.

The z-resolution is more or less determined by the accuracy of the beam deflection. Artifacts caused by the geometry of the tip are especially essential, when so-called high aspect ratio structures have to be exactly measured, which can be seen from Figure 9.9 and 9.10.

Figure 9.9 shows the impossibility of imaging steep structures with conventional tips. In the sense of metrology the width of a steep trench can be measured as precise as the scanners work. The depth however is more or less unpredictable. Even if one had a nearly ideal shaped probe as shown in Figure 9.10, the remaining width W_p of the probe will ever influence the result. By measuring the probe width and using deconvolution algorithms the result can be corrected to a certain extent.

9.4.2.2 Oxide Sharpened Tips

Oxide sharpened tips are made from ordinary pyramidal Si_3N_4 tips. Nonmodified pyramidal tips have an apex radius of about 50 nm and a pyramidal shape with a half-opening angle of 35°. If the tip is sharpened the curvature radius is reduced to about 20 nm and the pyramidal half-opening angle is only 18° over a region of some 100–200 nm at the end of the pyramid.

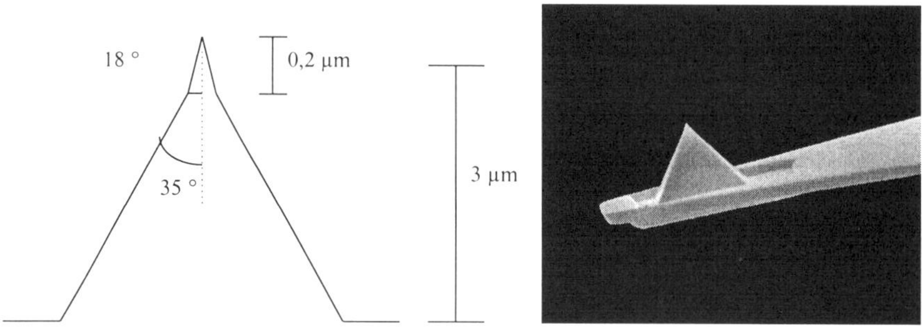

Figure 9.11. Typical shape of a sharpened tip. Radius of curvature is approx. 20 nm. (picture courtesy: Park Scientific Instruments, Sunnyvale, CA).

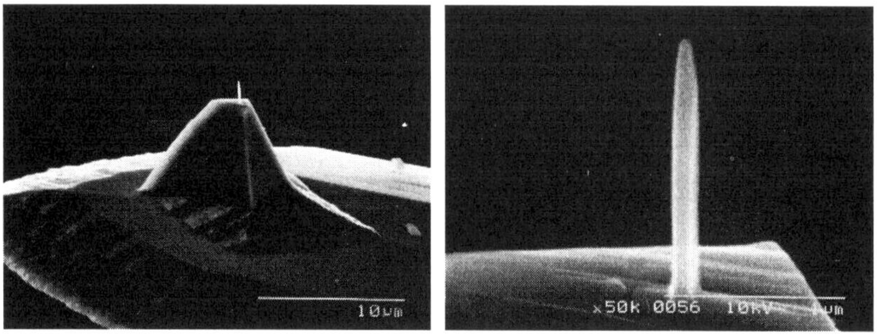

Figure 9.12. Typical shape of a FIB milled tip. (picture courtesy: Digital Instruments, Santa Barbara)

9.4.2.3 Focused Ion Beam Sharpened Tips

Another approach to get tips with a better aspect ratio is to sharpen ordinary Si or Si_3N_4 tips with a focused ion beam (FIB). An ion beam is capable of removing material from the surface of a solid by sputtering. A focused ion beam therefore can act as a very precise "mill". Unfortunately this fine technology is expensive.

By using this technique it is possible to achieve a 0.2–3 µm long region with a very high aspect ratio without compromising the tip stiffness in general. With FIB milled tips it is possible to resolve details of few nanometers size.

9.4.2.4 Sensors with Pear-like Tips For Sidewall Imaging

The ideal tip for precise 3D measurements (critical dimension measurements, CD measurements) is the miniaturized form of the well-known sensors from the 3D measuring machines: a rod with a ball at the end.

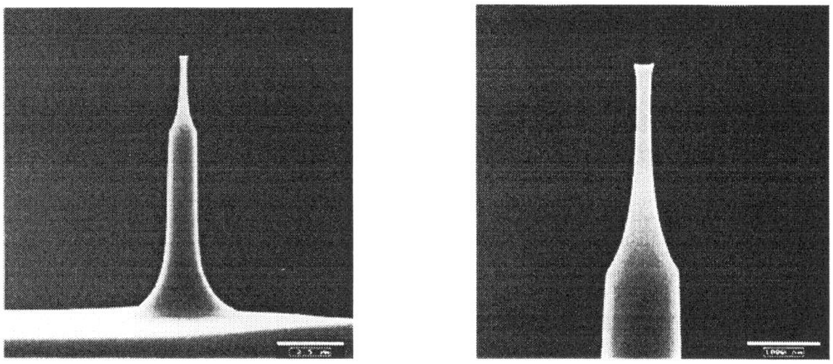

Figure 9.13. Special tip for sidewall imaging. (picture courtesy: IBM, Sindelfingen).

9.4.2.5 Contamination Tips Grown in the SEM

One way to get a tip suitable for imaging high aspect ratio structures is to grow a so-called "contamination needle" in a scanning electron microscope (SEM). The vacuum in the SEM column is often far from ideal. In many cases the residual gas might contain traces of oil, especially if the instrument is diffusion pumped and not equipped with a liquid nitrogen baffle. In this case the oil traces in the residual gas can be cracked by the electron beam of the SEM and a mixture of pure carbon and hydrocarbons will be deposited on the place where the electron beam is focused. If the instrument does not drift a needle with the diameter of the electron beam will grow.

These tips are often grown on tipless cantilevers or on top of pyramidal tips. They show a nearly unlimited aspect ratio, but are not very sharp. These tips are conductive since they consist mainly of carbon and consequently may also be used in the STM.

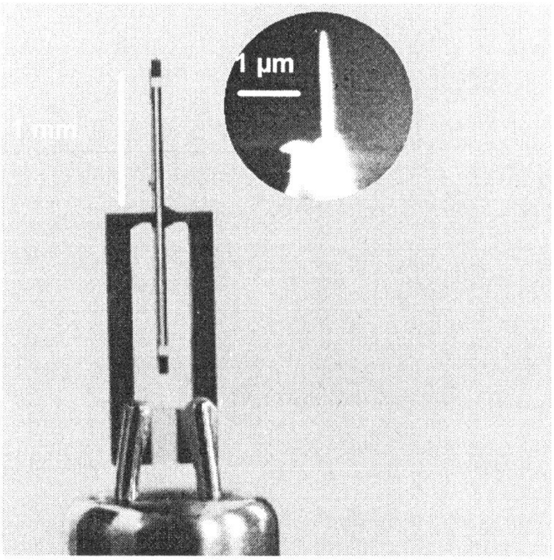

Figure 9.14. Typical shape of a contamination tip grown on top of the Carl Zeiss needle sensor.

Figure 9.15. Topometrix Supertip, a contamination tip grown on top of Si_3Ni_4 cantilever. (picture courtesy; of Topometrix Corp).

Because of their small diameter bending or breaking these cantilevers is very likely. Therefore cantilevers with these tips should be used in noncontact mode.

These cantilevers are commercially available from several manufacturers (see Appendix B) but they are rather expensive because of the necessity to handle each cantilever individually instead of batch processing them. Many SFM operators in academic institutions who have access to a SEM prefer to make contamination tips themselves, often on top of commercial cantilevers from well-known manufacturers.

9.4.3 Sensors for Derivative SFM Techniques

9.4.3.1 Cantilevers Without Tips At All

Some vendors offer cantilever chips whose cantilevers have no tip at all. The main purpose of these tips is to offer the possibility to add one's own tip at the end of the cantilever. This is often a contamination tip grown in the SEM as described earlier but other experiments can be done as well.

Tipless cantilevers are available in the same range of geometries, spring constants and resonance frequencies as normal cantilevers are.

9.4.3.2 Sensors for Magnetic Force Microscopy

Magnetic force microscopy (MFM) should consist – at least partially – of a ferromagnetic material. The first MFMs have been run with cantilevers made from chemically etched and bent Ni-wires. While these cantilevers can be produced relatively easily in almost every laboratory, they have several drawbacks:

- the production of these probes is time consuming and requires a certain experience;
- these probes produce a considerable magnetic stray field which may severely disturb the magnetic structures which should be visualized;

- the mechanical quality factor of bent wires is small (approx. 50) which limits the sensitivity if the probe is used in dynamic noncontact mode MFM and the hydrodynamic damping is removed by placing the set-up in a vacuum chamber.

Because of the drawbacks of bent Ni wires MFM users soon switched to microfabricated cantilevers coated with a ferromagnetic material. These cantilevers have several advantages:

- they can be batch fabricated and they are readily available to the customer;
- they fit to the scanning heads of most AFM vendors and do not require tinkering;
- a broader range of magnetic materials can be used, depending on the application. Tips can be covered with hard magnetic and high magnetic moment, hard magnetic and small magnetic moment and soft magnetic materials.
- because the magnetic material is only a thin film, its magnetic stray field decays much faster for coated tips then for massive wires. This reduces the influence of the tip on the sample to be imaged;
- microfabricated tips offer a wider range of spring constants and resonance frequencies compared to wires. Besides that, cantilevers made of monocrystalline silicon have a very big mechanical quality factor (Q approx. 10 000) which can be used for extremely sensitive measurements.
- while the MFM is run mostly in dynamic noncontact mode, the spring constant of MFM cantilevers (1–5 N/m) is considerably less than the spring constants commonly used for noncontact or intermediate contact topographic imaging (20–100 N/m). The reason for this is that the van der Waals force which is used in the topographic noncontact mode is much stronger than the magnetostatic interaction used in the MFM.

9.4.3.3 Sensors for Electric Force Microscopy

Sensors for Electric Force Microscopy (EFM) should have a tip on a defined potential. This requires the tip to be conductive which can be achieved by coating micromachined cantilevers of any kind with a metal, mostly gold or chromium, or using cantilevers made of doped and therefore conductive silicon.

Concerning useful spring constants for EFM the same applies as for MFM: the cantilevers often used have a spring constant of 1–5 N/m. In case of small potentials or charges to be detected, cantilevers with smaller spring constants might be helpful.

9.4.3.4 Sensors with Chemical Sensitivity

There are several approaches to gain some insight into the chemical structure of a surface by means of scanning force microscopy. Adding functional groups such as a

CH_3-, COOH- or NH_3- to the surface will cause the friction to be dependent on the surface chemistry of the sample which can be measured by lateral force microscopy. Similarly, making the cantilever hydrophilic allows one to distinguish hydrophilic and hydrophobic areas in the friction force image [16, 17].

Up to the time of writing (March 1997) no commercial sources of chemically modified cantilevers are known to the authors. It is expected however, that this will change in the future.

9.5 Brief Market Analysis and Trends

Just as it is essential to look at microscope vendors, when looking at probe manufacturers it is also important to analyze the SPM market before conducting the probe market.

From our evaluation, the SPM market is still expanding by 7–10% each year. Since the domain of SPMs is still basic and applied research, quite a big portion of this market is occupied by noncommercial self-made equipment, which however in most cases require the use of commercial probes. Only for very specific purposes, are noncommercial probes being used.

The probe market is also characterized by a delayed development compared to the SPM market, since with the first sales of an SPM only a certain number of probes are sold. This means that for a certain period of time the user will not need new probes. However together with new instrument sales, the market grows for spare probe sales for previously sold instruments.

From our evaluation, we see the SPM market share as in Figure 9.15.

We assume that more than 5000 SPMs are running worldwide and 1000 follow each year worldwide. Each SPM consumes 150 tips a year and we assume the average price of a tip to be 20 US$ so we can conclude that the probe market is in the order of 15 Mio US$ and grows by 20%.

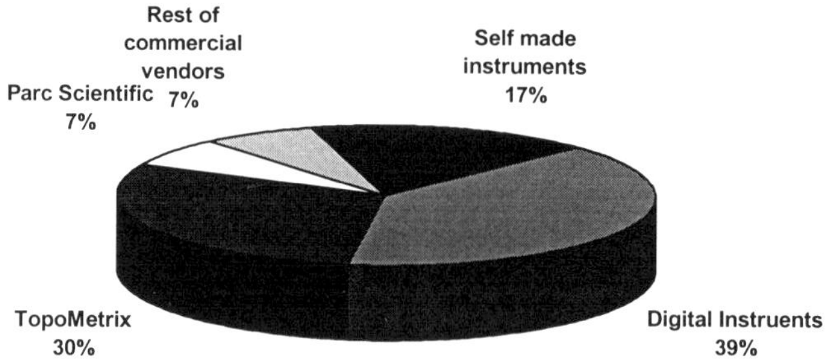

Figure 9.16. estimated market share.

Trends

Improvement of the aspect ratio
Current aspect ratios are 10:1. The goal is 20:1 to further minimize artifacts.

Very hard tips [16]

There are various methods published to produce a diamond tip on standard tips or to coat tips by means of CVD with diamond. Tip radii of 20 nm have been reached. Advantages and applications are very low wear, electrical conductivity through dotation (simultaneous STM/AFM) and possible nanolithography applications.

Chemical sensitive tips [17, 18]

Those tip are chemically active e.g. with OH-groups.

Multifunctional tips

Multifunctional tips offer the following advantages:

- new results from the combination of the results in different modes
- different results on exactly the same area
- time saving, since no expensive method is need changing.

Ultrasoft cantilever

Specifically for biological samples, contact mode may damage the samples. Ultrasoft cantilevers made of PMMA could perfectly be used especially under liquids.

9.6 References

[1] Binnig, G., Rohrer, H., *Rev. Mod. Phys.* **59** (1987), 3, 615–625.
[2] Synge, E. H., *Philos. Mag.* **6** (1928), 356–362.
[3] Heinzelmann, H., Pohl, D. W., *Appl. Physics A* **59** (1994), 89–101.
[4] Besocke, K.: *Surface Science* **181** (1987), 145.
[5] Fuchs, H., Schimmel, Th., *Speichern atomar kleiner "Zeichen"*, Philip Morris Research Award, 94, pp. 32–37.

[6] Meyer, G., Amer, N. M., *Appl. Phys. Lett* **53** (1988), 1045.
[7] Rugar, D., Mamin, H. J., Güthner, P., *Appl. Phys. Lett* **55** (1989), 2588.
[8] Yuan, C. W., Batalla, E., de Lozanne, A., Kirk, M., Tortonese, M., *Appl. Phys. Lett.* **65** (1994), 1308.
[9] Stahl, U., Yuan, C. W., de Lozanne, A. L., Tortonese, M., *Appl. Phys. Lett.* **65** (1994), 2878.
[10] McClelland, G. M., Erlandsson, R., and Chiang, S., in: *Review of Progress in Quantitative Nondestructive Evaluation*, Thompson, D. O., Chimenti, D. E., eds.; New York: 1987.
[11] Schönenberger , C., Alvarado, S. F., *Rev. Sci. Instr.* **60** (1989), 3131.
[12] Martin, Y., Wickramasinghe, H. K., *Appl. Phys. Lett.* **50** (1989), 1455.
[13] Göddenhenrich, T., Lemke, H., Hartmann, U., Heiden, C., *Appl. Phys. Lett.* **56** (1990), 2578.
[14] Shalom, S., Lieberman, K., Lewis, A., Cohen, S. R., *Rev. Sci. Instr.* **63**(9) (1992), 4061.
[15] Akmine, S., Albrecht, T. R., Zdeblick, M. J., Quate, C. F., *IEEE Electr. Dev. Lett.* **10** (1989), 489.
[16] Niedermann et al., *J. Vac. Sci. Technol. A* (May/June 1996).
[17] Frisbie, C. D., Rosznanyai, L. F., Noy, A., Wrighton, M. S., Lieber, C. M., *Science* **265** (1994), 2071.
[18] Noy, A., Frisbie, C. D., Rosznanyai, L. F., Wrighton, M. S., Lieber, C. M., *J. Am. Chem. Soc.* **117**(30) (1995) 7943.

9.7 Appendix A: Manufacturers of Microfabricated SFM-Sensors

Since the basic key to success when operating an SPM is to have appropriate and high-quality probes there are strong relations between microscope vendors and probe manufacturers. Some microscope vendors give such a high priority to the probes that they manufacture them on their own and some treat them somewhat proprietarily.

Based on interviews with more than 100 SPM users, we give a very brief estimate of some major microscope vendors/probe manufacturers.

Nanosensors GmbH

Fully specialized on the manufacturing of Si etched tips. Nanosensors deliver their probes via dealers or via microscope vendors or direct to customers.

Contact:

IMO-Building,
Im Amtmann 6
D-35578 Wetzlar, Germany
Tel.: ++49-6441-978 840, Fax.: ++49-6441-978 841
e-mail: nanosensors@compuserve.com
WWW: http://ourworld.compuserve.com/hompages/nanosensors

Digital Instruments Inc.

As market leader in the SPM field, DI has the biggest basket of probes. The Si_3Ni_4-probes are self manufactured. The "Oxide sharpened Silicon Nitride Probe" from Olympus are offered as well as Nanosensor probes. Some of NS-probes are further treated and re-offered for specific purposes. There are no restrictions known in selling the probes.

Contact:

520 E. Montecito St.
Santa Barbara, CA 93103 USA
Tel: ++1-805-899-3380, Fax: ++1-805-899-3392
e-mail: info@di.com
WWW: http://www.di.com

TopoMetrix Corp.

To is offering Nanosensors probes. Si_3Ni_4 probes are self-manufactured. To is also manufacturing contamination tips.
There are no restrictions known in selling the probes.

Contact:

5403 Betsy Ross Drive
Santa Clara, CA 95054-1162 USA
Tel: ++1-408-982-9700, Fax: ++1-408-982-9751
e-mail: service@topometrix.com
WWW: http://www.topometrix.com

Park Scientific Instruments (recently a part of ThermoSpectra Corporation)

PS controls both Si and SiNi process technology. Moreover PS is license holder of the so-called piezolevers, which do not requiring optical means for cantilever deflection sensing. Concerning some of their probes, restrictions in selling are known. Only PS customers can order Ultralever APUL-20AU-25/-10, Piezolever APPL, Piezolever VPPL

Contact:

1171 Borregas Ave.
Sunnyvale, CA 94089 USA
Tel: ++1-408-747-1600, Fax: ++1-408-747-1601
e-mail: info@park.com
WWW: http://www.park.com

Nanonics Ltd.

Na has focused on their so-called micro pipette technology, which allows mean to control the SNOM probe by means of a standard AFM set-up.

Contact:

The Manhat Technology Park
P.O. Box 48330
Jerusalem 91487, Israel
Tel: ++972-2-679-6601; Fax: ++972-2-679-6597

Olympus Optical Corp., Ltd.

Among SPM users we interviewed, the probes from Olympus are considered to be expensive but of according quality. No sales restrictions are known.

Contact:

2–43 Hatagaya Sibuyaku
Tokyo, Japan 151
Tel: ++81-3-3377-2111, Fax: ++81-3-3375-6550

Materials Analytical Services

Produces high-quality contamination and E-beam sharpened tips.
No sales restriction known

Contact:

616 Hutton Street, Suite 101
Raleigh, North Carolina 27606 USA
Tel: ++1-919-829 7041; Fax: ++1-919-829 5518

IBM/Veeco

Produces standard, conical and special by formed tips at the technology center Sindelfingen, Germany.
No sales restrictions known

Contact:

IBM German Manufacturing Technology Center
Sindelfingen, Germany
Tel: ++49-7031-16-5102; FAX: ++49-7031-16-5152

9.8 Appendix B: Some Important Sensor Products

Table 9.1. Overview.

Type	Manufacturer	Thickness [μm]	Width [μm]	Length [μm]	Spring constant [Nm^{-1}]
80-ESPW-8	LOT/NS	2	50	450	0.19
NP 20	DI	0,5		120/200	0,3
OTR4	DI/Olympus	0,4		100/200	0,022/0,094
OMCL-TR 400 PS-X	Olympus	0.4		100/200	0,022/0,094
Microlever APFL-06NO	Park Scientific	0,6	18	180	0,05
Ultralever APUL-06AU		0,8	25	180	0,16
APUL-20AU		1,8	25	180	0,16
Piezolever APPL-20NO		2,0/4,0			
1520-00	Topometrix		18	200	0,032
1680–00			18	200	0,032
1650-00			30	120	30 · · · 80
Supertip 1700-00			18	200	0,032
Ultrasharp SCS 10	NT-MDT				
SCH 10					
SCF 10					
Metris 3352	Burleigh Instr.	1,2· · ·2,5	40· · ·43	442	0,04· · ·0,85
Metris 3354		0,5· · ·5	10· · ·0	100· · ·50	0,02· · ·10
Super Sensor	Nanonics			100-500	0,1-100
milled Si	MAS				0,12
milled SiNi					0,12
HART-Probe					
AFM 9991-3	IBM/Veeco				
AFM 9991					

Resonance frequency [kHz]	Tip radius [nm]	Application	Price per unit [DEM]	Price per piece [DEM]
13	<10	AFM, LEM cont.Mode	8	61,00
			50	33,00
			385	22,17
5···50	20···40	AFM, LFM MMAFM cM	20	13,25
			approx. 500	7,96
13/40		AFM, LFM MMAFM, M	35	56,85
			105	50,55
			approx. 500	34,50
13/40		AFM, LFM MMAFM, M	70	34,64
22		AFM, cM	10	64,00
			50	45,60
25		AFM cM	10	82,00
			50	68,40
25		AFM ncM	10	91,00
			25	80,40
		AFM cM/non-cM	10	114,00
			25	91,20
5···15	1···10	AFM, LFM	20	44,00
5···15	1···10	AFM, LFM	10	264,00
240···420	20	AFM, ncM	8	110,00
5···15		AFM, LFM cM; for liquids	10	154,00
	<10	AFM, LFM	20	30,00
			200	15,00
			400	12,75
8···18		AFM	15	93,33
6···100	10	AFM, LFM	20	20,00
			500	10,20
10–400	5	AFM, LFM		95,00
	$\leq 5 \cdots 10$	AFM, LFM	5	200,00
	$\leq 5 \cdots 10$	AFM, LFM	5	130,00
	$\leq 2,5$	AFM, LFM	5	200,00
	≤ 10	AFM	12	73,00
	≤ 5	AFM	12	248,00

List of Symbols and Abbreviations

Symbol	Designation	Chapter
a	receiving area of thermoelectric IR radiation microsensor	1
	activation factor	2
	particle diameter	7
a_i	coefficient describing contribution of component i to a flavor	3
	ith-order temperature coefficient for quartz crystals	5
A	classical (crisp) set	2
	electrode area	7
$\tilde{A}, \tilde{A}_j$	fuzzy sets	2
$A_\mathrm{i}, A_\mathrm{r}$	particle displacements or velocities of incident and reflected waves	5
A_{ijk}	acoustic amplitude response for three-layered structure	5
A_0	incident amplitude	5
b_i	coefficient describing contribution of concentration X_i to a flavor	3
$\tilde{B}_l$	fuzzy value assigned to Y_k	2
C	velocity of sound	5
	concentration	7
$\vec{C}_i$	cluster center vector	2
C_L	longitudinal velocity of sound	5
C_P	thermal capacity	5
C_{Tr}	transverse velocity of sound	5
C	concentration	3
C	cylindrical refractive error	8
C_max	maximum permissible power dissipated in electrophoresis capillary	6
d	distance from sensor to object	5
	diameter of circular capillary	6
d_m	thickness of medium	5
d_q	desired output for layer q	2
D	detectivity of thermoelectric IR radiation microsensor	1
	particle diameter	5
	diffusion coefficient	7
	distance	8
	disk diameter	8
D_i	diffusion coefficient of component i	6
e	electronic charge	6
E	Young's modulus	5
	electrical field	6
f	function calculating net input	2
	analytical function of various parameters (equation (5.10))	5
	frequency of sound wave	5
	focal length	8

Symbol	Designation	Chapter
Δf	frequency change on mass loading of an acoustic microsensor	5
Δf	frequency change $(f - f_0)$ for quartz crystal	5,7
f_D	Doppler frequency	5
f_i	strength of receptive field i	2
	friction coefficient of species i	6
f_r	resonance frequency of transducer	5
f_R, f_T	received and transmitted frequencies	5
$f(x)$	activation function performed by artificial neuron	2
F	force	6
	Faraday constant	7
F_0	unperturbed fundamental frequency of oscillator	5
G	shear modulus	5
ΔG	change in free enthalpy	7
h'	distance between mire images (keratometry)	8
$h_{cj}(t)$	neighborhood function (neighborhood kernel) for neurons c and j	2
Δh	molar reaction enthalpy	7
H	number of receptive fields	2
	height equivalent to a theoretical plane	6
	distance between mires (keratometry)	8
ΔH	change in enthalpy	7
i	limiting current at planar electrode	7
$I(t)$	test pulse induced in nerve axon	3
J	error measure	2
J_p	summed square error over pattern P	2
k	concentration of disperse phase	5
k_i	heat coefficient	5
k_n	concentration of component n	5
k_0	propagation constant	5
k_1, k_{-1}	reaction on-rate (forward reaction) and off-rate (reverse reaction)	7
K	bulk modulus	5
	equilibrium binding constant	7
l	eye length	8
L	number of layers in a perceptron	2
	distance between transducers	5
	length of capillary	6
Δm	change in mass on quartz crystal microbalance	7
M_i	number of patterns in Θ_i	2
n	number of electrons transferred per molecule	7
	refractive index	7
n_c	peak capacity (maximum number of separated peaks on electrophoretogram)	6
N	number of theoretical plates	6
N_l	neuron count in layer l	2
O_L	node count in layer L	2
$O_b(t)$	output of bulbar neurons	3
O_i	output	2

Symbol	Designation	Chapter
$O_{pm}(t)$	response of periglomerular and mitral cells	3
p, p_0	sonic alternating pressure	5
$p(x, t)$	sonic alternating pressure of a plane elastic wave	5
P	Joule heating	6
r	radius of sphere	6
$\vec{r}_c, \vec{r}_j$	vectors to positions of neurons c and j	2
R	resistivity of thermoelectric IR radiation microsensor	1
	ohmic resistance of conductive layer of bolometric IR sensor	1
	particle radius	5
	reflection coefficient of acoustic wave	5
	resolution of electrophoretogram	6
	resistance of liquid column in electrophoretic capillary	6
	radius of curvature	8
R_i	output value of receptive field i	2
R_{ij}	reflection coefficient of medium i to medium j	5
$R_{xy}(\tau)$	correlation function	5
$\Re$	responsivity of thermoelectric IR radiation microsensor	1
s	path length of sound	5
S	Seebeck coefficient	1
	detection threshold of flavor in 50% of specified sample population	3
	spherical refractive error	8
S_{m}	sensitivity of acoustic microsensor	5
SMAX	softmaximum function	2
SMIN	softminimum function	2
ΔS	change in entropy	7
t	time	3,5
	analysis time for capillary electrophoresis	6
t_1, t_2	times between electronic pulses from transducers 1 and 2	5
T	absolute temperature	1,7
	transmission coefficient	5
ΔT	temperature difference $(T - T_0)$ between two quartz crystal plates	5
u_i	function performed by artificial neuron	2
$u_{l,j}$	output of jth neuron of layer l	2
U	set of all possible output values	2
	applied voltage	6
U, U_0	electrical voltages proportional to P, P_0	5
$U_{L,q}$	output of neuron q at layer L	2
v	velocity of flow	5
	wave velocity	7
v_{eo}	electroosmotic velocity of fluid	6
v_i	linear velocity of species i	6
v_{m}	velocity of medium	5
V	volume flow	5
	volume of annular Helmholtz layer	6
V_i	output of channel i	4
$V_i(t)$	time-dependent signal from sensor i	3

Symbol	Designation	Chapter
$\vec{w}$	weight vector of neuron j	2
$w_{l,j,i}$	weight connecting ith and jth neurons of layers $l-l$ and l in MLP, resp.	2
W	weight matrix, with components w	2
$\vec{x}$	input vector	2
x	ultrasonic signal received	5
$\vec{x}_i$	input to fuzzy system, e.g. artificial neuron	2
x_i	preprocessing metric	3
x'_i	array normalization function	3
Δx	path length of sound	5
X_i	concentration of chemical species i	3
Y	ultrasonic signal received	5
$\tilde{Y}_k$	fuzzy output	2
$z_i e$	net charge on species i	6
Z	thermoelectric figure of merit	1
	acoustic impedance	5
Z_0	free-space impedance of vacuum	1
α	attenuation of sound	5
	angle between directions of flow and of sound propagation	5
	asymptotic divergence angle	8
$\bar{\alpha}$	thermal expansion coefficient	5
α_{relax}	relaxation losses of sound in a liquid	5
α_{sc}	scattering losses of sound in a liquid	5
α_{therm}	thermal losses of sound in a liquid	5
α_{vis}	viscous losses of sound in a liquid	5
α_0	basic absorption of emulsion or dispersion medium	5
α_1	attenuation of sound in a liquid	5
β	compressibility	5
Δ	refractive error	8
ε	dielectric constant of buffer and Helmholtz layer	6
ε_0	permittivity of a vacuum	1
ζ	zeta potential	6
η	viscosity of homogeneous medium	6
η	solvent viscosity	7
η_s	shear viscosity	5
η_v	bulk viscosity	5
$\theta_{\text{b}}, \theta_{\text{f}}$	return pipe and feed pipe temperatures	5
$\theta_{\text{x}}, \theta_{\text{y}}$	deflection angles in x and y directions	8
Θ_i	set of training patterns	2
λ	thermal conductivity	1
	wavelength of sound	5
	resonance wavelength	8
λ_0	wavelength at which 100% absorption occurs	1
μ	learning parameter	2
	set of membership values of U	2
	friction coefficient	2

Symbol	Designation	Chapter
μ_{avg}	average mobility of components in electrophoresis	6
μ_A	characteristic function of set A	2
$\mu_{\tilde{A}}$	membership function	2
μ_i	electrophoretic mobility of species i	6
μ_0	permeability of a vacuum	1
$\Delta\mu_{app}$	difference in apparent mobility of separate components in electrophoresis	6
ν	kinematic viscosity	5
ν_t	time-dependent learning rate	2
ρ	density	5
ρ	rotational relaxation time	7
ρ, φ	polar coordinates at pupil	8
ρ_e	charge density	6
σ	electrical conductivity	1,6
σ_i^2	normalization parameter	2
σ_x^2	variance of migrating zone width in electrophoresis	6
τ	time constant of thermoelectric IR radiation microsensor	1
τ_m	time at which correlation function has a maximum	5
φ	phase of sound wave	5
	angular displacement	8
Φ	cylindrical refractive power	8
χ	thermal conductivity	5
ω_i	output of artificial receptive field i	2
$\mathscr{X}$	domain	2
$\wedge_t$	t-norm	2
$\vee_t$	t-conorm	2
$\oslash$	diameter of thermoelectric IR radiation microsensor	1

Abbreviation	Explanation	Chapter
ABS	antilock braking system	2
AC	alternating current	3
ACE	angiotensinogen converting enzyme	7
AFM	atomic force microscope, atomic force microscopy	9
Arg	arginine	6
ASBC	American Society of Brewing Chemists	3
ASEN	application-specific electronic nose	3
Asn	asparagine	6
AVC	automatic volume control	3
BADD	basic defuzzification distribution	2
BEM	boundary element method	5
BIA	biomolecular interaction analysis	7
BPN	backpropagative neural network	3
CA	cluster analysis	3
CAM	content-addressable memory	3
cAMP	cyclic adenosine monophosphate	3

Abbreviation	Explanation	Chapter
CAPA	coupled acoustic and piezoelectric analysis	5
CBADD	customizable basic defuzzification distribution	2
CCA	canonical correlation analysis	3
CCD	charge-coupled device	8
CD	critical dimension	9
CE	capillary electrophoresis	5
cGMP	cyclic guanosine monophosphate	3
ch.	channel	4
cm	contact mode	9
CNS	central nervous system	3
DC	direct current	3
DOP	dioctyl hydrogen phosphate	4
EFM	electric force microscopy	9
EOG	electro-olfactogram	3
EOG	electro-oculography	8
EPL	external plexiform layer	3
ERG	electroretinography	8
FCS	fluorescence correlation spectroscopy	7
FEM	finite element method	1,5
FIA	flow injection analysis	6
FIB	focused ion beam	9
FID	flame ionization detector	3
FIR	far infrared	1,2
FITC	fluorescein isothiocyanate	6
FPA	flavor profile analysis	3
FPA	focal plane array	1
FPIA	fluorescence polarization immunoassay	7
FPL	fuzzy programming language	2
FPM	flexural plate mode	5
FT	Fourier transform	3
FU	flavor unit	3
GC	gas chromatography, gas chromatograph	3,6
GC-EN	gas chromatography–electronic nose	3
GCL	granule cell layer	3
Gln	glutamine	6
Gly	glycine	6
GMP	disodium guanylate	4
G_{olf}	G-protein	3
HETP	height equivalent to a theoretical plane	6
HOMO	highest occupied molecular orbital	3
HPLC	high-performance liquid chromatography	3,6
HTS	high throughput screening	7
HTSC	high-temperature superconducting	1
HV	high voltage	6
IDT	interdigital transducer	5
IMP	disodium inosate	4

Abbreviation	Explanation	Chapter
IOP	intraocular pressure	8
IR	infrared	1,3
ISE	ion-sensitive electrode	3
LC	liquid chromatography	3
LIF	laser-induced fluorescence	6
LMS	least mean squares	2
LOD	limit of detection	7
LOT	lateral olfactory tract	3
LUMO	lowest unoccupied molecular orbital	3
MBL	mitral cell body layer	3
MFM	magnetic force microscopy	9
MLP	multilayer perceptron	2
MOB	main olfactory bulb	3
MOS	metal oxide sensor	5
MRA	multiple regression analysis	3
MRR	molecular receptive range	3
MS	mass spectrometry, mass spectrometer	3,6
MSG	monosodium glutamate	4
M/T	mitral/tufted (cells)	3
MV	multivariate (analysis)	3
NA	numerical aperture	8
ncm	noncontact mode	9
NDT	nondestructive testing	5
NETD	noise equivalent temperature difference	1
NFL	nerve fiber layer	8
NMR	nuclar magnetic resonance (spectrometry)	3
OBP	odorant-binding protein	3
OC	olfactory cortex	3
OCT	optical coherence tomography	8
OG	original gravity	5
OPA	*o*-phthaldialdehyde	6
OPD	optical path deviation	8
PARC	pattern recognition	3
PCA	principal component analysis	3
PCR	principal component regression	3
PCR	polymerase chain reaction	7
PDMS	polydimethylsiloxane	3
PDP	parallel distributed processing	3
Phe	phenylalanine	6
PHB	poly(hydroxybutyrate)	5
PLS	partial least-squares regression	3
POC	primary olfactory complex	3
PVDF	poly (vinylidene fluoride)	5
QCM	quartz crystal microbalance	5,7
RBF	radial basis function	2
RBFN	radial basis function network	2

Abbreviation	Explanation	Chapter
RIE	reactive ion etching	6
RIFS	reflectometric interference spectroscopy	7
SAW	surface acoustic wave	5,7
SCCE	synchronized cyclic capillary electrophoresis	6
SEM	scanning electron microscope	9
Ser	serine	6
SFM	scanning force microscope	9
SH-APM	shear horizontal acoustic plate mode	5
SPM	scanning probe microscope	8
SPR	surface plasmon resonance	3,7
STM	scanning tunneling microscope	9
STP	standard temperature and pressure	3
TAS	total analysis system	6
TC	temperature coefficient	5
TCR	temperature coefficient of resistance	1
TOC	total organic carbon	6
TOMA	trioctylmethylammonium chloride	4
TSK	Takagi–Sugeno–Kang	2
TSM	thickness shear mode	5
UHV	ultrahigh vacuum	3
UV	ultraviolet	1
VECP	visually evoked cortical potential	8
VF	vinyl fluoride	5
VF_2	difluoroethylene	5
VF_3	trifluoroethylene	5
VF_4	tetrafluoroethylene	5
VOC	volatile organic compound	3
WPNI	whey protein nitrogen index	4
YBCO	$YBa_2Cu_3O_{7-x}$	1
YSZ	yttria-stabilized ZrO_2	1
μ-TAS	miniaturized total analysis system	6
3-AFC	three-alternative forced choice test	3
7-TD	7-transmembrane domain	3

Index

Note: Index entries referring to Figures and Tables are indicating by *italic page numbers.*

Index prepared by Paul Nash